AF356649

Biocomposite Inks for 3D Printing

Editor

Gary Chinga Carrasco

MDPI • Basel • Beijing • Wuhan • Barcelona • Belgrade • Manchester • Tokyo • Cluj • Tianjin

Editor
Gary Chinga Carrasco
RISE PFI
Trondheim
Norway

Editorial Office
MDPI
St. Alban-Anlage 66
4052 Basel, Switzerland

This is a reprint of articles from the Special Issue published online in the open access journal *Bioengineering* (ISSN 2306-5354) (available at: www.mdpi.com/journal/bioengineering/special_issues/3D_ink).

For citation purposes, cite each article independently as indicated on the article page online and as indicated below:

LastName, A.A.; LastName, B.B.; LastName, C.C. Article Title. *Journal Name* **Year**, *Volume Number*, Page Range.

ISBN 978-3-0365-1738-4 (Hbk)
ISBN 978-3-0365-1737-7 (PDF)

Contents

About the Editor

Gary Chinga Carrasco

Dr. Gary Chinga Carrasco graduated from the Norwegian University of Science and Technology (NTNU) with background in Cell Biology (Cand. scient. 1997) and Chemical Engineering (Dr. ing. 2002). He is currently lead scientist at RISE PFI in the area of biocomposites. He has published more than 110 peer-reviewed papers (including 7 critical reviews). Dr. Chinga Carrasco was one of two recipients of the Norwegian award "Treforedlingsprisen 2018"–awarded for his nanocellulose research and winner of the 2021 TAPPI Nanotechnology Division Mid-Career Award for his research contributions related to the nanotechnology of renewable materials and contributions to the technical community. He has years of experience as coordinator of national and international projects, working closely with national and international R&D partners and industry. Currently his research focuses on tailor-made nanocelluloses, characterization, biomedical devices, and novel biocomposite materials, with a main focus on 3D.

bioengineering

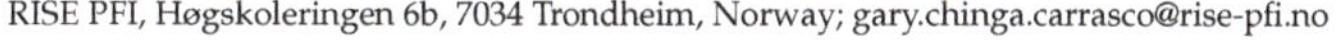

Editorial

Biocomposite Inks for 3D Printing

Gary Chinga-Carrasco

RISE PFI, Høgskoleringen 6b, 7034 Trondheim, Norway; gary.chinga.carrasco@rise-pfi.no

Citation: Chinga-Carrasco, G. Biocomposite Inks for 3D Printing. *Bioengineering* **2021**, *8*, 102. https://doi.org/10.3390/bioengineering8080102

Received: 9 July 2021
Accepted: 19 July 2021
Published: 22 July 2021

Publisher's Note: MDPI stays neutral with regard to jurisdictional claims in published maps and institutional affiliations.

Three-dimensional (3D) printing has evolved massively during the last years and is demonstrating its potential in tissue engineering, wound dressings, cell culture models for drug testing, and prosthesis, to name a few. One important factor is the optimized composition of inks that can facilitate the deposition of cells, fabrication of vascularized tissue and the structuring of complex constructs that are similar to the human micro-environment or functional organs. Some of the key aspects regarding the formulation of bioinks (inks biocompatible with cells) are, e.g., the tailoring of mechanical properties of the supporting matrix, biocompatibility considering the targeted tissue and the rheological behavior of the ink which may affect the cell viability, proliferation and cell differentiation. Biocomposite inks can include several polymers, such as polyhydroxyalkanoates, polylactic acid, collagen, agarose, alginate, nanocellulose, and may be complemented with cross-linkers to stabilize the constructs and with bioactive molecules to add functionality. Hence, these topics were covered by this Special Issue, which was supported by international groups with key competence in these areas of research and development.

The advances regarding the regeneration of functional tubular tissues and organs were explored by Jeong et al. [1]. The authors described several technologies such as extrusion-based, inkjet, laser-assisted and stereolithography-based bioprinting, and considering relevant inks, based on collagen, gelatin, alginate and synthetic polymers. The limitations of traditional methods to fabricate shape-free structures were mentioned, emphasizing the applicability of free-shape constructs which are based on indirect 3D printing, i.e., a hydride system where a 3D mold is printed to indirectly form 3D constructs. According to the authors, extrusion-based systems (also called direct-ink writing) are flexible and appropriate for fabrication of tubular structures. However, tubular structures such as the esophagus, blood vessels, and trachea are still demanding to fabricate and apply as clinical substitutes due to various physiological aspects [1].

Fused deposition modelling (FDM) is an extrusion-based technology, where a melted polymer is deposited layer by layer in predefined x,y,z locations. There are various polymers that can be applied for FDM 3D printing and the most applied is polylactic acid (PLA). However, natural polymers such as polyhydroxyalkanoate (PHA) are becoming an interesting but still limited alternative within the area of 3D printing and in the biomedical field [2]. PHA is naturally produced by microorganisms such as bacteria and archaea. Due to the versatility of the polymer various application areas were mentioned by Giubilini et al., including drug delivery, vessel stenting, tissue engineering, emphasizing also the opportunities offered by 3D printing such as the fabrication of non-toxic, resorbable scaffolds for tissue regeneration [2].

Collagen, which is found in the extracellular matrix (ECM), is another natural polymer that has been utilized for decades to enhance cell cultures, and more recently as a biomaterial for 3D bioprinting and tissue engineering [3]. The popularity of collagen is exemplified by the great number of commercial products currently available. However, the poor mechanical properties of collagen are mentioned as a limitation of the biomaterial, and the authors provide some strategies for chemical and physical cross-linking to counteract this limitation [3]. The authors also dedicated a section to regulatory considerations which is interesting and valid also for other biomaterials and products. Furthermore, the hydrolysis of collagen leads to the formation of gelatin that can be combined with alginate to form a biocomposite ink for the 3D printing of scaffolding material. This was successfully demonstrated by Somasekharan et al. [4], where a hydrogel composed of alginate dialdehyde, gelatin and platelet-rich-plasma was formulated to 3D bioprint cross-linked cell-laden constructs with 80% cell viability. Agarose is another linear polysaccharide that has been reported to form constructs with high stiffness [5]. The agarose was modified (by TEMPO mediated oxidation) and combined with minor amounts of native agarose to form a biocomposite ink for 3D printing by micro-extrusion. The authors demonstrated the potential of the biocomposite ink by printing a series of complex and impressive 3D self-standing shapes.

The application of bacteria in the area of 3D printing and biofabrication was reviewed by Shavandi and Jalalvandi [6]. The authors mentioned the printing of bacteria aided by polymers such as gelatin and alginate and cross-linked with calcium to stabilize the constructs. Such systems may find their application area in the biofabrication of model biofilms containing a predefined distribution of bacteria such as *Pseudomonas aeruginosa* and *Staphylococcus aureus* which may be used to test antibacterial agents for, e.g., treatment of infected wounds where these bacteria are common pathogens.

Cellulose fibers have been explored for years as reinforcement of polymers to form biocomposites. Cellulose fibers can also be mechanically processed to obtain microfibrillated cellulose. In this Special Issue, a timely application of a polypropylene/microfibrillated cellulose biocomposite was demonstrated in the 3D printing of prosthetic products. The main advantage of this approach is the possibility to 3D print medical devices that are tailor-made for individual patients. This was successfully demonstrated by Stenvall et al. [7], where a transtibial prosthesis was 3D printed by FDM. The prosthesis was tested by patients and clinicians, and positive feedbacks were obtained regarding the use of a biocomposite product instead of conventional prosthesis [7]. Cellulose can also be processed chemically to obtain a printable gel [8]. The cellulose gel (20 wt%) was tested extensively following statistical approaches to find the best parameters for optimal 3D printing. The authors demonstrated the printability of the gel by firstly printing simple cubes and then more complex structures such as ear models [8].

Nanocellulose is one of the most recent biomaterials that have entered the 3D printing space. The shear-thinning property of nanocellulose is most appropriate for 3D printing by micro-extrusion systems. The stiffness of pre-defined 3D constructs can be tailored for the targeted tissue [9]. Wang et al. [9] reviewed several aspects of nanocelluloses, including the impact of surface charge and modification on, e.g., cell survival, cell attachment and proliferation. However, according to the authors, aspects that still require attention are the control of biodegradability in the human body and potential nanotoxicity, which are also considered major topics of research by the scientific community.

There are several types of nanocelluloses that can be obtained by various pre-treatments [9]. Enzymes are also applied in the pre-treatment step in order to facilitate the nanofibrillation. Kangas et al. [10] demonstrated the production of unbleached and delignified nanocelluloses based on an enzymatic pre-treatment, and their potential suitability as ink for 3D printing. The study demonstrated that the enzymatic pre-treatment was more effective on the delignified pulp and that an additional fluidization step was required to secure a nanocellulose grade with adequate morphology and rheology for 3D printing by micro-extrusion systems. The authors proved that the material was not cytotoxic and could be used to print self-standing 3D constructs.

Espinosa et al. [11] demonstrated the application of biocomposite inks (containing TEMPO nanocellulose, varying amounts of alginate and cross-linked with Ca^{2+}), for wound dressings. Wound care causes a significant economic burden on patients and healthcare systems; thus, research on advanced wound dressings that could be tailor-made by 3D printing has been a major area of research during the last years. In this specific study, TEMPO nanocellulose-based inks performed well in 3D printing operations and the 3D printed constructs had in addition great capacity to maintain water [11], which are considered beneficial characteristics for novel and personalized wound-dressing devices.

Research on cancer, one of the most abundant diseases worldwide, would benefit from developing relevant tissue-mimicking micro-environments in order to test new drug candidates. Rosendahl et al. [12] reported on an extensive study, including gene expression analysis, and showed the effect of 3D printed TEMPO nanocellulose scaffolds on cancer cells, as a step to develop novel tumor model systems. The analysis demonstrated that 3D printed nanocellulose scaffolds induced cancer stem cell characteristics on both genetic and cellular levels [12]. In addition, a heterogenous cell population was revealed, growing in multiple layers mimicking the in vivo situation in contrast to conventional 2D cell cultures where cells grow in a monolayer with a homogeneous cell population. The authors concluded that carboxylated nanocellulose represents a promising material for 3D cell culture models for cancer applications and drug screening.

In summary, the studies included in this Special Issue cover a vast area of research and provide clear examples of 3D printing technologies and applications, also emphasizing the benefit of additional converging technologies, such as chemical engineering, nanotechnology, biotechnology and gene sequencing. It is expected that these technologies, combined with artificial intelligence and advances in gene editing will lead to exponential growth and further disruption of this fascinating area of research, with a main focus on the manufacturing of physiologically relevant and functional bioconstructs.

Funding: This work received no external funding.

Conflicts of Interest: The author declares no conflict of interest.

References

1. Jeong, H.-J.; Nam, H.; Jang, J.; Lee, S.-J. 3D Bioprinting Strategies for the Regeneration of Functional Tubular Tissues and Organs. *Bioengineering* **2020**, *7*, 32. [CrossRef] [PubMed]
2. Giubilini, A.; Bondioli, F.; Messori, M.; Nyström, G.; Siqueira, G. Advantages of Additive Manufacturing for Biomedical Applications of Polyhydroxyalkanoates. *Bioengineering* **2021**, *8*, 29. [CrossRef]
3. Chan, W.W.; Yeo, D.C.L.; Tan, V.; Singh, S.; Choudhury, D.; Naing, M.W. Additive Biomanufacturing with Collagen Inks. *Bioengineering* **2020**, *7*, 66. [CrossRef] [PubMed]
4. Somasekharan, L.; Kasoju, N.; Raju, R.; Bhatt, A. Formulation and Characterization of Alginate Dialdehyde, Gelatin, and Platelet-Rich Plasma-Based Bioink for Bioprinting Applications. *Bioengineering* **2020**, *7*, 108. [CrossRef]
5. Gu, Y.; Schwarz, B.; Forget, A.; Barbero, A.; Martin, I.; Shastri, V.P. Advanced Bioink for 3D Bioprinting of Complex Free-Standing Structures with High Stiffness. *Bioengineering* **2020**, *7*, 141. [CrossRef] [PubMed]
6. Chavandi, A.; Jalalvandi, E. Biofabrication of Bacterial Constructs: New Three-Dimensional Biomaterials. *Bioengineering* **2019**, *6*, 44. [CrossRef] [PubMed]
7. Stenvall, E.; Flodberg, G.; Pettersson, H.; Hellberg, K.; Hermansson, L.; Wallin, M.; Yang, L. Additive Manufacturing of Prostheses Using Forest-Based Composites. *Bioengineering* **2020**, *7*, 103. [CrossRef] [PubMed]
8. Huber, T.; Najaf Zadeh, H.; Feast, S.; Roughan, T.; Fee, C. 3D Printing of Gelled and Cross-Linked Cellulose Solutions; an Exploration of Printing Parameters and Gel Behaviour. *Bioengineering* **2020**, *7*, 30. [CrossRef] [PubMed]
9. Wang, X.; Wang, Q.; Xu, C. Nanocellulose-Based Inks for 3D Bioprinting: Key Aspects in Research Development and Challenging Perspectives in Applications—A Mini Review. *Bioengineering* **2020**, *7*, 40. [CrossRef] [PubMed]
10. Kangas, H.; Felissia, F.E.; Filgueira, D.; Ehman, N.V.; Vallejos, M.E.; Imlauer, C.M.; Lahtinen, P.; Area, M.C.; Chinga-Carrasco, G. 3D Printing High-Consistency Enzymatic Nanocellulose Obtained from a Soda-Ethanol-O2 Pine Sawdust Pulp. *Bioengineering* **2019**, *6*, 60. [CrossRef] [PubMed]
11. Espinosa, E.; Filgueira, D.; Rodríguez, A.; Chinga-Carrasco, G. Nanocellulose-Based Inks—Effect of Alginate Content on the Water Absorption of 3D Printed Constructs. *Bioengineering* **2019**, *6*, 65. [CrossRef] [PubMed]
12. Rosendahl, J.; Svanström, A.; Berglin, M.; Petronis, S.; Bogestål, Y.; Stenlund, P.; Standoft, S.; Ståhlberg, A.; Landberg, G.; Chinga-Carrasco, G.; et al. 3D Printed Nanocellulose Scaffolds as a Cancer Cell Culture Model System. *Bioengineering* **2021**, *8*, 97. [CrossRef]

 *bioengineering*

Review

3D Bioprinting Strategies for the Regeneration of Functional Tubular Tissues and Organs

Hun-Jin Jeong [1] , Hyoryung Nam [2], Jinah Jang [2,3,4,5,]* and Seung-Jae Lee [1,6,]*

[1] Department of Mechanical Engineering, Wonkwang University, 460, Iksan-daero, Iksan-si, Jeollabuk-do 54538, Korea; hunjinjeong312@gmail.com

[2] Department of Creative IT Engineering, Pohang University of Science and Technology, 77 Cheongam-Ro, Nam-Gu, Pohang, Gyeongbuk 37673, Korea; alfzmtea@postech.ac.kr

[3] School of Interdisciplinary Bioscience and Bioengineering, Pohang University of Science and Technology, 77 Cheongam-Ro, Nam-Gu, Pohang, Gyeongbuk 37673, Korea

[4] Department of Mechanical Engineering, Pohang University of Science and Technology, 77 Cheongam-Ro, Nam-Gu, Pohang, Gyeongbuk 37673, Korea

[5] Institute of Convergence Science, Yonsei University, 50, Yonsei-ro, Seodaemun-gu, Seoul 03722, Korea

[6] Department of Mechanical and Design Engineering, Wonkwang University, 460, Iksan-daero, Iksan-si, Jeollabuk-do 54538, Korea

* Correspondence: jinahjang@postech.ac.kr (J.J.); sjlee411@wku.ac.kr (S.-J.L.)

Received: 29 February 2020; Accepted: 30 March 2020; Published: 31 March 2020

Abstract: It is difficult to fabricate tubular-shaped tissues and organs (e.g., trachea, blood vessel, and esophagus tissue) with traditional biofabrication techniques (e.g., electrospinning, cell-sheet engineering, and mold-casting) because these have complicated multiple processes. In addition, the tubular-shaped tissues and organs have their own design with target-specific mechanical and biological properties. Therefore, the customized geometrical and physiological environment is required as one of the most critical factors for functional tissue regeneration. 3D bioprinting technology has been receiving attention for the fabrication of patient-tailored and complex-shaped free-form architecture with high reproducibility and versatility. Printable biocomposite inks that can facilitate to build tissue constructs with polymeric frameworks and biochemical microenvironmental cues are also being actively developed for the reconstruction of functional tissue. In this review, we delineated the state-of-the-art of 3D bioprinting techniques specifically for tubular tissue and organ regeneration. In addition, this review described biocomposite inks, such as natural and synthetic polymers. Several described engineering approaches using 3D bioprinting techniques and biocomposite inks may offer beneficial characteristics for the physiological mimicry of human tubular tissues and organs.

Keywords: 3D bioprinting; biocomposite ink; tubular tissue; tubular organ

1. Introduction

Tubular tissues and organs exist with various forms and functions in the gastrointestinal (esophagus, intestines), respiratory (trachea), vascular (veins, arteries), and urinary (bladder, urethra) systems [1]. These tubular tissues have various diseases and malfunctions requiring appropriate therapeutic interventions, such as donor tissue transplantation, autologous implant, and replacement with a synthetic prosthesis. Autologous transplantation is considered as one of the best therapeutic methods; however, in the case of the trachea and esophagus tissue with little redundancy and non-existent autologous tissues, therapeutic approaches using donor tissues or synthetic prosthesis are required [2–4]. Donor tissue transplantation is an ideal option, but there remains a disparity between the number of the appropriate donors and the high demand for the therapeutic use of donor tissue [5–7]. In addition, finding a suitable donor tissue is not easy since most of the tubular tissues are associated with poor prognosis after surgery [4,8]. For these reasons, many tissue engineering approaches have been researched for the manufacture of suitable tubular tissues and organs.

Successful bladder tissue engineering using tissue-engineered hollow spherical biodegradable structures was first reported in 1996 [9]. Since then, many studies have been reported to create artificial functional tubular tissue, such as tracheal tissues. However, these tubular tissues generally have similar morphological features, but the fabrication of tubular tissues requires high-level microfabrication-techniques due to their complex hierarchical macro- and micro-structure containing the different cell types and extracellular matrix (ECM) [10–13].

Various microfabrication techniques for tissue engineering, such as electrospinning [14–19], cell-sheet engineering [20–22], and mold-casting [23–26], have been widely studied to make complex multi-layered architecture for artificial functional tubular tissues. These approaches would use additional substrates (e.g., rotating rod, sacrificial mold) to create the tubular architecture, as well as requiring complex and multiple manufacturing processes. Besides, these approaches are not only insufficient to create the tubular structure with a target-specific mechanical property but also restrict shape-freedom due to the technical limitations.

3D bioprinting technique is emerging as an alternative to overcome the limitations of fabrication in terms of building tubular tissues and organs [27,28]. 3D bioprinting has also been utilized for higher-complexity structures with printable biocomposite inks containing the living cells and natural and synthetic polymers [29]. The greatest benefit of 3D bioprinting technology in tubular tissue engineering is the ability to fabricate tubular structures with multi-layer free-form constructs, as well as allowing the placement of biomaterials in a cell and printable ink containing the biochemical microenvironmental cues [10]. In particular, this technology has unlimited possibilities that are feasible for producing complex tissues and organs. Additionally, it can be applied anatomically and clinically since it should facilitate the manufacture of patient-tailored 3D structures [30–32].

Therefore, this review dealt with the state-of-the-art of 3D bioprinting technologies, various biocomposite inks, and their applications to tubular tissue engineering focused on a blood vessel, trachea, and esophagus tissue regeneration.

2. 3D Bioprinting Techniques

To achieve the building of 3D-engineered human tissue and organ analogs, it is necessary to use accurate and well-controlled fabrication methods for suitable biomaterials and living cells. Due to these functional 3D fabrication requirements, four types of 3D bioprinting techniques have been developed based on the principle of releasing the printable biomaterials from the printing head.

2.1. Extrusion-Based Bioprinting Systems

The principles of extrusion-based bioprinting are dispensing biomaterials through the nozzle by physical force (e.g., pneumatic pressure, piston, or metal screw) and pneumatic pressure (Figure 1a). The extrusion head moves in the x, y, and z directions under the instruction of the CAD-CAM software to produce a 3D architecture by staked biomaterial onto the substrate. Even if this technique has a lower accuracy than the other 3D bioprinting methods (ink-jet, laser-based), it can be capable of various biomaterials, such as cell-laden bioink, cell-spheroids, hydrogels, and high-viscosity thermoplastic polymers. This bioprinting method allows the extrusion of an extensive range of viscous materials $(6–30 \times 10^7 \text{ mPa·s})$, and the resolution of the extruding in the exit of the nozzle is in the range of 100 μm–millimeter [33,34]. Among them, thermoplastic polymer, such as polycaprolactone (PCL) [35–37], poly (lactide-co-glycolic-acid, PLGA) [38,39], poly (L-lactic acid, PLLA) [40,41], have been widely applied to fabricate hard-tissue and sturdy supporting constructs. The distinct advantage of this bioprinting technique is that it can be installed in a multi-head system, allowing the simultaneous use of one or more biomaterials, such as synthetic polymer and cell-laden bioink. Therefore, given the ability to quickly manufacturing complex 3D tissue structures that morphologically and biologically mimic the human body, the extrusion technique is regarded as a promising clinical approach [30].

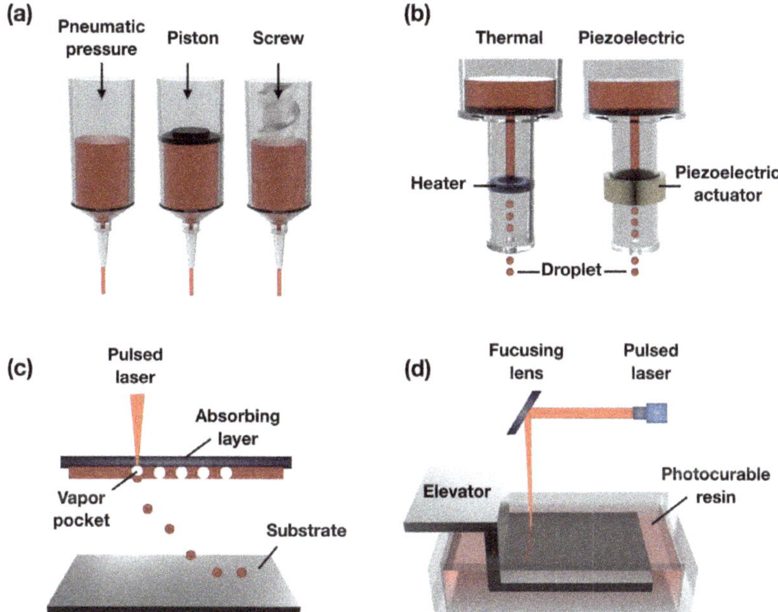

Figure 1. Schematic illustration of the (**a**) extrusion-based; (**b**) ink-jet; (**c**) assistant laser; and (**d**) stereolithography-based 3D bioprinting systems.

2.2. Ink-Jet Bioprinting Systems

Ink-jet bioprinting makes use of mechanical pulses, such as piezoelectric and thermal, to manufacture small-sized bioink droplets (spot size resolution of around 50–75 μm) and is referred to as the drop-on-demand method (Figure 1b). In the case of the piezoelectric ink-jet, break piezoelectrical materials are made by generating acoustic waves using an actuator to create droplets [42]. The thermal ink-jet uses the generated small air bubbles by applying electrical heat within a heated printhead to make droplets. These air bubbles are able to control the material being extruded from the exit of the print nozzle [43]. Cell viability in ink-jet bioprinting may differ depending on the applied mechanical pulses.

The main advantage of the ink-jet technique is that it enables to print with picoliter-volume droplets to build micro-structures because it can control the desired ejected droplet size as a variation of ultrasound parameters, such as amplitude, pulse, and time. However, there are drawbacks, including the need for low-viscosity material (3.5–12 mPa·s) to avoid clogging, and bioprinting material should be quickly gelated in post-print for 3D build constructs. In addition, the mechanical property of the post-printed construct has weak solidity, and the drying of a printed droplet on the substrate during bioprinting is a problem to be solved.

2.3. Laser-Assisted Bioprinting

Laser-assisted bioprinting uses a laser source to fabricate at high precision onto substrates. There are two separate approaches: laser-guided direct writing and laser-induced forward transfer (LIFT) [44,45]. Compared to other printing approaches, laser-assisted printing was rarely used in the old days; however, it has recently become increasingly popular as the 3D printing method for microfabrication.

This technique consists of a focusing system (to align and focus laser), an absorbing layer (ribbon), a pulsed laser beam (to induce the transfer of bioink), and a substrate for the bioink layer (Figure 1c) [46].

In brief, the laser pulse is induced on the absorbing layer to create a high-pressure bubble from the bioink layer and then drop the bioink onto the substrate. This bioprinting method can also use bioink with the high-viscosity and high-concentration of the cell (1–300 mPa·s) since it is a nozzle-free system [47]. Therefore, the formation of delicate shaping and arrangement of the cell patterning can be achieved while bioprinting without affecting cell viability. However, the fast gelation of bioink is essential for achieving a highly precise shape. For this reason, the laser-assisted bioprinting is a time-consuming process because of the relatively low flow rate for the crosslinking of the material. Therefore, the challenge of building 3D constructs of clinically relevant size remains [48].

2.4. Stereolithography-Based Bioprinting

Stereolithography (SLS or SL) is a form of 3D bioprinting technology using a laser source (ultraviolet, infrared radiation) for building 3D structures. An SLS bioprinting system consists of the light source, a reservoir with the liquid photocurable resin, an elevator system, and a digital mirror device (DMD) [49]. This technique was developed commercially in 1986 by Chuck Hill and then widely used for creating prototypes and complex production parts (Figure 1d). It can allow relatively rapid manufacturing time since the whole layers of the 2D slicing pattern of the 3D constructs are irradiated in a photopolymer reservoir. Also, this system can fabricate the sub-micron structure with highly precise 3D shapes (~1.2 μm) because a laser source can be focused on a small spot in the photocurable resin [50].

As this stereolithography bioprinting technology has been applied in the field of tissue engineering, various materials have been developed to contain bioink and cells with photo-initiators. In particular, commonly used photocurable materials in stereolithography bioprinting technique include the acrylics and epoxies. However, to apply the tissue-engineered approaches, the biocompatible photocurable resin needs to contain propylene fumarate (PPF) and trimethylene carbonate (TMC). These biocompatible photocurable materials have been widely used to manufacture with complex architecture and sacrificial mold [51]. However, SLS 3D bioprinting is still challenging due to cytotoxicity of the photocurable resins and high cost for system installation.

3. Printable Biocomposite Inks for Various 3D Bioprinting Techniques

Printable biocomposite inks are generally classified as natural, synthetic, and functional polymers. Natural polymers (e.g., collagen, gelatin, alginate, ECM-based ink) have been widely used in the field of the tissue-engineering and have been considered promising biomaterials with similar components of native tissue or organs in the human body [52–54]. In particular, protein-based natural polymers, such as collagen, gelatin, and ECM-based ink, have a remarkable capacity to help regenerate the epithelial layer, which is essential for creating the functional tubular tissue. Alginate bioink has an inferior biological activity compared to protein-based natural polymers. However, it has been widely utilized as bioprinting material to build the tubular constructs due to the easily controllable printability and excellent biocompatibility.

Ideally, synthetic polymers support the structure of the 3D printed target-tissue and degrade completely after implantation without side effects. Also, synthetic polymers have to pass the verification of strict criteria to be applied in clinical settings. In this section, we have described representative biocompatible synthetic polymers, such as thermoplastic polycaprolactone (PCL), polyethylene glycol (PEG), and polylactic acid (PLA), which have been recognized by the Food and Drug Administration (FDA) or widely applied in the field of tubular tissue-engineering [55–58]. In addition, we previously reported the combinations of natural polymer and synthetic polymer or nanocellulose that have been tried for enhancing mechanical properties and positive biochemical factors [59].

3.1. Natural Polymers

3.1.1. Collagen

Collagen is a representative natural polymer applied to bioprinting, and it consists of proline and glycine, with a triple-helix arrangement of polypeptides, as the most abundant protein in the human body [52,60]. In human organs (e.g., skin, bone, cartilage, vessel) and connective tissues, various collagen types exist, such as collagen I, III, and IV [61]. Among them, collagen type I is the most abundant and also the most commonly used in 3D bioprinting [62]. Collagen has characteristics of little cross-species immunological reaction and low toxicity, as well as allowing enhanced cell attachment and proliferation due to the presence of asparagine-glycine-aspartic acid (RGD) residues [11]. For this reason, collagen-based bioinks are regarded as highly promising biomimetic materials.

The main advantage of the collagen-based bioink is that it enables the embedding of living cells with ECM components and biochemical materials. However, because it has a crosslinking property, the use of a crosslinker or gelation process by temperature is essential for the construction of the 3D structure. In addition, the mechanical strength and bioprinting property of the collagen-based bioinks are dependent on viscosity due to collagen contents. For this control, many studies have combined collagen with biocomposite materials, such as fibrin [63,64], alginate [65], chitosan [60], and agarose [66], for improving printability and mechanical properties. Collagen-based bioink is undoubtedly an excellent biomaterial, but there remains scope for the improvement for use as bioprinting materials.

3.1.2. Gelatin

Gelatin is a type of protein obtained from collagen as a partially hydrolyzed form, and it is a biodegradable and biocompatible natural polymer [67]. Likewise, as with collagen, gelatin-based bioink can enhance cell attachment and proliferation because it has an RGD sequence with abundant integrin-binding motifs. Gelatin is dissolved in water to maintain the thermo-sensitivity property, but it reversibly forms a low-viscosity soluble state at human body temperature [68]. Because of these limitations in maintaining the form, using only the gelatin-based bioink as a printable biomaterial is not suitable to build sturdy 3D tissue structures. Therefore, many studies have attempted the development of the printable gelatin-based composite ink mixed with other polymer materials, such as PCL [69,70], chitosan hydrogel [71], hyaluronic acid [72], fibrin [73], alginate [74,75], and silk [76,77], for improving structure's stability.

Gelatin methacrylate (GelMA) has been widely used as an advanced bioink that modifies photocrosslinkable polymers [78]. The mixture of GelMA with photoinitiator (e.g., 2-hydroxy-1-(4-(hydroxyethoxy) phenyl)-2-methyl-1-propanone (Irgacure 2959)) undergoes rapid crosslinking after extrusion through exposure to UV light (360–480 nm in wavelength). This crosslinkable property enables the structural stabilization after bioprinting. Also, GelMA has excellent biological characteristics of cell adhesion, biodegradability, and cell migration because it involves collagen and gelatin components, such as integrin-binding motifs and RGD. Due to these promising properties, many studies have applied the combined material for improving the desired quality.

3.1.3. dECM Ink

dECM-based ink has been regarded as a promising material for 3D bioprinting [79]. dECM-based ink is fabricated by the decellularization process of the target-tissue. It has an inherent component of tissue-specific microenvironment cues, such as proteoglycans, glycoprotein, and collagenous protein. To date, various dECM-based inks have been reported for target-specific tissues, such as derived skin [80], bone [39], vessel [59], liver [81], kidney [82], and so on. Each derived tissue has different printability properties, but all have the distinguishing feature of temperature-responsive gelation under the physiological environment [83].

3.1.4. Alginate

Alginic acid, also called alginate, is an anionic polysaccharide distributed in the cell walls of brown algae [84]. It has a hydrophilic property and forms a viscous gel when hydrated. Alginate hydrogel has been applied as a wound dressing material because it has good biocompatibility and is structurally similar to natural ECM with a bioinert property [85]. In addition, alginate has been extensively used as an ink to fabricate the 3D structure in the field of tissue engineering because it can robustly form a cell-compatible hydrogel by instantly polymerizing using multivalent cations (e.g., Ca^{2+}, Ba^{2+}). Given their facilitation of tissue formation, hydrogel inks have been modified for a variety of tissue-engineered approaches, such as bone [86], cartilage [87], and vascular tissue [88]. In addition, because the alginate has no cell-adhesive site, the bioactive component enables the addition of the signal trigger, such as RGD, for cell viability and differentiation [89].

3.2. Synthetic Polymer

3.2.1. Polycaprolactone

PCL is one of the aliphatic polyesters; it is the most frequently used biomaterial for 3D bioprinting in the field of tissue engineering. PCL has superior printability due to its low melting temperature and glass-transition temperature. In addition, it is well known as a clinically applicable biomaterial approved by the FDA as a biocompatible and biodegradable polymer [90].

The degradation rate of biomaterials must be carefully considered before the fabrication of the target-specific tissue-engineered structure. If using the 3D scaffold with quickly degradation materials, there is a possibility of the mechanical property rapidly degrading after implantation in the body. In this regard, PCL has a great benefit as it can control the degradation rate by blending of different ratios of the polymer and copolymers [91,92]. The degradation mechanism of the PCL has a bulk erosion process by hydrolysis, and, in this process, PCL does not release toxic components [93,94]. Because of these convenient advantages, PCL is actively utilized as various bioprinting materials.

3.2.2. Polylactic Acid

Polylactide or polylactic acid (PLA) is most widely used for creating tissue-engineered architecture [95]. Also, PLA has been approved by the USA FDA for human clinical applications. The PLA has been used as a biomaterial for frequency 3D bioprinting because of its readily available thermoplastic properties [96]. Although there are differences depending on molecular weight (MW), PLA has relatively high mechanical properties, with an approximate tensile modulus of 3 GPa and tensile strength of 50–70 MPa [97]. The MW has a significant effect on biodegradability, but high-MW PLA is likely to cause inflammation and infection in vivo [98]. Therefore, before 3D bioprinting, the MW property must be considered for the mechanical properties of the target tissue.

3.2.3. Polyglycolic Acid

Polyglycolic acid (PGA) is a thermoplastic material with a high melting point and glass transition temperature, and it is more acidic and hydrophilic than PLA [99]. In addition, it is used as a surgical suture fiber because of its high mechanical strength and biocompatibility [100]. In the field of tissue engineering, solvent casting and compression molding are used to create PGA-based porous scaffolds [56]. However, PGA requires precise control as it is highly sensitive to degradation. Additionally, glycolic acid produced during the biodegradation process can be absorbed into the body, but the increased acid concentrations in the surrounding tissues may cause tissue damage.

3.3. Functional Polymer

As mentioned above, generally, biocomposite inks are classified as natural- and synthetic-based polymers, and these have been attempted to be used for complex and cell-compatible 3D constructs as tissue-engineering approaches [101]. Among them, hydrogel-type inks (i.e., alginate, collagen, dECM ink) have been considered as attractive materials because these can provide an optimized environment to a living cell. However, to be suitable for 3D bioprinting, these hydrate materials require adequate rheological properties to keep shape during bioprinting and must have cross-linking abilities, allowing to retain the 3D structure fidelity after bioprinting. Recently, the importance of versatile bioink materials in the field of tissue engineering has led to the development of functional polymers with improved biocompatibility, rheological behavior, and mechanical properties [102]. In this section, functional polymers that improve the bioprinting stability and fidelity when combined with nanocellulose biomaterials are introduced.

Nanocellulose refers to cellulosic nanomaterials, including cellulose nanocrystals (CNC) and cellulose nanofibrils (CNF) [103–105]. Gary Chinga-Carrasco et al. developed printable ink with bagasse, which is an underutilized agro-industrial residue [106]. This functional polymer has demonstrated non-cytotoxicity, stable bioprinting property, and shape fidelity, as well as potential that a low-value agro-industrial residue (bagasse) can be converted into a high-value product as disposable bioinks for 3D bioprinting. However, further evaluation is required for clinical applications. Kajsa Markstedt et al. developed functional bioink that combines the outstanding shear thinning properties of nanofibrillated cellulose (NFC) with the alginate [105]. The nanocellulose-based bioink enables fidelity bioprinting of 2D structure as well as 3D construct, which is anatomically the shape of a human ear and sheep meniscus. Also, nanocellulose-based bioink exhibits excellent cell viability. Therefore, these functional polymers using cellulose nanofibrils have shown promising potential as 3D bioprinting materials.

4. Recent Design Approaches for Engineering Tubular Structures

Tubular tissues and organs, such as the gastrointestinal tract, urinary tract, and respiratory tract, exist everywhere in the human body, serving major functions, including distributing fluids and air through the organs [107]. Almost all of the tubular tissues have a multilayer cellular structure from the innermost to the outermost, and the inner structure has an endothelium cell layer [6]. There are no existent pioneering fabrication techniques to fabricate tubular-shaped tissues and organs in the field of tissue engineering [5]. However, various fabrication methodologies have been suggested for constructing tubular structures with mimicking the inherent multilayer cellular constructs.

Traditional methods, such as casting, cell sheet assembly, and dip coating, have been attempted to create the tubular structures. The casting method creates the tubular structure by the biomaterials filling in the sacrificial mold and then demolding after appropriate chemical processes, such as gelation or crosslinking (Figure 2a). This method was proposed in 1986 by Weinberg and Bell, who made artificial vascular structures using the collagen-containing fibroblast and smooth muscle cells [23]. Since then, cell sheet assembly technology has been reported for reproducing hierarchical multi-layered cellular structures (Figure 2b) [20–22]. This method has been facilitated for creating the multilayer tubular structure by rolling on the rod using the stacked monolayer fabricated by biological functional materials containing extracellular matrix and target-specific cell components. The dip-coating method can also produce multiple tubular structures using rods by repeatedly dipping in the hydrogel and cross-linker agent (Figure 2c) [108–110]. These traditional methods have shown the promising ability to mimicking the cellular arrangement of native tubular tissues. However, there has been an unmet challenge in implementing physiological and mechanical properties suitable to tissue-specific complex environments. Also, these methods have unavoidable hurdles for fabricating shape-free forms and controllable structures. To overcome these challenges, hybrid-type technology, combining the traditional method with 3D bioprinting, has been tried to fabricate a free-form tubular structure [111]. Hybrid-type approaches have shown the possibility of creating a free-form tubular construct, although the structure has not been directly printed.

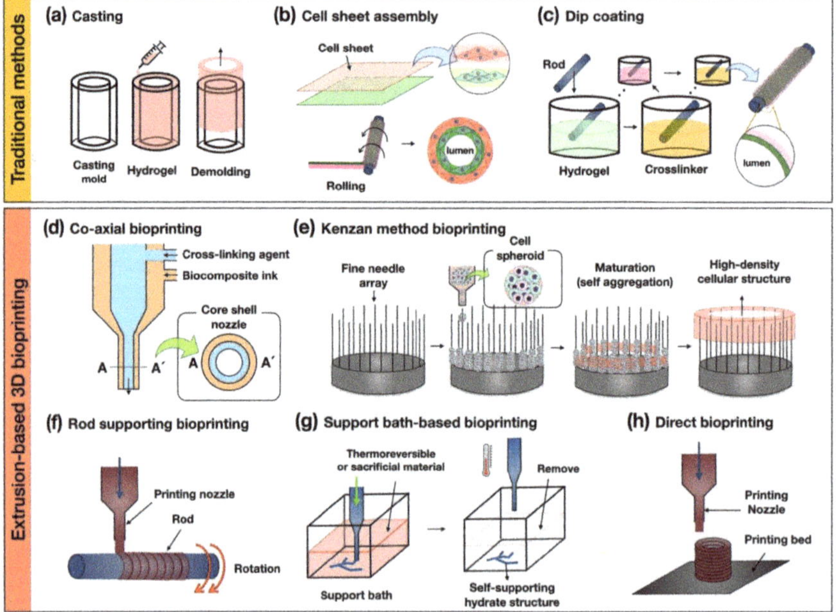

Figure 2. Schematic illustrations of the traditional methods of (**a**) casting; (**b**) cell sheet assembly; (**c**) dip coating and the extrusion-based 3D (**d**) co-axial; (**e**) kenzan method; (**f**) rod supporting; (**g**) support bath-based; and (**h**) direct bioprinting for fabricating tubular structures.

3D bioprinting technology has been emerging as a promising approach to facilitate complex structures and spatial cell positioning in tubular tissue engineering [112,113]. Among the various 3D bioprinting techniques (e.g., extrusion-based, ink-jet, laser-assisted, stereolithography-based 3D bioprinting), extrusion-based 3D bioprinting has been one of the most utilized for creating the tubular structure because it is relatively convenient for the installation of the system and availability of a wide range of biomaterials. In this article, we summarized and classified several categories of the extrusion-based 3D bioprinting for building a multilayer tubular structure (co-axial-, kenzan method-, rod supporting-, support bath-, direct bioprinting) (Table 1).

The co-axial bioprinting method is capable of creating a tubular structure using a core–shell nozzle that is capable of extruding two or more biomaterials (Figure 2d). Several research teams have been employing this method to print complex tubular structures with biocomposite inks. Gao et al. developed the printable hybrid bioink containing a mixture of vascular tissue-derived decellularized extracellular matrix (VdECM), alginate, and human umbilical vein endothelial cells [59]. Subsequently, they fabricated a perfusable multilayer blood vessel by co-axial bioprinting [88]. Yongxiang Luo et al. printed a 3D porous scaffold with regular macropores and a network of a controllable hollow structure as an embedded vasculature-like system using co-axial bioprinting [114]. This method can allow not only the building of a hollow construct with functional biological components but is also capable of fabricating permeable vascular-embedded 3D constructs. Moreover, it can manufacture small-diameter vascular structures with endothelial and smooth muscle layers, as well as being able to print long-length warping vascular structures with a minimal amount of time. However, this method has limitations in terms of making the anatomical bifurcate structure and stacking hierarchical constructs.

Table 1. Advantages and disadvantages of the extrusion-based 3D bioprinting.

Extrusion-Based 3D Bioprinting	Advantages	Disadvantages	Reference
Co-axial bioprinting	- The building of a hollow construct with functional biological components - Fabricating permeable vascular-embedded 3D constructs - Manufacturing small-diameter vascular structures with endothelial and smooth muscle layers - Printing long-length warping vascular structures in a short amount of time	- Difficulty in making anatomical bifurcate structures - Difficulty in stacking hierarchical constructs	[59,88,114]
Kenzan method bioprinting	- High cell density - Reduced printing time on spheroid bioprinting	- The requirement of the additional cell spheroid fabrication process - Fixed interneedle distance	[115–118]
Rod supporting bioprinting	- Manufacturing ability of the self-supporting multi-layer hollow structure with target-specific cell components - Sequential printing using two or more printable biomaterials	- Dependence on rotating rod shape	[31,119]
Support bath-based bioprinting	- Enabling 3D bioprinting of the hydrate biomaterials - Creating a complex 3D anatomical architecture	- Limitation of available biomaterials - High cost	[120,121]
Direct bioprinting	- High freedom of shape	- Long production time	[122,123]

Kenzan method bioprinting was invented by Koich Nakayama. It can create a high-density cellular structure by locating cell spheroids on a fine needle array (Figure 2e) [115–118]. The main principle of this method uses the natural and intrinsic feature of cell-to-cell self-aggregation. Recently, this research group fabricated the esophagus-like tubular structure without scaffold using the multicellular spheroids that matured during several periods in the bioreactor to create the rigid organoids.

The rod supporting bioprinting method produces a hollow construct by dispensing printable biomaterials on a rotating rod (Figure 2f). The rotating rod is provided as temporary support to the printed biomaterial for keeping a 3D shape and is removed when the printed structure is considered to be self-supporting. Sang-Woo Bae et al. printed the artificial tracheal structure with a synthetic polymer (i.e., PCL) and cell-laden bioink (epithelial cells and bone-marrow stem cells) by sequential extruding on the rotating rod [119]. Qing Gao et al. fabricated a hydrogel-based vascular structure with multilevel fluidic channels using a combination with co-axial bioprinting [31]. The main advantage of this bioprinting method is the manufacturing ability of the self-supporting multi-layer hollow structure with target-specific cell components though sequential bioprinting using above two or more printable biomaterials, such as polymer-based and cell-laden bioink. However, there is a disadvantage in that the figuration of the printed structure is dependent on the rotating rod shape.

The support bath-based bioprinting method refers to using a thermo-sensitivity gel bath or sacrificial materials for supporting the bioprinting biomaterials (Figure 2g). This method obtains a self-supporting structure, keeping the reversible condition for removing the biomaterials after bioprinting in the gel bath. Thomas J. Hinton et al. introduced this bioprinting method as the freeform reversible embedding of suspended hydrogel (FRESH), which uses this technique to print the biomimetic section of the human right coronary arterial tree with alginate-based bioink [120]. The research team, Tal Dvir et al. printed the endothelial cell-laden hydrogel to create the blood vessel-embedded cardiac tissue of the rabbit scale in the supporting bath with an aqueous solution containing sodium alginate, xanthan gum, calcium carbonate [121]. This method enables 3D bioprinting of the hydrate biomaterials, including alginate, collagen, and fibrin. Also, it can create complex 3D anatomical architectures, including branched coronary arteries, embedded vascular system organoids, etc. These results have demonstrated the potential of the approach for engineering personalized tissues and the bioprinting of patent-tailored biochemical microenvironment.

Direct bioprinting can be used to stack the printable biomaterials layer-by-layer for building the 3D structure (Figure 2h). In order to use this bioprinting technique, biomaterials must have sufficient solidity properties to maintain the 3D structure. For bioprinting hydrate bioink, in particular, it is essential to consider the rheological properties of materials during the bioprinting process. Anthony Atala et al. used direct bioprinting to build a 3D urethra tubular scaffold with a polymeric framework and cell-laden fibrin hydrogel [122]. The polymeric framework consisting of the PCL and PLCL (polylactic acid-co-ε-caprolactone) stably supports the printed hydrogel bioink for the desirable fabrication of the 3D tubular structure. Recently, Yifei Jin et al. developed self-supporting hydrogel by mixing laponite nanoclay and then successfully printed sturdy architecture without supporting the gel bath and polymeric structures [123].

5. Application of the 3D Printed Tubular-Organs with Various Biocomposite Inks

5.1. Esophagus

The esophagus is one of the gastrointestinal tracts and a 20–25 cm hollow structure, connecting the oropharynx and the stomach [124–126]. It allows the transport of food to the stomach by peristalsis and contractions of the muscle layer. Because the esophagus has a complex hierarchical structure (mucosa, submucosa, muscle layers), this must be considered when fabricated using engineering approaches [127–129].

Every year, 5000 to 10,000 patients are diagnosed with an esophageal disease requiring partial repair or full-thickness circumferential replacement, such as esophageal cancer, malignancy, congenital long-gap atresia, and esophageal achalasia. The strategy of the esophageal treatment is normally a gastric pull-up or autotransplantation using intestine or skin. In the case of autograft, using the gastrointestinal tract is an unavoidable strategy to achieve circumferential full-thickness repair since there is no substitute for esophagus tissue. However, although the autograft might allow the transfer of liquids or solid matter, complete restoration of the native tissue is compromised. Therefore, several approaches using 3D bioprinting technology have been researched to achieve the esophageal substitution, which replicates primary histological features of hierarchical cellular structures (Table 2).

Table 2. The 3D bioprinting technique and biocomposite ink for the esophageal tubular structure.

3D Bioprinting Technique	Biocomposite Ink	Reference
Kenzan method bioprinting	Cell spheroids with human dermal fibroblasts, human esophageal smooth muscle cells, human bone marrow-derived mesenchymal stem cells, human umbilical vein endothelial cells	[115]
Rod supporting bioprinting and electrospinning	Polyurethane (PU), polycaprolactone (PCL)	[15]
Rod supporting bioprinting and electrospinning	Polycaprolactone (PCL)	[14]
Direct bioprinting	Thermal polyurethane (TPU), polylactic acid (PLA)	[130]

Yosuke Takeoka et al. developed a scaffold-free biomimetic structure for the regeneration of the esophagus using the kenzan method bioprinting (Figure 3) [115]. This team used the maturated cell spheroids of the normal human dermal fibroblasts (NHDFs), human esophageal smooth muscle cells (HESMCs), human bone marrow-derived mesenchymal stem cells (MSCs), and human umbilical vein endothelial cells (HUVECs) to print the tubular multicellular structures. Mechanical and histochemical assessment of the printed esophagus-like tubular structure had been done with the content ratio of those cell sources. The high proportion of mesenchymal stem cell groups tended to give greater mechanical strength as well as the expressed α-smooth muscle actin and vascular endothelial growth factor (VEGF) on immunohistochemistry. After bioprinting of esophagus-like scaffold-free tubular structures with demonstrated multicellular proportion, it was matured in bioreactor and then transplanted into rats as esophageal grafts. The esophageal grafts were implanted between the stomach and esophagus with a silicone tube. Results showed the grafts were maintained in vivo for 30 days, and the epithelium extended and covered the inner lumen and was able to pass food as well. The epithelialization of the inner surface of the esophageal lumen should be considered as the key regenerative factor because it must be done postoperatively with non-sterilized solid matter. In this respect, this research result has been promising as a potential substitute for esophageal transplantation using bioprinting.

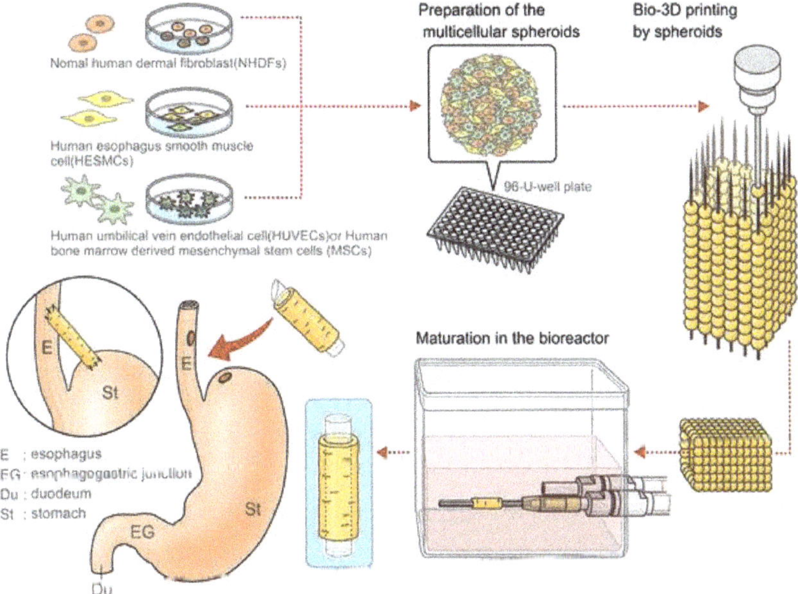

Figure 3. Schematic illustration of the esophageal tubular structure by kenzan method bioprinting [115].

In Gul Kim et al. employed both techniques of the supporting rod bioprinting and electrospinning to build the enhanced tubular structure with two-layer [15]. That study evaluated the 3D printed esophageal graft and the effect of bioreactor cultivation on cell maturity for muscle regeneration and epithelialization. To fabricate the tubular framework, the membrane was manufactured by electrospun polyurethane (PU) on the rotating rod (diameter: 2 mm), and then to improve the mechanical stability, the PCL strand was squeezed using the extrusion-based system. Cell-seeded (hMSCs) tubular frameworks were matured in the customized bioreactor system, and shear stress of the 0.1 dyne/cm^2 flow-induced with a pattern of 1 min/2 min for engagement/resting was applied to invigorate the frameworks. In comparison results of the histological analysis in the circumferential esophageal defects in a rat model from bioreactor cultivation and the omentum-cultured groups, both the groups showed over 80% mucosal regeneration without a fistula. The follow-up study by this research group suggested the further extended bioreactor culture system that could apply the different mechanical stimuli and biochemical reagents at the inner lumen and outside of the scaffold [16]. Among the mechanical stimuli, in particular, the intermittent shear flow by hydrostatic pressure and the shear stress by flow media were relevant to improving efficacy for differentiation of the epithelial and muscle lineage compared to steady shear flow.

Similarly, Eun-Jae Chung et al. utilized both supporting rod bioprinting and electrospinning and developed an esophageal scaffold reinforced by a 3D-printed PCL ring [14]. After bioprinting the reinforcing ring on the rod, a thin PCL layer was formed by electrospinning to form a nano-structured tubular structure. The printed tubular structures were wrapped into the omentum of rats for 2 weeks and then orthotopically transplanted for a circumferential esophageal defect. When macroscopically observed in the in vivo study, no fistulas, necrosis of the anastomosis, or abscess formation was found in the surrounding of the operating sections.

Maohua Lin et al. used direct bioprinting to a fabricated esophageal tubular stent with spiral patterns that applied the optimized design by computational simulation [130]. The printed esophageal tubular stent consisted of a mixed biodegradable polymer of medical-grade thermal polyurethane (TPU) and PLA in optimum proportion to achieve appropriate mechanical stiffness and flexibility. The group of the tubular stent with 10% PLA was investigated as a remarkable condition in the anti-migration force, self-expansion force, and human esophagus epithelial cell viability.

5.2. Blood Vessel

Blood vessels are components of the circulatory systems with hollow tube structures in the tissue and organs [131]. These transport blood cells, oxygen, and nutrients throughout the body and receive the CO_2 and waste from the metabolic activity of the peripheral cells and tissues. Blood vessels are divided into arteries, veins, and capillaries according to their structural characteristic and biological functions. In general, the artery and vein walls consist of three layers: tunica intima (squamous endothelium), tunica media (smooth muscle cells), and tunica adventitia (fibrous collagen) [132,133].

Generally, blood vessel disorder refers to the hardening, enlargement, and narrowing of arteries and veins. These health problems trigger arterial diseases, which can cause death, such as coronary artery heart disease, cardiovascular disease, peripheral artery disease. Worldwide annual mortalities related to cardiovascular disease are expected to rise to 23.3 million by 2030 [134]. To date, revascularization strategies have included the stent, surgical bypass grafting, and angioplasty. Also, commercialized off-the-shelf alternatives, such as polytetrafluoroethylene (PTFE), gore-tex, and dacron, are also being proved clinically effective when replacing large-diameter vessels ($\geq$6 mm). When using a small-diameter ($\leq$6 mm) vascular graft, however, it has caused a thrombosis event with the closing lumen and the lack of long-term patency as well as intimal hyperplasia. Considering the limitations of current vascular grafts, tissue-engineered vascular graft (TEVG) has been developed using 3D bioprinting technology. In particular, for the clinical applications of the TEVG, anti-thrombosis and long-term patency overcome the essential issues. Recently, to achieve this goal, several interesting studies have reported generating a tubular structure with a biochemical component capable of physiological remodeling (Table 3).

Gao et al. successfully fabricated the tubular bio-blood-vessel (BBV) with hybrid bioink (a mixture of VdECM and alginate) using the versatile 3D co-axial bioprinting method (Figure 4) [59]. The VdECM/alginate hybrid bioink containing the atorvastatin-loaded PLGA microspheres (APMS) and endothelial progenitor cells (EPCs) provides a favorable environment to promote the proliferation and neovascularization. When bioink encapsulating APMS/EPCs is used with a core–shell nozzle, the inner shell is filled with $CaCl_2$ solution (CPF127) for ionically crosslinking by releasing the calcium ion. The co-axial cell-printed tubular structure has been estimated in an ischemia model in nude mouse hind limb. It has induced an increased rate of neovascularization and the remarkable regeneration of ischemic limbs. The noteworthy point is that this research has achieved the creation of tubular structures with a broad range of diameters by controlling the core–shell nozzle. In addition, functional encapsulation of the cell/drug-laden bioink has shown potential for expansion as the printed BBV with the carrier that enables anti-thrombosis and long-term patency.

Table 3. The 3D bioprinting technique and biocomposite ink for the vascular tubular structure.

3D Bioprinting Technique	Biocomposite Ink	Reference
Co-axial bioprinting	Vascular-tissue-derived decellularized extracellular matrix (VdECM) with alginate	[59]
Rod supporting bioprinting	Fibrinogen and gelatin	[135]
Rod supporting bioprinting and electrospinning	Polycaprolactone (PCL)	[136]
Co-axial bioprinting and rod supporting bioprinting	Alginate	[31]
Support bath-based bioprinting	Alginate and gelatin slurry support bath	[120]
Co-axial printing and support bath-based bioprinting	Photocrosslinkable bioelastomer prepolymers ink (dimethyl itaconate. 1,8-ictanediol and triethyl citrate) and carbomer gel bath	[137]
Direct bioprinting	Pluronic 127 and gelatin methacrylate (GelMA)	[138]

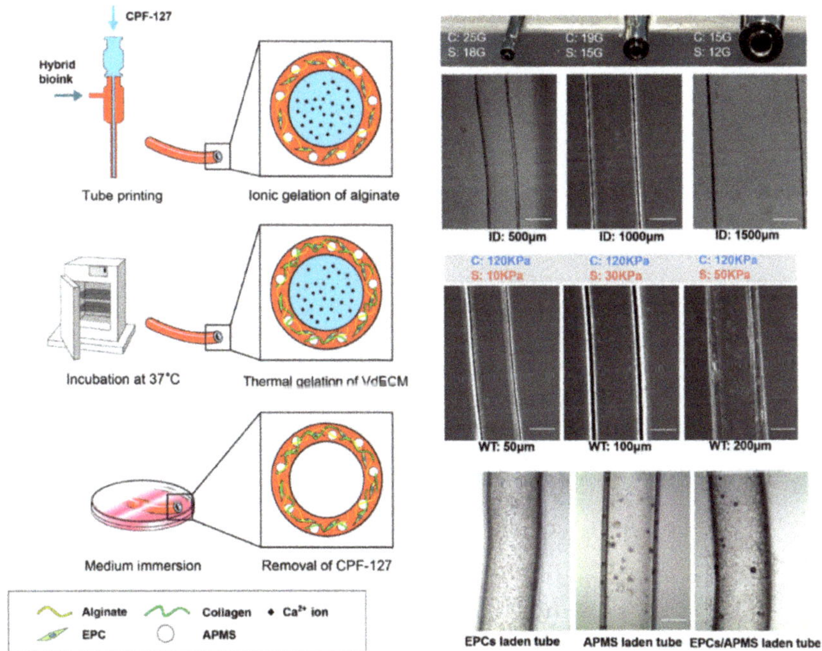

Figure 4. Schematic illustration and structural images of the vascular tubular structure by co-axial bioprinting [59].

Sebastian Freeman et al. developed fibrin-based vascular constructs using rod supporting bioprinting [135]. The printable bioink consists of the fibrinogen with gelatin to achieve the desired shear-thinning property for self-standing. Unprintable fibrinogen was used as a printable biomaterial by blending the favorable rheological properties with heat-treated gelatin. During two months of the cultures after bioprinting, the burst pressure of the tubular structure reached 1110 mm Hg, and the remarkable improvement of the tensile mechanical properties was achieved in both the circumferential and axial elastic moduli.

Sang Jin Lee et al. combined rod supporting bioprinting and electrospinning for mechanical robustness and to build multi-layered structures using synthetic polymers (e.g., PCL) [136]. To induce potent angiogenic activity, the printed tubular structure was coated with polydopamine (PDA) and vascular endothelial growth factor (VEGF) on the surface. The coated-PDA layer enhanced the ability of the hydrophilicity; it also remarkably increased the vascular cell proliferation and angiogenic differentiation during in vivo/in vitro.

Quing Gao et al. fabricated hydrogel-based perfusable vascular structure with multilevel fluidic channels in tubular tissue approaches by using the 3D bioprinting that combined with co-axial and rod supporting bioprinting [31]. Partially cross-linked hollow alginate containing the fibroblasts and smooth muscle cells were extruded through a two co-axial nozzle and then printed along with a rotating supporting rod. The printed tubular structure exhibited a sufficiently strong mechanical strength (ultimate strength: 0.148 MPa) for the implantation due to the fusion of adjacent crosslinking reaction. Encapsulation of the fibroblast in the tubular structure showed over 90% survival within 1 week in vivo. This research has shown the ability to directly fabricating a perfusable vessel-like structure by cell-laden biomaterials through a coupled co-axial bioprinting and rod supporting method.

Thomas J. Hinton et al. 3D bioprinted a more complex structure than the perfusable arterial tree with alginate bioink and embedded it in the gelatin slurry support bath using a support bath-based bioprinting technology [120]. Perfusion structures mimicking a portion of the right arterial tree obtained through MRI data were printed in multiple branches with 3D tortuosity. Houman Savoji et al., similarly, printed vascular tubes (using core–shell nozzle) via freeform reversible embedding of photocrosslinkable bioelastomer prepolymers within a carbomer hydrogel bath by co-axial bioprinting and support bath-based bioprinting [137]. This tubular tissue-engineered approach to create a further advanced tubular structure made the significant achievement of mechanical robustness and recreated complex 3D anatomical architectures.

Kolesky et al. fabricated embedded vasculature constructs, repleted with multiple types of cells and an extracellular matrix (ECM), using direct bioprinting [138]. An aqueous fugitive ink composed of pluronic 127 was used for easy printing and removing under mild conditions to create vascular channels. In addition, gelatin methacrylate (GelMA) was used as a bulk matrix and cell carrier. After infilling and photopolymerizing the GelMA matrix on the fugitive pluronic 127 ink, the fugitive ink was removed by cooling the printed constructs below 4 °C, yielding open channels to fabricate the embedded vasculature constructs. Using this 3D bioprinting process, the potential of 3D vascular embedded constructs with human neonatal dermal fibroblasts (HNDFs) and human umbilical vein endothelial cells (HUVECs) was convincingly demonstrated. This result showed the development possibility for remodeling heterogeneous tissue constructs containing vasculature and multiple cell types.

5.3. Trachea

The trachea is a tubular structure in which the lower respiratory tract begins and refers to the pathway that begins immediately running between the larynx and the bronchi. The trachea is a composite tubular structure consisting of epithelium, basement membrane, connective tissue, smooth muscle, and cartilaginous layer. The tubular shape is about 2 cm in diameter and 11 cm in length with a flat posterior. The trachea acts as an airway to enter and exit the air during respiration. In addition, when debris, such as dust, enters into the trachea with air, it functions to move and remove the debris using ciliary movement and mucus [139,140].

Tracheas are becoming increasingly damaged due to severe environmental pollution. In addition, damage to the trachea has become a serious problem due to the increased use of ventilators for the treatment of patients [141]. In order to solve this problem, the transplantation of donor tissue from a deceased person to an injured organ has been reported. However, not only is it difficult to obtain donor tissue but even if it is obtained, there is a disadvantage that it can be transplanted through a complex pretreatment process over a long period of time [142]. The method of treating organ damage depends on the extent and length of the involvement of the site of injury. End-to-end anastomosis is a common treatment for circumferential injuries. However, end-to-end anastomosis has disadvantages, such as continuous endotracheal intubation, rupture, or stenosis of anastomosis after surgery. In addition, end-to-end anastomosis cannot be applied if more than 50% of the trachea needs to be excised. Such cases are difficult to treat clinically [143,144]. Tissue engineering is an appropriate approach to solve these problems; in addition, recent advances in 3D bioprinting technology have enabled the production of more sophisticated and systematic artificial structures (Table 4) [145,146].

Table 4. The 3D bioprinting techniques and biocomposite inks for the tracheal tubular structure.

3D Bioprinting Technique	Biocomposite Ink	Reference
Kenzan method bioprinting	Cell spheroids with chondrocytes, endothelial cells, and mesenchymal stem cells	[147]
Direct bioprinting and rod-supporting bioprinting	Polycaprolactone (PCL), silicone	[148]
Direct bioprinting	Polyurethane (PU)	[149]
Direct bioprinting	Polycaprolactone (PCL)	[150]

Daisuke Taniguchi et al. developed an artificial trachea using kenzan method bioprinting [147]. They assessed the circumferential tracheal replacement using scaffold-free trachea-like grafts generated from spheroids consisting of several types of cells—chondrocytes, endothelial cells, and mesenchymal stem cells—to build 3D structures. This artificial trachea from spheroids was matured in a bioreactor and transplanted into a rat. In the transplantation, they used silicone stents to prevent collapse. As a result, chondrogenesis and vasculogenesis could be observed in this artificial trachea.

Manchen Gao et al. printed a biodegradable reticular PCL scaffold with similar morphology to the rabbits' native trachea by direct bioprinting [148]. Chondrocytes were cultured in this 3D scaffold and conducted into the subcutaneous of nude mice. The scaffold showed the successful reconstruction and the proper supporting force to maintain the lumen as well as presenting remarkable cartilaginous properties both in vitro and in vivo.

Cheng-Tien Hsieh et al. fabricated a tissue-engineered trachea with structural similarity to the native trachea from water-based biocomposite ink at low temperature using direct bioprinting [149].

In that research, two kinds of water-based biodegradable polyurethanes with different physicochemical properties were used as biocomposite ink. The human MSCs were seeded into this tracheal construct, and then the construct was implanted in nude mice. After 6 weeks, the results showed dynamic compression moduli of the scaffolds that were 0.3–0.8 MPa under the force of 0.1–0.8 N, which was similar to the native trachea. It also confirmed gas-tightness by airflow test at positive and negative air pressures. Moreover, MSCs seeded in the tracheal scaffolds were grown into cartilage-like tissue. It expressed chondrogenic potential and secreted glycosaminoglycans (GAGs) and collagen after 14 days in vitro culture without any exogenous growth factors, such as bioactive factors or small molecular drugs. They showed that the tracheal scaffold of biocomposite inks and 3D bioprinting techniques might be used to fabricate personalized artificial tracheas for clinical applications. It also showed the possibility of incorporating exogenous growth factors in the water-based biocomposite ink to enhance the chondrogenesis of the MSCs.

Jae Yeon Lee et al. developed an artificial tracheal structure PCL framework by extrusion bioprinting and silicone band by direct bioprinting and rod-supporting bioprinting (Figure 5) [150]. In particular, the states of the PCL extrusion were precisely controlled to create dotted circular patterns so that the bellows framework had about 300 μm pores in the wall except for groove parts. Then, they used a rod supporting bioprinting to print ring-shaped bands into the outer grooves of the PCL framework using a medical-grade silicone elastomer. The PCL framework was put around the rotating rods and rotated. They proved the potential of this artificial scaffold to be applied immediately in emergencies.

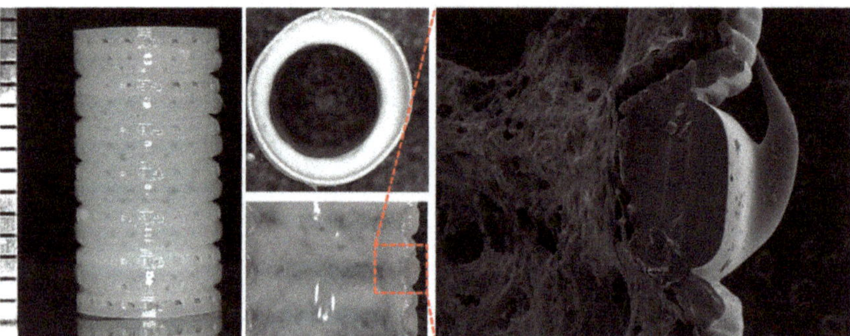

Figure 5. Structural images of the tracheal tubular structure by direct and rod-supporting bioprinting [150].

6. Future Perspectives and Concluding Remark

The biofabricated tissue-engineered tubular constructs require particular features of target-specific mechanical properties, anatomical accuracy, autoimmune acceptance, long-term patency, and similar cell arrangement for creating and mimicking of native tissues. This represents a significant technical challenge, and, to date, clinically meaningful tubular structure bioprinting approaches have been reported, utilizing versatile additive manufacturing techniques and biocomposite inks.

Various 3D bioprinting methodologies have emerged in the field of tissue engineering, and advanced technologies have been secured through rapid technological developments. Among them, the extrusion-based bioprinting system has been actively used to fabricate tubular structures with advantages of ease construction and flexibility in the use of various biomaterials. The fabrication of multi-layered tubular structures using supporting rods has been actively used, and approaches with more histologically close multi-cellular components using cell-spheroids have been developed. Co-axial bioprinting has also established itself as a promising approach that allows easy fabrication of freely adjustable perfusable tubular structures. In addition, this technology can be loaded with a variety of functional drugs, as well as cells that help bioenvironment cues, which are used in expandability platforms, such as vascular-embedded organoids and drug-screening devices. Support bath-based bioprinting has the advantage of being able to produce free- and multi-branched self-supporting forms using hydrate materials.

Despite the breakthrough in 3D bioprinting technology, artificial tubular structures of the esophagus, blood vessels, and trachea still face challenges for application as clinical substitutes [151–153]. This is because these tissues are exposed to high clinical needs, such as contraction, expansion by peristalsis, and blood pressure. In particular, the esophagus and trachea inevitably contact with external contaminants, such as liquid, food, and air, after insertion of the artificial tubular structure, thus hindering the growth of the functional endothelial layer. In addition, in the case of blood vessels, the absence of the endothelium layer may induce thrombus and stenosis. Therefore, pre-maturation, such as omentum culture, of bioreactor is considered a significant factor in the growth of artificial tissues before surgical approaches. In addition, many researchers are working to rebuild tubular organs, such as the stomach, intestine [154], bladder [9], and urethra [122], as well as the mentioned ones in this review. Thus, future developments of artificial tubular tissue should simultaneously entail the promising benefits provided by 3D bioprinting as well as the development of functional biocomposite inks and optimal cell culture techniques for target-tissues.

Author Contributions: Investigation, H.-J.J. and H.N.; writing—original draft preparation, H.-J.J. and H.N.; writing—review and editing, H.-J.J. and H.N.; supervision, J.J. and S.-J.L.; All authors have read and agreed to the published version of the manuscript.

Funding: This research was supported by National Research Foundation of Korea (NRF) grant funded by the Ministry of Education, Science, and Technology (NRF 2016R1D1A1B01006658) and by Basic Science Research Program through the National Research Foundation of Korea (NRF) funded by the Ministry of Education (2015R1A6A3A04059015) and by the MSI T(Ministry of Science and ICT), Korea, under the ICT Consilience Creative program (IITP-2019-2011-1-00783) supervised by the IITP (Institute for Information and communications Technology Planning and Evaluation).

Conflicts of Interest: The authors declare no conflict of interest.

References

1. Saksena, R.; Gao, C.Y.; Wicox, M.; de Mel, A. Tubular organ epithelialisation. *J. Tissue Eng.* **2016**, *7*, 2041731416683950. [CrossRef]

2. Alcantara, P.S.; Spencer-Netto, F.A.; Silva-Junior, J.F.; Soares, L.A.; Pollara, W.M.; Bevilacqua, R.G. Gastro-esophageal isoperistaltic bypass in the palliation of irresectable thoracic esophageal cancer. *Int. Surg.* **1997**, *82*, 249–253.

3. Badylak, S.F.; Vorp, D.A.; Spievack, A.R.; Simmons-Byrd, A.; Hanke, J.; Freytes, D.O.; Thapa, A.; Gilbert, T.W.; Nieponice, A. Esophageal reconstruction with ECM and muscle tissue in a dog model. *J. Surg. Res.* **2005**, *128*, 87–97. [CrossRef]

4. Gawad, K.A.; Hosch, S.B.; Bumann, D.; Lubeck, M.; Moneke, L.C.; Bloechle, C.; Knoefel, W.T.; Busch, C.; Kuchler, T.; Izbicki, J.R. How important is the route of reconstruction after esophagectomy: A prospective randomized study. *Am. J. Gastroenterol.* **1999**, *94*, 1490–1496. [CrossRef]

5. Galliger, Z.; Vogt, C.D.; Panoskaltsis-Mortari, A. 3D bioprinting for lungs and hollow organs. *Transl. Res.* **2019**, *211*, 19–34. [CrossRef]

6. Gora, A.; Pliszka, D.; Mukherjee, S.; Ramakrishna, S. Tubular Tissues and Organs of Human Body-Challenges in Regenerative Medicine. *J. Nanosci. Nanotechnol.* **2016**, *16*, 19–39. [CrossRef]

7. Gangemi, A.; Tzvetanov, I.G.; Beatty, E.; Oberholzer, J.; Testa, G.; Sankary, H.N.; Kaplan, B.; Benedetti, E. Lessons Learned in Pediatric Small Bowel and Liver Transplantation From Living-Related Donors. *Transplantation* **2009**, *87*, 1027–1030. [CrossRef]

8. Napier, K.J.; Scheerer, M.; Misra, S. Esophageal cancer: A Review of epidemiology, pathogenesis, staging workup and treatment modalities. *World J. Gastrointest. Oncol.* **2014**, *6*, 112–120. [CrossRef]

9. Atala, A. Tissue engineering of human bladder. *Br. Med. Bull.* **2011**, *97*, 81–104. [CrossRef]

10. Murphy, S.V.; Atala, A. 3D bioprinting of tissues and organs. *Nat. Biotechnol.* **2014**, *32*, 773–785. [CrossRef]

11. Kim, J.E.; Kim, S.H.; Jung, Y. Current status of three-dimensional printing inks for soft tissue regeneration. *Tissue Eng. Regen. Med.* **2016**, *13*, 636–646. [CrossRef] [PubMed]

12. Pati, F.; Jang, J.; Ha, D.H.; Won Kim, S.; Rhie, J.W.; Shim, J.H.; Kim, D.H.; Cho, D.W. Printing three-dimensional tissue analogues with decellularized extracellular matrix bioink. *Nat. Commun.* **2014**, *5*, 3935. [CrossRef] [PubMed]

13. Jang, J.; Park, H.J.; Kim, S.W.; Kim, H.; Park, J.Y.; Na, S.J.; Kim, H.J.; Park, M.N.; Choi, S.H.; Park, S.H.; et al. 3D printed complex tissue construct using stem cell-laden decellularized extracellular matrix bioinks for cardiac repair. *Biomaterials* **2017**, *112*, 264–274. [CrossRef] [PubMed]

14. Chung, E.J.; Ju, H.W.; Yeon, Y.K.; Lee, J.S.; Lee, Y.J.; Seo, Y.B.; Hum, P.C. Development of an omentum-cultured oesophageal scaffold reinforced by a 3D-printed ring: Feasibility of an in vivo bioreactor. *Artif. Cell Nanomed. Biotechnol.* **2018**, *46*, S885–S895. [CrossRef]

15. Kim, I.G.; Wu, Y.; Park, S.A.; Cho, H.; Choi, J.J.; Kwon, S.K.; Shin, J.W.; Chung, E.J. Tissue-Engineered Esophagus via Bioreactor Cultivation for Circumferential Esophageal Reconstruction. *Tissue Eng. Part A* **2019**, *25*, 1478–1492. [CrossRef]

16. Wu, Y.; Kang, Y.G.; Kim, I.G.; Kim, J.E.; Lee, E.J.; Chung, E.J.; Shin, J.W. Mechanical stimuli enhance simultaneous differentiation into oesophageal cell lineages in a double-layered tubular scaffold. *J. Tissue Eng. Regen. Med.* **2019**, *13*, 1394–1405. [CrossRef]

17. Zhu, Y.B.; Leong, M.F.; Ong, W.F.; Chan-Park, M.B.; Chian, K.S. Esophageal epithelium regeneration on fibronectin grafted poly(L-lactide-co-caprolactone) (PLLC) nanofiber scaffold. *Biomaterials* **2007**, *28*, 861–868. [CrossRef]

18. Kuppan, P.; Sethuraman, S.; Krishnan, U.M. PCL and PCL-Gelatin Nanofibers as Esophageal Tissue Scaffolds: Optimization, Characterization and Cell-Matrix Interactions. *J. Biomed. Nanotechnol.* **2013**, *9*, 1540–1555. [CrossRef]

19. Soliman, S.; Laurent, J.; Kalenjian, L.; Burnette, K.; Hedberg, B.; La Francesca, S. A multilayer scaffold design with spatial arrangement of cells to modulate esophageal tissue growth. *J. Biomed. Mater. Res. B* **2019**, *107*, 324–331. [CrossRef]

20. L'Heureux, N.; Paquet, S.; Labbe, R.; Germain, L.; Auger, F.A. A completely biological tissue-engineered human blood vessel. *FASEB J.* **1998**, *12*, 47–56.

21. Masuda, S.; Shimizu, T.; Yamato, M.; Okano, T. Cell sheet engineering for heart tissue repair. *Adv. Drug Deliv. Rev.* **2008**, *60*, 277–285. [CrossRef]

22. Geng, W.X.; Ma, D.Y.; Yan, X.R.; Liu, L.Q.; Cui, J.H.; Xie, X.; Li, H.M.; Chen, F.L. Engineering tubular bone using mesenchymal stem cell sheets and coral particles. *Biochem. Biophys. Res. Commun.* **2013**, *433*, 595–601. [CrossRef]

23. Weinberg, C.B.; Bell, E. A blood vessel model constructed from collagen and cultured vascular cells. *Science* **1986**, *231*, 397–400. [CrossRef]

24. Carvalho, C.R.; Costa, J.B.; da Silva Morais, A.; Lopez-Cebral, R.; Silva-Correia, J.; Reis, R.L.; Oliveira, J.M. Tunable Enzymatically Cross-Linked Silk Fibroin Tubular Conduits for Guided Tissue Regeneration. *Adv. Healthc. Mater.* **2018**, *7*, e1800186. [CrossRef]

25. Schutte, S.C.; Chen, Z.; Brockbank, K.G.; Nerem, R.M. Cyclic strain improves strength and function of a collagen-based tissue-engineered vascular media. *Tissue Eng. Part A* **2010**, *16*, 3149–3157. [CrossRef]

26. Boccafoschi, F.; Rajan, N.; Habermehl, J.; Mantovani, D. Preparation and characterization of a scaffold for vascular tissue engineering by direct-assembling of collagen and cells in a cylindrical geometry. *Macromol. Biosci.* **2007**, *7*, 719–726. [CrossRef]

27. Park, J.H.; Jang, J.; Lee, J.S.; Cho, D.W. Current advances in three-dimensional tissue/organ printing. *Tissue Eng. Regen. Med.* **2016**, *13*, 612–621. [CrossRef]

28. Jang, J.; Yi, H.G.; Cho, D.W. 3D Printed Tissue Models: Present and Future. *ACS Biomater. Sci. Eng.* **2016**, *2*, 1722–1731. [CrossRef]

29. Park, J.Y.; Jang, J.; Kang, H.W. 3D Bioprinting and its application to organ-on-a-chip. *Microelectron. Eng.* **2018**, *200*, 1–11. [CrossRef]

30. Gao, G.; Kim, B.S.; Jang, J.; Cho, D.W. Recent Strategies in Extrusion-Based Three-Dimensional Cell Printing toward Organ Biofabrication. *ACS Biomater. Sci. Eng.* **2019**, *5*, 1150–1169. [CrossRef]

31. Gao, Q.; Liu, Z.J.; Lin, Z.W.; Qiu, J.J.; Liu, Y.; Liu, A.; Wang, Y.D.; Xiang, M.X.; Chen, B.; Fu, J.Z.; et al. 3D Bioprinting of Vessel-like Structures with Multilevel Fluidic Channels. *ACS Biomater. Sci. Eng.* **2017**, *3*, 399–408. [CrossRef]

32. Jang, J. 3D Bioprinting and In Vitro Cardiovascular Tissue Modeling. *Bioengineering* **2017**, *4*, 71. [CrossRef]

33. Shim, J.H.; Lee, J.S.; Kim, J.Y.; Cho, D.W. Bioprinting of a mechanically enhanced three-dimensional dual cell-laden construct for osteochondral tissue engineering using a multi-head tissue/organ building system. *J. Micromech. Microeng.* **2012**, *22*, 08501410. [CrossRef]

34. Chang, C.C.; Boland, E.D.; Williams, S.K.; Hoying, J.B. Direct-write bioprinting three-dimensional biohybrid systems for future regenerative therapies. *J. Biomed. Mater. Res. B* **2011**, *98*, 160–170. [CrossRef]

35. Shor, L.; Guceri, S.; Chang, R.; Gordon, J.; Kang, Q.; Hartsock, L.; An, Y.H.; Sun, W. Precision extruding deposition (PED) fabrication of polycaprolactone (PCL) scaffolds for bone tissue engineering. *Biofabrication* **2009**, *1*, 015003. [CrossRef]

36. Neufurth, M.; Wang, X.H.; Wang, S.F.; Steffen, R.; Ackermann, M.; Haep, N.D.; Schroder, H.C.; Muller, W.E.G. 3D printing of hybrid biomaterials for bone tissue engineering: Calcium-polyphosphate microparticles encapsulated by polycaprolactone. *Acta Biomater.* **2017**, *64*, 377–388. [CrossRef]

37. Jeong, H.J.; Gwak, S.J.; Kang, N.U.; Hong, M.W.; Kim, Y.Y.; Cho, Y.S.; Lee, S.J. Bioreactor mimicking knee-joint movement for the regeneration of tissue-engineered cartilage. *J. Mech. Sci. Technol.* **2019**, *33*, 1841–1850. [CrossRef]

38. Mironov, A.V.; Grigoryev, A.M.; Krotova, L.I.; Skaletsky, N.N.; Popov, V.K.; Sevastianov, V.I. 3D printing of PLGA scaffolds for tissue engineering. *J. Biomed. Mater. Res. Part A* **2017**, *105*, 104–109. [CrossRef]

39. La, W.G.; Jang, J.; Kim, B.S.; Lee, M.S.; Cho, D.W.; Yang, H.S. Systemically replicated organic and inorganic bony microenvironment for new bone formation generated by a 3D printing technology. *RSC Adv.* **2016**, *6*, 11546–11553. [CrossRef]

40. Serra, T.; Planell, J.A.; Navarro, M. High-resolution PLA-based composite scaffolds via 3-D printing technology. *Acta Biomater.* **2013**, *9*, 5521–5530. [CrossRef]

41. Song, Y.; Li, Y.; Song, W.; Yee, K.; Lee, K.Y.; Tagarielli, V.L. Measurements of the mechanical response of unidirectional 3D-printed PLA. *Mater. Des.* **2017**, *123*, 154–164. [CrossRef]

42. Li, K.; Liu, J.K.; Chen, W.S.; Zhang, L. Controllable printing droplets on demand by piezoelectric inkjet: Applications and methods. *Microsyst. Technol.* **2018**, *24*, 879–889. [CrossRef]

43. Derby, B. Bioprinting: Inkjet printing proteins and hybrid cell-containing materials and structures. *J. Mater. Chem.* **2008**, *18*, 5717–5721. [CrossRef]

44. Odde, D.J.; Renn, M.J. Laser-guided direct writing for applications in biotechnology. *Trends Biotechnol.* **1999**, *17*, 385–389. [CrossRef]

45. Michael, S.; Sorg, H.; Peck, C.T.; Koch, L.; Deiwick, A.; Chichkov, B.; Vogt, P.M.; Reimers, K. Tissue engineered skin substitutes created by laser-assisted bioprinting form skin-like structures in the dorsal skin fold chamber in mice. *PLoS ONE* **2013**, *8*, e57741. [CrossRef]

46. Schiele, N.R.; Corr, D.T.; Huang, Y.; Raof, N.A.; Xie, Y.; Chrisey, D.B. Laser-based direct-write techniques for cell printing. *Biofabrication* **2010**, *2*, 032001. [CrossRef]

47. Wust, S.; Muller, R.; Hofmann, S. Controlled Positioning of Cells in Biomaterials-Approaches Towards 3D Tissue Printing. *J. Funct. Biomater.* **2011**, *2*, 119–154. [CrossRef]

48. Melchels, F.P.; Feijen, J.; Grijpma, D.W. A review on stereolithography and its applications in biomedical engineering. *Biomaterials* **2010**, *31*, 6121–6130. [CrossRef]

49. Wang, Z.J.; Abdulla, R.; Parker, B.; Samanipour, R.; Ghosh, S.; Kim, K. A simple and high-resolution stereolithography-based 3D bioprinting system using visible light crosslinkable bioinks. *Biofabrication* **2015**, *7*, 045009. [CrossRef]

50. Heller, C.; Schwentenwein, M.; Russmueller, G.; Varga, F.; Stampfl, J.; Liska, R. Vinyl Esters: Low Cytotoxicity Monomers for the Fabrication of Biocompatible 3D Scaffolds by Lithography Based Additive Manufacturing. *J. Polym. Sci. Polym. Chem.* **2009**, *47*, 6941–6954. [CrossRef]

51. Lee, K.W.; Wang, S.F.; Fox, B.C.; Ritman, E.L.; Yaszemski, M.J.; Lu, L.C. Poly(propylene fumarate) bone tissue engineering scaffold fabrication using stereolithography: Effects of resin formulations and laser parameters. *Biomacromolecules* **2007**, *8*, 1077–1084. [CrossRef] [PubMed]

52. Hunt, N.C.; Grover, L.M. Cell encapsulation using biopolymer gels for regenerative medicine. *Biotechnol. Lett.* **2010**, *32*, 733–742. [CrossRef] [PubMed]

53. Sell, S.A.; Wolfe, P.S.; Garg, K.; McCool, J.M.; Rodriguez, I.A.; Bowlin, G.L. The Use of Natural Polymers in Tissue Engineering: A Focus on Electrospun Extracellular Matrix Analogues. *Polymers* **2010**, *2*, 522–553. [CrossRef]

54. Malafaya, P.B.; Silva, G.A.; Rcis, R.L. Natural-origin polymers as carriers and scaffolds for biomolecules and cell delivery in tissue engineering applications. *Adv. Drug Deliv. Rev.* **2007**, *59*, 207–233. [CrossRef] [PubMed]

55. Milazzo, M.; Negrini, N.C.; Scialla, S.; Marelli, B.; Fare, S.; Danti, S.; Buehler, M. Additive Manufacturing Approaches for Hydroxyapatite-Reinforced Composites. *Adv. Funct. Mater.* **2019**, *29*, 1903055. [CrossRef]

56. Sabir, M.I.; Xu, X.X.; Li, L. A review on biodegradable polymeric materials for bone tissue engineering applications. *J. Mater. Sci.* **2009**, *44*, 5713–5724. [CrossRef]

57. Wu, W.; DeConinck, A.; Lewis, J.A. Omnidirectional Printing of 3D Microvascular Networks. *Adv. Mater.* **2011**, *23*, H178–H183. [CrossRef]

58. Kim, B.S.; Jang, J.; Chae, S.; Gao, G.; Kong, J.S.; Ahn, M.; Cho, D.W. Three-dimensional bioprinting of cell-laden constructs with polycaprolactone protective layers for using various thermoplastic polymers. *Biofabrication* **2016**, *8*, 035013. [CrossRef]

59. Gao, G.; Lee, J.H.; Jang, J.; Lee, D.H.; Kong, J.S.; Kim, B.S.; Choi, Y.J.; Jang, W.B.; Hong, Y.J.; Kwon, S.M.; et al. Tissue Engineered Bio-Blood-Vessels Constructed Using a Tissue-Specific Bioink and 3D Coaxial Cell Printing Technique: A Novel Therapy for Ischemic Disease. *Adv. Funct. Mater.* **2017**, *27*, 1700798. [CrossRef]

60. Ma, L.; Gao, C.Y.; Mao, Z.W.; Zhou, J.; Shen, J.C.; Hu, X.Q.; Han, C.M. Collagen/chitosan porous scaffolds with improved biostability for skin tissue engineering. *Biomaterials* **2003**, *24*, 4833–4841. [CrossRef]

61. Skardal, A.; Mack, D.; Kapetanovic, E.; Atala, A.; Jackson, J.D.; Yoo, J.; Soker, S. Bioprinted Amniotic Fluid-Derived Stem Cells Accelerate Healing of Large Skin Wounds. *Stem Cell Transl. Med.* **2012**, *1*, 792–802. [CrossRef] [PubMed]

62. Dalgleish, R. The human type I collagen mutation database. *Nucleic Acids Res.* **1997**, *25*, 181–187. [CrossRef] [PubMed]

63. Cummings, C.L.; Gawlitta, D.; Nerem, R.M.; Stegemann, J.P. Properties of engineered vascular constructs made from collagen, fibrin, and collagen-fibrin mixtures. *Biomaterials* **2004**, *25*, 3699–3706. [CrossRef]

64. Han, C.M.; Zhang, L.P.; Sun, J.Z.; Shi, H.F.; Zhou, J.; Gao, C.Y. Application of collagen-chitosan/fibrin glue asymmetric scaffolds in skin tissue engineering. *J. Zhejiang Univ.-Sci. B* **2010**, *11*, 524–530. [CrossRef] [PubMed]

65. Kim, G.; Ahn, S.; Kim, Y.; Cho, Y.; Chun, W. Coaxial structured collagen-alginate scaffolds: Fabrication, physical properties, and biomedical application for skin tissue regeneration. *J. Mater. Chem.* **2011**, *21*, 6165–6172. [CrossRef]

66. Ulrich, T.A.; Jain, A.; Tanner, K.; MacKay, J.L.; Kumar, S. Probing cellular mechanobiology in three-dimensional culture with collagen-agarose matrices. *Biomaterials* **2010**, *31*, 1875–1884. [CrossRef] [PubMed]

67. Shin, J.Y.; Jeong, S.J.; Lee, W.K. Fabrication of porous scaffold by ternary combination of chitosan, gelatin, and calcium phosphate for tissue engineering. *J. Ind. Eng. Chem.* **2019**, *80*, 862–869. [CrossRef]

68. Wang, X.H.; Yan, Y.N.; Pan, Y.Q.; Xiong, Z.; Liu, H.X.; Cheng, B.; Liu, F.; Lin, F.; Wu, R.D.; Zhang, R.J.; et al. Generation of three-dimensional hepatocyte/gelatin structures with rapid prototyping system. *Tissue Eng.* **2006**, *12*, 83–90. [CrossRef]

69. Dulnik, J.; Denis, P.; Sajkiewicz, P.; Kolbuk, D.; Choinska, E. Biodegradation of bicomponent PCL/gelatin and PCL/collagen nanofibers electrospun from alternative solvent system. *Polym. Degrad. Stab.* **2016**, *130*, 10–21. [CrossRef]

70. Fu, W.; Liu, Z.L.; Feng, B.; Hu, R.J.; He, X.M.; Wang, H.; Yin, M.; Huang, H.M.; Zhang, H.B.; Wang, W. Electrospun gelatin/PCL and collagen/PLCL scaffolds for vascular tissue engineering. *Int. J. Nanomed.* **2014**, *9*, 2335–2344. [CrossRef]

71. Roehm, K.D.; Madihally, S.V. Bioprinted chitosan-gelatin thermosensitive hydrogels using an inexpensive 3D printer. *Biofabrication* **2018**, *10*, 015002. [CrossRef] [PubMed]

72. Shin, J.; Kang, H.W. The Development of Gelatin-Based Bio-Ink for Use in 3D Hybrid Bioprinting. *Int. J. Precis. Eng. Manuf.* **2018**, *19*, 767–771. [CrossRef]

73. Sharma, R.; Smits, I.P.M.; De La Vega, L.; Lee, C.; Willerth, S.M. 3D Bioprinting Pluripotent Stem Cell Derived Neural Tissues Using a Novel Fibrin Bioink Containing Drug Releasing Microspheres. *Front. Bioeng. Biotechnol.* **2020**, *8*, 57. [CrossRef] [PubMed]

74. Ouyang, L.; Yao, R.; Zhao, Y.; Sun, W. Effect of bioink properties on printability and cell viability for 3D bioplotting of embryonic stem cells. *Biofabrication* **2016**, *8*, 035020. [CrossRef] [PubMed]

75. Ouyang, L.; Yao, R.; Mao, S.; Chen, X.; Na, J.; Sun, W. Three-dimensional bioprinting of embryonic stem cells directs highly uniform embryoid body formation. *Biofabrication* **2015**, *7*, 044101. [CrossRef] [PubMed]

76. Das, S.; Pati, F.; Choi, Y.J.; Rijal, G.; Shim, J.H.; Kim, S.W.; Ray, A.R.; Cho, D.W.; Ghosh, S. Bioprintable, cell-laden silk fibroin-gelatin hydrogel supporting multilineage differentiation of stem cells for fabrication of three-dimensional tissue constructs. *Acta Biomater.* **2015**, *11*, 233–246. [CrossRef]

77. Xiong, S.; Zhang, X.; Lu, P.; Wu, Y.; Wang, Q.; Sun, H.; Heng, B.C.; Bunpetch, V.; Zhang, S.; Ouyang, H. A Gelatin-sulfonated Silk Composite Scaffold based on 3D Printing Technology Enhances Skin Regeneration by Stimulating Epidermal Growth and Dermal Neovascularization. *Sci. Rep.* **2017**, *7*, 4288. [CrossRef]

78. Gauvin, R.; Chen, Y.C.; Lee, J.W.; Soman, P.; Zorlutuna, P.; Nichol, J.W.; Bae, H.; Chen, S.C.; Khademhosseini, A. Microfabrication of complex porous tissue engineering scaffolds using 3D projection stereolithography. *Biomaterials* **2012**, *33*, 3824–3834. [CrossRef]

79. Kim, B.S.; Kim, H.; Gao, G.; Jang, J.; Cho, D.W. Decellularized extracellular matrix: A step towards the next generation source for bioink manufacturing. *Biofabrication* **2017**, *9*, 034104. [CrossRef]

80. Kim, B.S.; Gao, G.; Kim, J.Y.; Cho, D.W. 3D Cell Printing of Perfusable Vascularized Human Skin Equivalent Composed of Epidermis, Dermis, and Hypodermis for Better Structural Recapitulation of Native Skin. *Adv. Healthc. Mater.* **2019**, *8*, 1801019. [CrossRef]

81. Lee, H.; Han, W.; Kim, H.; Ha, D.H.; Jang, J.; Kim, B.S.; Cho, D.W. Development of Liver Decellularized Extracellular Matrix Bioink for Three-Dimensional Cell Printing-Based Liver Tissue Engineering. *Biomacromolecules* **2017**, *18*, 1229–1237. [CrossRef] [PubMed]

82. Ali, M.; Kumar, P.R.A.; Yoo, J.J.; Zahran, F.; Atala, A.; Lee, S.J. A Photo-Crosslinkable Kidney ECM-Derived Bioink Accelerates Renal Tissue Formation. *Adv. Healthc. Mater.* **2019**, *8*, 1800992. [CrossRef]

83. Jang, J.; Park, J.Y.; Gao, G.; Cho, D.W. Biomaterials-based 3D cell printing for next-generation therapeutics and diagnostics. *Biomaterials* **2018**, *156*, 88–106. [CrossRef] [PubMed]

84. Colosi, C.; Shin, S.R.; Manoharan, V.; Massa, S.; Costantini, M.; Barbetta, A.; Dokmeci, M.R.; Dentini, M.; Khademhosseini, A. Microfluidic Bioprinting of Heterogeneous 3D Tissue Constructs Using Low-Viscosity Bioink. *Adv. Mater.* **2016**, *28*, 677–684. [CrossRef] [PubMed]

85. Aderibigbe, B.A.; Buyana, B. Alginate in Wound Dressings. *Pharmaceutics* **2018**, *10*, 42. [CrossRef]

86. Cunniffe, G.M.; Gonzalez-Fernandez, T.; Daly, A.; Sathy, B.N.; Jeon, O.; Alsberg, E.; Kelly, D.J. Three-Dimensional Bioprinting of Polycaprolactone Reinforced Gene Activated Bioinks for Bone Tissue Engineering. *Tissue Eng. Part A* **2017**, *23*, 891–900. [CrossRef]

87. Narayanan, L.K.; Huebner, P.; Fisher, M.B.; Spang, J.T.; Starly, B.; Shirwaiker, R.A. 3D-Bioprinting of Polylactic Acid (PLA) Nanofiber-Alginate Hydrogel Bioink Containing Human Adipose-Derived Stem Cells. *ACS Biomater. Sci. Eng.* **2016**, *2*, 1732–1742. [CrossRef]

88. Gao, G.; Park, J.Y.; Kim, B.S.; Jang, J.; Cho, D.W. Coaxial Cell Printing of Freestanding, Perfusable, and Functional In Vitro Vascular Models for Recapitulation of Native Vascular Endothelium Pathophysiology. *Adv. Healthc. Mater.* **2018**, *7*, 1801102. [CrossRef]

89. Wang, L.; Shelton, R.M.; Cooper, P.R.; Lawson, M.; Triffitt, J.T.; Barralet, J.E. Evaluation of sodium alginate for bone marrow cell tissue engineering. *Biomaterials* **2003**, *24*, 3475–3481. [CrossRef]

90. Li, Z.; Tan, B.H. Towards the development of polycaprolactone based amphiphilic block copolymers: Molecular design, self-assembly and biomedical applications. *Mater. Sci. Eng. C Mater. Biol. Appl.* **2014**, *45*, 620–634. [CrossRef]

91. Woodruff, M.A.; Hutmacher, D.W. The return of a forgotten polymer-Polycaprolactone in the 21st century. *Prog. Polym. Sci.* **2010**, *35*, 1217–1256. [CrossRef]

92. Guo, B.L.; Ma, P.X. Synthetic biodegradable functional polymers for tissue engineering: A brief review. *Sci. China Chem.* **2014**, *57*, 490–500. [CrossRef] [PubMed]

93. Lam, C.X.F.; Hutmacher, D.W.; Schantz, J.T.; Woodruff, M.A.; Teoh, S.H. Evaluation of polycaprolactone scaffold degradation for 6 months in vitro and in vivo. *J. Biomed. Mater. Res. Part A* **2009**, *90*, 906–919. [CrossRef] [PubMed]

94. Lam, C.X.F.; Savalani, M.M.; Teoh, S.H.; Hutmacher, D.W. Dynamics of in vitro polymer degradation of polycaprolactone-based scaffolds: Accelerated versus simulated physiological conditions. *Biomed. Mater.* **2008**, *3*, 034108. [CrossRef] [PubMed]

95. Patricio, T.; Domingos, M.; Gloria, A.; D'Amora, U.; Coelho, J.F.; Bartolo, P.J. Fabrication and characterisation of PCL and PCL/PLA scaffolds for tissue engineering. *Rapid Prototyp. J.* **2014**, *20*, 145–156. [CrossRef]

96. Zhang, B.; Seong, B.; Nguyen, V.; Byun, D. 3D printing of high-resolution PLA-based structures by hybrid electrohydrodynamic and fused deposition modeling techniques. *J. Micromech. Microeng.* **2016**, *26*, 025015. [CrossRef]

97. Farah, S.; Anderson, D.G.; Langer, R. Physical and mechanical properties of PLA, and their functions in widespread applications—A comprehensive review. *Adv. Drug Deliv. Rev.* **2016**, *107*, 367–392. [CrossRef]

98. Andreopoulos, A.G.; Hatzi, E.C.; Doxastakis, M. Controlled release systems based on poly(lactic acid). An in vitro and in vivo study. *J. Mater. Sci.-Mater. Med.* **2000**, *11*, 393–397. [CrossRef]

99. Stock, U.A.; Mayer, J.E., Jr. Tissue engineering of cardiac valves on the basis of PGA/PLA Co-polymers. *J. Long Term Eff. Med. Implant.* **2001**, *11*, 249–260. [CrossRef]

100. Rouhollahi, F.; Hosseini, S.A.; Alihosseini, F.; Allafchian, A.; Haghighat, F. Investigation on the Biodegradability and Antibacterial Properties of Nanohybrid Suture Based on Silver Incorporated PGA-PLGA Nanofibers. *Fiber Polym.* **2018**, *19*, 2056–2065. [CrossRef]

101. Park, J.Y.; Gao, G.; Jang, J.; Cho, D.W. 3D printed structures for delivery of biomolecules and cells: Tissue repair and regeneration. *J. Mater. Chem. B* **2016**, *4*, 7521–7539. [CrossRef]

102. Angelopoulos, I.; Allenby, M.C.; Lim, M.; Zamorano, M. Engineering inkjet bioprinting processes toward translational therapies. *Biotechnol. Bioeng.* **2020**, *117*, 272–284. [CrossRef] [PubMed]

103. Wagberg, L.; Decher, G.; Norgren, M.; Lindstrom, T.; Ankerfors, M.; Axnas, K. The build-up of polyelectrolyte multilayers of microfibrillated cellulose and cationic polyelectrolytes. *Langmuir* **2008**, *24*, 784–795. [CrossRef] [PubMed]

104. Chinga-Carrasco, G.; Yu, Y.D.; Diserud, O. Quantitative Electron Microscopy of Cellulose Nanofibril Structures from Eucalyptus and Pinus radiata Kraft Pulp Fibers. *Microsc. Microanal.* **2011**, *17*, 563–571. [CrossRef]

105. Markstedt, K.; Mantas, A.; Tournier, I.; Avila, H.M.; Hagg, D.; Gatenholm, P. 3D Bioprinting Human Chondrocytes with Nanocellulose-Alginate Bioink for Cartilage Tissue Eng. Applications. *Biomacromolecules* **2015**, *16*, 1489–1496. [CrossRef]

106. Chinga-Carrasco, G.; Ehman, N.V.; Pettersson, J.; Vallejos, M.E.; Brodin, M.W.; Felissia, F.E.; Hakansson, J.; Area, M.C. Pulping and Pretreatment Affect the Characteristics of Bagasse Inks for Three-dimensional Printing. *ACS Sustain. Chem. Eng.* **2018**, *6*, 4068–4075. [CrossRef]

107. Basu, J.; Ludlow, J.W. Platform technologies for tubular organ regeneration. *Trends Biotechnol.* **2010**, *28*, 526–533. [CrossRef]

108. Wilkens, C.A.; Rivet, C.J.; Akentjew, T.L.; Alverio, J.; Khoury, M.; Acevedo, J.P. Layer-by-layer approach for a uniformed fabrication of a cell patterned vessel-like construct. *Biofabrication* **2016**, *9*, 015001. [CrossRef]

109. Harding, S.I.; Afoke, A.; Brown, R.A.; MacLeod, A.; Shamlou, P.A.; Dunnill, P. Engineering and cell attachment properties of human fibronectin-fibrinogen scaffolds for use in tissue engineered blood vessels. *Bioprocess Biosyst. Eng.* **2002**, *25*, 53–59. [CrossRef]

110. Liu, Y.; Jiang, C.; Li, S.; Hu, Q. Composite vascular scaffold combining electrospun fibers and physically-crosslinked hydrogel with copper wire-induced grooves structure. *J. Mech. Behav. Biomed. Mater.* **2016**, *61*, 12–25. [CrossRef]

111. Ouyang, L.; Burdick, J.A.; Sun, W. Facile Biofabrication of Heterogeneous Multilayer Tubular Hydrogels by Fast Diffusion-Induced Gelation. *ACS Appl. Mater. Interfaces* **2018**, *10*, 12424–12430. [CrossRef]

112. Park, J.H.; Jang, J.; Lee, J.S.; Cho, D.W. Three-Dimensional Printing of Tissue/Organ Analogues Containing Living Cells. *Ann. Biomed. Eng.* **2017**, *45*, 180–194. [CrossRef]

113. Sun, W.; Starly, B.; Daly, A.C.; Burdick, J.A.; Groll, J.; Skeldon, G.; Shu, W.; Sakai, Y.; Shinohara, M.; Nishikawa, M.; et al. The bioprinting roadmap. *Biofabrication* **2020**, *12*, 022002. [CrossRef]

114. Luo, Y.; Lode, A.; Gelinsky, M. Direct plotting of three-dimensional hollow fiber scaffolds based on concentrated alginate pastes for tissue engineering. *Adv. Healthc. Mater.* **2013**, *2*, 777–783. [CrossRef]

115. Takeoka, Y.; Matsumoto, K.; Taniguchi, D.; Tsuchiya, T.; Machino, R.; Moriyama, M.; Oyama, S.; Tetsuo, T.; Taura, Y.; Takagi, K.; et al. Regeneration of esophagus using a scaffold-free biomimetic structure created with bio-three-dimensional printing. *PLoS ONE* **2019**, *14*, e0211339. [CrossRef]

116. Moldovan, N.I.; Hibino, N.; Nakayama, K. Principles of the Kenzan Method for Robotic Cell Spheroid-Based Three-Dimensional Bioprinting. *Tissue Eng. Part B-Rev.* **2017**, *23*, 237–244. [CrossRef]

117. Moldovan, L.; Barnard, A.; Gil, C.H.; Lin, Y.; Grant, M.B.; Yoder, M.C.; Prasain, N.; Moldovan, N.I. iPSC-Derived Vascular Cell Spheroids as Building Blocks for Scaffold-Free Biofabrication. *Biotechnol. J.* **2017**, *12*, 1700444. [CrossRef]

118. Zhang, X.Y.; Yanagi, Y.; Sheng, Z.J.; Nagata, K.; Nakayama, K.; Taguchi, T. Regeneration of diaphragm with bio-3D cellular patch. *Biomaterials* **2018**, *167*, 1–14. [CrossRef]

119. Bae, S.W.; Lee, K.W.; Park, J.H.; Lee, J.; Jung, C.R.; Yu, J.; Kim, H.Y.; Kim, D.H. 3D Bioprinted Artificial Trachea with Epithelial Cells and Chondrogenic-Differentiated Bone Marrow-Derived Mesenchymal Stem Cells. *Int. J. Mol. Sci.* **2018**, *19*, 1624. [CrossRef]

120. Hinton, T.J.; Jallerat, Q.; Palchesko, R.N.; Park, J.H.; Grodzicki, M.S.; Shue, H.J.; Ramadan, M.H.; Hudson, A.R.; Feinberg, A.W. Three-dimensional printing of complex biological structures by freeform reversible embedding of suspended hydrogels. *Sci. Adv.* **2015**, *1*, e1500758. [CrossRef]

121. Noor, N.; Shapira, A.; Edri, R.; Gal, I.; Wertheim, L.; Dvir, T. 3D Printing of Personalized Thick and Perfusable Cardiac Patches and Hearts. *Adv. Sci.* **2019**, *6*, 1900344. [CrossRef]

122. Zhang, K.; Fu, Q.; Yoo, J.; Chen, X.; Chandra, P.; Mo, X.; Song, L.; Atala, A.; Zhao, W. 3D bioprinting of urethra with PCL/PLCL blend and dual autologous cells in fibrin hydrogel: An in vitro evaluation of biomimetic mechanical property and cell growth environment. *Acta Biomater.* **2017**, *50*, 154–164. [CrossRef]

123. Jin, Y.F.; Liu, C.C.; Chai, W.X.; Compaan, A.; Huang, Y. Self-Supporting Nanoclay as Internal Scaffold Material for Direct Printing of Soft Hydrogel Composite Structures in Air. *ACS Appl. Mater. Interfaces* **2017**, *9*, 17457–17466. [CrossRef]

124. Talley, N.J. Introducing Expert Review of Gastroenterology and Hepatology. *Expert Rev. Gastroenterol. Hepatol.* **2007**, *1*, 1–2. [CrossRef]

125. Yoshida, N.; Watanabe, M.; Baba, Y.; Iwagami, S.; Ishimoto, T.; Iwatsuki, M.; Sakamoto, Y.; Miyamoto, Y.; Ozaki, N.; Baba, H. Risk factors for pulmonary complications after esophagectomy for esophageal cancer. *Surg. Today* **2014**, *44*, 526–532. [CrossRef]

126. Kulkarni, A.A.; Chauhan, V.; Sharma, V.; Singh, H. Gastric Conduit Perforation: A Late Fatal Complication after Esophagectomy. *Cureus* **2019**, *11*, e4987. [CrossRef]

127. Saxena, A.K.; Kofler, K.; Ainodhofer, H.; Hollwarth, M.E. Esophagus tissue engineering: Hybrid approach with esophageal epithelium and unidirectional smooth muscle tissue component generation in vitro. *J. Gastrointest. Surg.* **2009**, *13*, 1037–1043. [CrossRef]

128. Saxena, A.K.; Ainoedhofer, H.; Hollwarth, M.E. Esophagus tissue engineering: In vitro generation of esophageal epithelial cell sheets and viability on scaffold. *J. Pediatr. Surg.* **2009**, *44*, 896–901. [CrossRef]

129. Gong, C.; Hou, L.; Zhu, Y.; Lv, J.; Liu, Y.; Luo, L. In vitro constitution of esophageal muscle tissue with endocyclic and exolongitudinal patterns. *ACS Appl. Mater. Interfaces* **2013**, *5*, 6549–6555. [CrossRef]

130. Lin, M.H.; Firoozi, N.; Tsai, C.T.; Wallace, M.B.; Kang, Y.Q. 3D-printed flexible polymer stents for potential applications in inoperable esophageal malignancies. *Acta Biomater.* **2019**, *83*, 119–129. [CrossRef]

131. Dimitrievska, S.; Niklason, L.E. Historical Perspective and Future Direction of Blood Vessel Developments. *CSH Perspect. Med.* **2018**, *8*, a025742. [CrossRef]

132. Neufurth, M.; Wang, X.H.; Tolba, E.; Dorweiler, B.; Schroder, H.C.; Link, T.; Diehl-Seifert, B.; Muller, W.E.G. Modular Small Diameter Vascular Grafts with Bioactive Functionalities. *PLoS ONE* **2015**, *10*, e0133632. [CrossRef]

133. Wenger, R.; Giraud, M.N. 3D Printing Applied to Tissue Engineered Vascular Grafts. *Appl. Sci.* **2018**, *8*, 2631. [CrossRef]

134. Carrabba, M.; Madeddu, P. Current Strategies for the Manufacture of Small Size Tissue Eng. Vascular Grafts. *Front. Bioeng. Biotechnol.* **2018**, *6*, 41. [CrossRef]

135. Freeman, S.; Ramos, R.; Chando, P.A.; Zhou, L.X.; Reeser, K.; Jin, S.; Soman, P.; Ye, K.M. A bioink blend for rotary 3D bioprinting tissue engineered small-diameter vascular constructs. *Acta Biomater.* **2019**, *95*, 152–164. [CrossRef]

136. Lee, S.J.; Kim, M.E.; Nah, H.; Seok, J.M.; Jeong, M.H.; Park, K.; Kwon, I.K.; Lee, J.S.; Park, S.A. Vascular endothelial growth factor immobilized on mussel-inspired three-dimensional bilayered scaffold for artificial vascular graft application: In vitro and in vivo evaluations. *J. Colloid Interface Sci.* **2019**, *537*, 333–344. [CrossRef]

137. Savoji, H.; Davenport Huyer, L.; Mohammadi, M.H.; Lai, B.F.L.; Rafatian, N.; Bannerman, D.; Shoaib, M.; Bobicki, E.R.; Ramachandran, A.; Radisic, M. 3D Printing of Vascular Tubes Using Bioelastomer Prepolymers byFreeform Reversible Embedding. *ACS Biomater. Sci. Eng.* **2020**, *6*, 1333–1343. [CrossRef]

138. Kolesky, D.B.; Truby, R.L.; Gladman, A.S.; Busbee, T.A.; Homan, K.A.; Lewis, J.A. 3D bioprinting of vascularized, heterogeneous cell-laden tissue constructs. *Adv. Mater.* **2014**, *26*, 3124–3130. [CrossRef]

139. Rock, J.R.; Randell, S.H.; Hogan, B.L.M. Airway basal stem cells: A perspective on their roles in epithelial homeostasis and remodeling. *Dis. Model. Mech.* **2010**, *3*, 545–556. [CrossRef]

140. Nam, H.; Choi, Y.; Jang, J. Vascularized Lower Respiratory-Physiology-On-A-Chip. *Appl. Sci.* **2020**, *10*, 900. [CrossRef]

141. Foltinová, J.; Schrott-Fischer, A.; Zilínek, V.; Foltin, V.; Freysinger, W. Is the Trachea a Marker of the Type of Environmental Pollution? *Larygoscope* **2009**, *112*, 713–720. [CrossRef] [PubMed]

142. Fillinger, M.F.; Kerns, D.B.; Bruch, D.; Reinitz, E.R.; Schwartz, R.A. Does the end-to-end venous anastomosis offer a functional advantage over the end-to-side venous anastomosis in high-output arteriovenous grafts? *J. Vasc. Surg.* **1990**, *12*, 676–688, discussion 688–690. [CrossRef] [PubMed]

143. Kaihara, S.; Kim, S.; Benvenuto, M.; Kim, B.S.; Mooney, D.J.; Tanaka, K.; Vacanti, J.P. End-to-End Anastomosis between Tissue-Engineered Intestine and Native Small Bowel. *Tissue Eng. Part A* **2007**, *5*, 339–346. [CrossRef] [PubMed]

144. Ten Hallers, E.J.O.; Rakhorst, G.; Marres, H.A.M.; Jansen, J.A.; van Kooten, T.G.; Schutte, H.K.; van Loon, J.P.; van der Houwen, E.B.; Verkerke, G.J. Animal models for tracheal research. *Biomaterials* **2004**, *25*, 1533–1543. [CrossRef]

145. Kaye, R.; Goldstein, T.; Zeltsman, D.; Grande, D.A.; Smith, L.P. Three dimensional printing: A review on the utility within medicine and otolaryngology. *Int. J. Pediatr. Otorhinolaryngol.* **2016**, *89*, 145–148. [CrossRef]

146. Zhong, N.P.; Zhao, X. 3D printing for clinical application in otorhinolaryngology. *Eur. Arch. Oto-Rhino-Laryngol.* **2017**, *274*, 4079–4089. [CrossRef]

147. Taniguchi, D.; Matsumoto, K.; Tsuchiya, T.; Machino, R.; Takeoka, Y.; Elgalad, A.; Gunge, K.; Takagi, K.; Taura, Y.; Hatachi, G.; et al. Scaffold-free trachea regeneration by tissue engineering with bio-3D printing. *Interact. Cardiovasc. Thorac.* **2018**, *26*, 745–752. [CrossRef]

148. Lee, J.Y.; Park, J.H.; Ahn, M.J.; Kim, S.W.; Cho, D.W. Long-term study on off-the-shelf tracheal graft: A conceptual approach for urgent implantation. *Mater. Des.* **2020**, *185*, 108218. [CrossRef]

149. Gao, M.C.; Zhang, H.Y.; Dong, W.; Bai, J.; Gao, B.T.; Xia, D.K.; Feng, B.; Chen, M.L.; He, X.M.; Yin, M.; et al. Tissue-engineered trachea from a 3D-printed scaffold enhances whole-segment tracheal repair. *Sci. Rep.-UK* **2017**, *7*, 5246. [CrossRef]

150. Hsieh, C.T.; Liao, C.Y.; Dai, N.T.; Tseng, C.S.; Yen, B.L.J.; Hsu, S.H. 3D printing of tubular scaffolds with elasticity and complex structure from multiple waterborne polyurethanes for tracheal tissue engineering. *Appl. Mater. Today* **2018**, *12*, 330–341. [CrossRef]

151. Brozovich, F.V.; Nicholson, C.J.; Degen, C.V.; Gao, Y.Z.; Aggarwal, M.; Morgan, K.G. Mechanisms of Vascular Smooth Muscle Contraction and the Basis for Pharmacologic Treatment of Smooth Muscle Disorders. *Pharmacol. Rev.* **2016**, *68*, 476–532. [CrossRef] [PubMed]

152. Carmeliet, P.; Moons, L.; Collen, D. Mouse models of angiogenesis, arterial stenosis, atherosclerosis and hemostasis. *Cardiovasc. Res.* **1998**, *39*, 8–33. [CrossRef]

153. Rumbaut, R.E.; Slaff, D.W.; Burns, A.R. Microvascular thrombosis models in venules and arterioles in vivo. *Microcirculation* **2005**, *12*, 259–274. [CrossRef] [PubMed]

154. Wengerter, B.C.; Emre, G.; Park, J.Y.; Geibel, J. Three-dimensional Printing in the Intestine. *Clin. Gastroenterol. Hepatol.* **2016**, *14*, 1081–1085. [CrossRef]

bioengineering

Review

Advantages of Additive Manufacturing for Biomedical Applications of Polyhydroxyalkanoates

Alberto Giubilini [1], Federica Bondioli [2], Massimo Messori [3], Gustav Nyström [4,5] and Gilberto Siqueira [4,*]

1   Department of Engineering and Architecture, University of Parma, 43124 Parma, Italy; alberto.giubilini@studenti.unipr.it
2   Department of Applied Science and Technology, Politecnico di Torino, 10129 Torino, Italy; federica.bondioli@polito.it
3   Department of Engineering "Enzo Ferrari", University of Modena and Reggio Emilia, 41125 Modena, Italy; massimo.messori@unimore.it
4   Cellulose & Wood Materials Laboratory, Empa—Swiss Federal Laboratories for Materials Science and Technology, 8600 Dübendorf, Switzerland; gustav.nystroem@empa.ch
5   Department of Health Sciences and Technology, ETH Zürich, 8092 Zürich, Switzerland
*   Correspondence: gilberto.siqueira@empa.ch

Abstract: In recent years, biopolymers have been attracting the attention of researchers and specialists from different fields, including biotechnology, material science, engineering, and medicine. The reason is the possibility of combining sustainability with scientific and technological progress. This is an extremely broad research topic, and a distinction has to be made among different classes and types of biopolymers. Polyhydroxyalkanoate (PHA) is a particular family of polyesters, synthetized by microorganisms under unbalanced growth conditions, making them both bio-based and biodegradable polymers with a thermoplastic behavior. Recently, PHAs were used more intensively in biomedical applications because of their tunable mechanical properties, cytocompatibility, adhesion for cells, and controllable biodegradability. Similarly, the 3D-printing technologies show increasing potential in this particular field of application, due to their advantages in tailor-made design, rapid prototyping, and manufacturing of complex structures. In this review, first, the synthesis and the production of PHAs are described, and different production techniques of medical implants are compared. Then, an overview is given on the most recent and relevant medical applications of PHA for drug delivery, vessel stenting, and tissue engineering. A special focus is reserved for the innovations brought by the introduction of additive manufacturing in this field, as compared to the traditional techniques. All of these advances are expected to have important scientific and commercial applications in the near future.

Keywords: polyhydroxyalkanoates; scaffolds; biomedicine; additive manufacturing; 3D printing; drug delivery; vessel stenting; tissue engineering

Citation: Giubilini, A.; Bondioli, F.; Messori, M.; Nyström, G.; Siqueira, G. Advantages of Additive Manufacturing for Biomedical Applications of Polyhydroxyalkanoates. *Bioengineering* 2021, 8, 29. https://doi.org/10.3390/bioengineering8020029

Received: 11 November 2020
Accepted: 16 February 2021
Published: 23 February 2021

Publisher's Note: MDPI stays neutral with regard to jurisdictional claims in published maps and institutional affiliations.

1. Introduction

The term "biopolymer" is nowadays very common and widely spread in different fields of application. However, it is sometimes improperly used, due to the fact that there is not a brief and comprehensive definition of this word. To clarify the meaning of "biopolymer", it is important to define the concepts of "bio-based" and "biodegradable", and if the former is strictly connected with the origin of the material, at the opposite, the latter is related to its end-of-life.

A material can be defined as bio-based if it derives in whole or in part from biomass resources, i.e., organic materials that are renewable [1].

A material can be properly defined as biodegradable if it can be used as a carbon source by microorganisms and converted safely into CO_2, biomass and water [2]. Besides, if the material undergoes a biodegradation and a physical disintegration level of at least 90%, in less than six months, then it can also be defined as "compostable" [3].

Hence, the family of biopolymers can be divided into three main groups:

1. Biopolymers coming from renewable resources but not being biodegradable, e.g., bio-based polyethylene terephthalate (bio-PET), bio-based polypropylene (bio-PP), and bio-based polyethylene (bio-PE);
2. Biopolymers coming from not-renewable resources but being biodegradable, e.g., polybutylene adipate terephthalate (PBAT);
3. Biopolymers coming from renewable resources and being biodegradable, e.g., polyhydroxyalkanoate (PHA), poly(lactic acid) (PLA), and polybutylene succinate (PBS).

In this review, the PHA family is taken into consideration, and particular interest is reserved to its application in the biomedical field. Since the beginning of the twenty-first century, an increasing number of scientific studies and clinical trials have been published about PHA medical devices for different final applications, such as tissue engineering, drug delivery, or as vascular stents [4]. Therefore, first this review is aimed to present and discuss the results obtained with PHA and traditional techniques, like solvent casting, phase separation, salt leaching, or electrospinning. Furthermore, a great importance is given to the introduction of additive manufacturing in this research field, and particularly to the innovations and advantages introduced by 3D printing, which allowed us to overcome some of the greatest limitations of traditional approaches. For example, thanks to additive manufacturing, it was possible to obtain a finer control over the porosity, a true development of the devices in all three dimensions, and even the reproduction of complex structures, which are able to mimic natural tissues and which are highly tailored to the physical requirements of each patient [5,6]. Finally, this review is concluded with a discussion, in the authors' opinion, of the most likely future biomedical perspectives for this promising class of biopolymer, and of the new targets that can be achieved, thanks to 3D printing, in a new way of considering medicine, with a high customization of medical care. In Figure 1 shows a schematic representation of the overall topic and structure of this review work.

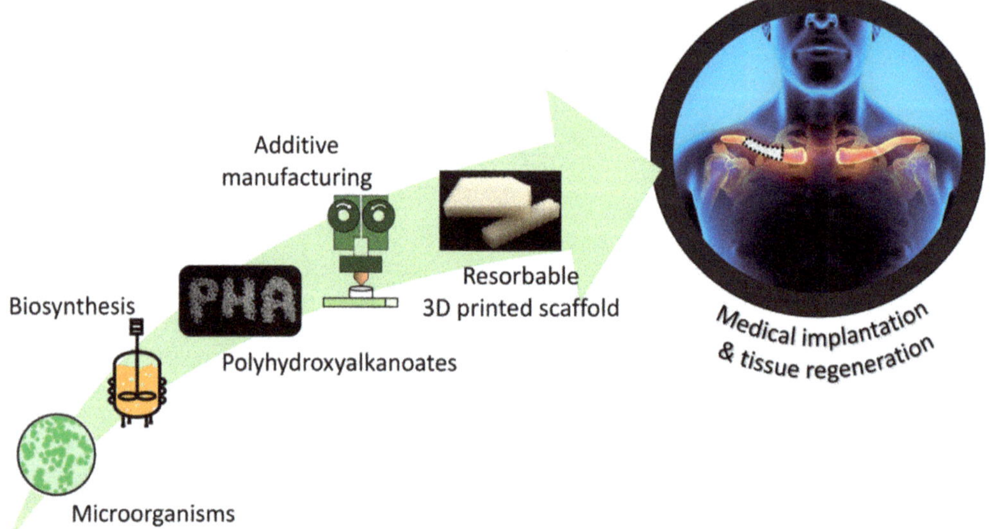

Figure 1. Schematic representation of the production, technological transformation, and biomedical applications of polyhydroxyalkanoate (PHA)-based devices.

The methodology carried out for the analysis of the literature started with searching published reviews on two of the most widespread databases, i.e., Scopus and ScienceDirect. Keywords selected for the literature search included PHA, additive manufacturing, biomedical application, biopolymer medical device, and PHA biosynthesis. These reviews were scanned, all parts related to PHA were highlighted, and the cited original research articles were acquired. After that, all references' abstracts were examined, and a first category clustering was performed according to this filtering system: (1) PHA production; (2) traditional PHA medical devices (solvent casting, salt leaching, thermally induced phase separation, non-solvent induced phase separation, emulsification, and electrospinning); (3) innovative PHA medical devices (Direct Ink Writing, Fused Deposition Modeling, Selective Laser Sintering, and Computer Aided Wet-Spinning). Afterwards, a second classification was implemented, to order all the references in accordance with the final medical application: (1) drug delivery; (2) vessel stenting; (3) bone tissue engineering, and (4) cartilage tissue engineering. Eventually, a combination of the two former groups was completed, and this synthesis was used as starting point for the manuscript development.

2. PHA: Biosynthesis and Properties

Due to the global awareness of the environmental impact of fossil-based polymers [7], the main goal of plastic industry is nowadays to tackle plastic pollution and its sociopolitical and economic challenges by developing new materials that can combine the advantages of traditional plastics with a sustainable production and disposal. In this research field, biopolymers play a central role due to their great benefits, such as carbon footprint reduction, saving of fossil resources and landfill decrease [8].

PHA is a large family of thermoplastic aliphatic polyesters mainly produced by prokaryotic organisms, such as bacteria, most prevalently Gram-negative [9], and archaea under conditions of nutrient depletion and in the presence of an excess of carbon source [10]. It is noteworthy to consider that, although only at a preliminary scientific research level, the production of PHAs from plants was achieved [11]. The general structure of PHAs is reported in Figure 2, where m can be equal or greater than one and R can be a hydrogen atom or an alkyl substituent, depending on the type of PHA [12]. Maurice Lemoigne, a French microbiologist, was the first researcher who identified the synthesis of PHAs from bacteria in 1926 by using a culture of *Bacillus megaterium* to isolate poly(3-hydroxybutyrate) (PHB) [13].

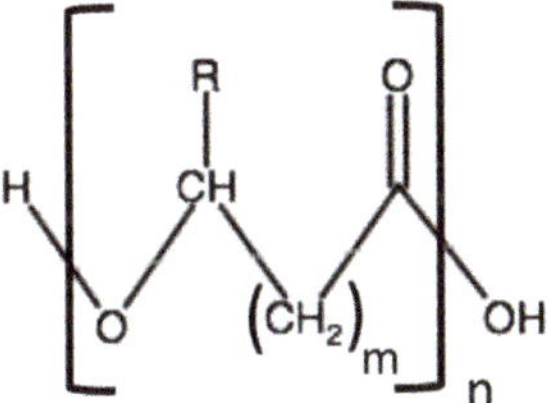

Figure 2. General chemical structure of polyhydroxyalkanoates (PHAs); "m" varies from 1 to 4 and "n" ranges from 100 to 30,000; R denotes a hydrogen atom or an alkyl side chain [12].

As several biopolymers belong to the PHA family, their classification is important, and they can be sorted depending on their chain-length monomeric composition, according to the number of carbon atoms per monomer, and here a great importance is played by the composition of the monomer side chain R [14]:

- Short-chain-length PHA (scl-PHA) has three to five carbon atoms;
- Medium-chain-length PHA (mcl-PHA) has 6 to 14 carbon atoms;
- Long-chain-length PHA (lcl-PHA) has more than 14 carbon atoms.

Generally, scl-PHAs, containing mainly 3-hydroxybutyrate (3HB) or 3-hydroxyvalerate (3HV) units, have a higher degree of crystallinity, a higher glass transition temperature, and a higher molecular mass compared to mcl-PHAs [15–17], containing 3-hydroxyhexanoate (3HH), 3-hydroxyoctanoate (3HO), 3-hydroxydecanoate (3HD), or 3-hydroxydodecanoate (3HHD) monomers.

Another possible distinction can be made between homopolymer, of which the most famous and widespread example is PHB, and copolymers, such as poly(3-hydroxybutyrate-*co*-3-hydroxyvalerate) (PHBV), poly(3-hydroxybutyrate-*co*-4-hydroxybutyrate) (P3HB-4HB), or poly(3-hydroxybutyrate-*co*-3-hydroxyhexanoate) (PHBH). In this latter case, also the monomer arrangement can define a further method of classification. In fact, the difference between block copolymers and random copolymers is due to the ordered succession of similar monomers, unlike a random distribution, distinctive of the second type of copolymers [18]. The physical blending or the chemical copolymerization allow us to obtain a final material with tuned properties, which directly depend on the structures of the single-constituent monomers [10]. For example, PHB has a high crystallinity and brittleness, which can be reduced by introducing a new monomer unit, such as 3HV or 3HH [19]. The molar composition ratio of the copolymers is a key factor to tune the final properties, such as elongation at break and degree of crystallinity, which increase with the increase of 3HV [20] or 3HH [21] molar content in the structure. Figure 3 shows a schematic representation and categorization of the PHA family according to the chain-length and the composition of the structural units.

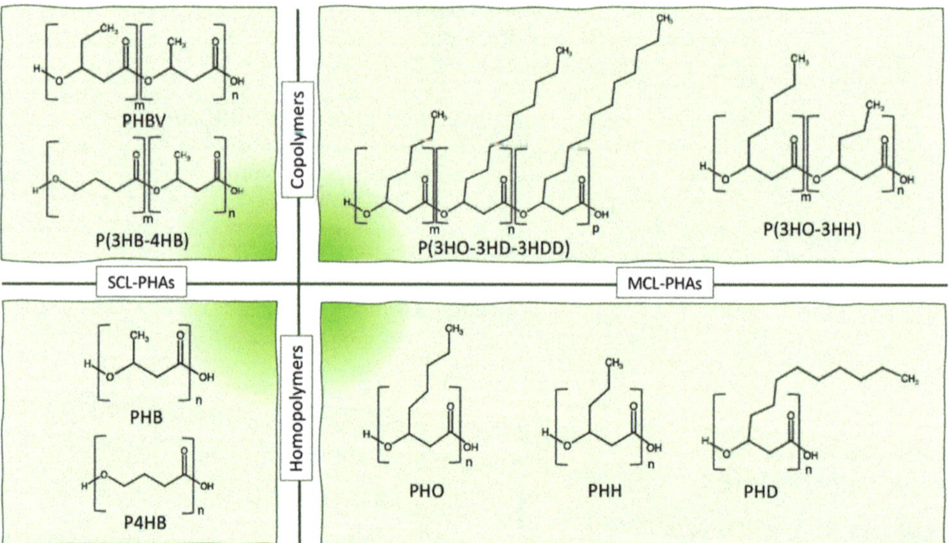

Figure 3. PHAs classification depending on the chain length and the chemical structure of the monomers. PHBV—poly(3-hydroxybutyrate-*co*-3-hydroxyvalerate); P(3HB-4HB)—poly(3-hydroxybutyrate-*co*-4-hydroxybutyrate); PHB—poly(3-hydroxybutyrate); P4HB—poly(4-hydroxybutyrate); P(3HO-3HD-3HDD)—poly(3-hydroxyoctanoate-*co*-3-hydroxydecanoate-*co*-3-hydroxydodecanoate); P(3HO-3HH)—poly(3-hydroxyoctanoate-*co*-3-hydroxyhexanoate); PHO—poly(3-hydroxyoctanoate); PHH—poly(3-hydroxyhexanoate); PHD—poly(3-hydroxydecanoate).

As already reported, PHAs are bio-based polymers, whose origin derives from bacterial and archaeal fermentation. Some microorganisms, when they are subjected to an environmental stress, such as a depletion of essential nutrients, can start a conversion of the carbon sources in hydroxyalkanoate units, such as carbon and energy reserve, which are further polymerized into PHA granules through a biosynthetic pathway and stored in the bacterial cell cytoplasm [22]. The average size of the PHA granules is approximately 0.2–0.5 µm [23,24]. In Figure 4, a transmission electron micrograph of *Rhodovulum visakhapatnamense* cells containing PHA granules is reported.

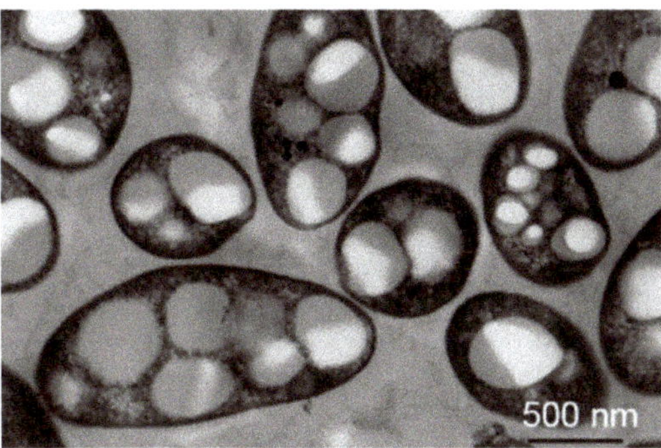

Figure 4. TEM image of *Rhodovulum visakhapatnamense* accumulating intracellular PHA granules, appearing as whitish and bright areas (adapted from Reference [25]).

The biosynthetic pathway of PHB consists of three enzymatic reactions catalyzed by three different enzymes: phbA, phbB, and phbC. The first reaction is a condensation of two acetyl coenzyme A (acetyl-CoA) molecules into acetoacetyl-CoA by β-ketoacyl-CoA thiolase (encoded by phbA). The second reaction is the reduction of acetoacetyl-CoA to (R)-3-hydroxybutyryl-CoA by an NADPH-dependent acetoacetyl-CoA dehydrogenase (encoded by phbB). Lastly, the (R)-3-hydroxybutyryl-CoA monomers are polymerized into PHB by PHB polymerase (encoded by phbC) [26,27]. The scheme in Figure 5 synthesizes the fundamental enzymatic biosynthetic pathway.

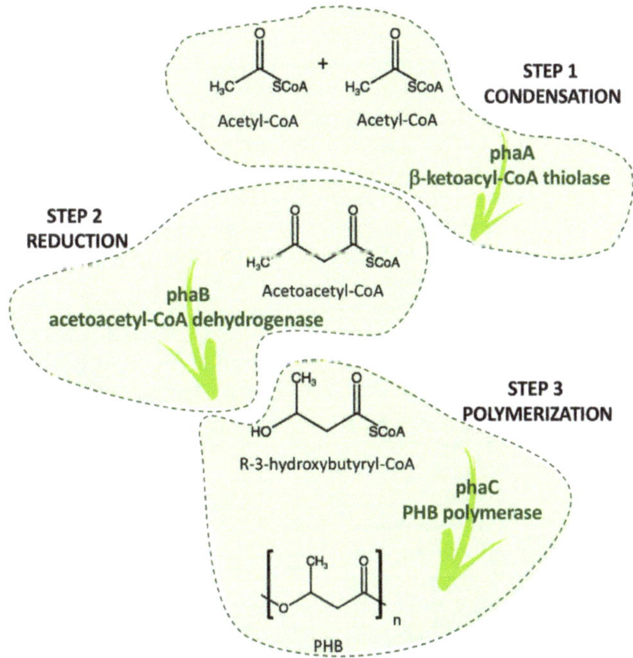

Figure 5. Biosynthetic pathway of poly(3-hydroxybutyrate) production within the bacterial cytoplasm. PHB is synthesized by the successive action of three enzymes: β-ketoacyl-CoA thiolase (phbA), acetoacetyl-CoA dehydrogenase (phbB), and PHB polymerase (phbC) in a three-step pathway.

Nowadays, the number of bacteria that is able to produce PHA is remarkable, i.e., more than eighty different genera [22]. The most commonly used bacteria species able to produce PHAs belong to the genera of *Alcaligenes, Azotobacter, Bacillus, Cupriavidus, Chromobacterium, Delftia, Pseudomonas, Ralstonia,* and *Staphylococcus* [28]. Different microorganisms own different polymerase enzymes, and this leads to the fact that every single microorganism is capable of producing small differences in the final biopolymer [29]. For example, *Ralstonia* bacteria have a particular polymerase enzyme that prioritizes the synthesis of scl-PHA [30]; on the opposite, *Pseudomonas* bacteria produce mcl-PHA [31]. Moreover, the PHA production yield can vary significantly from 0.25 g/L, using, for example, terephthalic acid as carbon source for *Pseudomonas putida* GO16, to 51.2 g/L, using commercial glycerol as carbon source for *Cupriavidus necator* DSM 545 [10].

Carbon is at the basis of organic chemistry and the fundamental element for all biomasses. There are different possible carbon sources that can be used to feed the microorganisms during PHA production and they can be classified in three different substrate groups: carbohydrates (e.g., sucrose, lactose, starch, or lignocellulose) [32–34], triacylglycerols (e.g., animal fats or plant oils) [35,36], and hydrocarbons. The last group is not economically significant since only few species of bacteria are capable to synthetize PHAs from this source and the process tends to have a low efficiency [37]. Apart from the carbon, other chemical compounds are required such as nitrogen sources, and some of the most used are $(NH_4)_2SO_4$, NH_4Cl, or NH_4NO_3 [22]. Variation in carbon to nitrogen ratios led to a different amount of PHA concentration in bacterial cells [38], and most of the studies showed that limiting nitrogen concentration while increasing carbon substrates had a positive effect on the PHA production rate [39,40]. Since the biosynthesis process ends with the storage of PHA granules into the cell cytoplasm, a further crucial step is required, the extraction of the PHAs granules from the bacterial cell. The approaches for biopolymer recovery can be different, and they are here synthesized:

- Solvent dissolution: The extraction is performed on pretreated cells, where PHA granules were made accessible by rupture of the cell membrane, and halogenated solvents are then used to dissolve the granules and then precipitate them in a non-solvent solution [41]. The biggest limitation of this method is the need of a high amount of harmful solvents, which hinders the environmental benefits of PHA biosynthesis [42]. In order to overcome this drawback, the use of non-halogenated solvents or supercritical CO_2 are being investigated as alternatives [43].
- Enzymatic digestion: This method consists of a digestion of the cell membrane by action of enzymes, followed by filtration, floatation, or centrifugation recovery of the PHA granules [44].
- Chemical digestion: The procedure consists, as in the previous procedure, of the digestion of the cell membrane by the chemical action of sodium hypochlorite at high pH values, which makes most of the cellular components soluble in water, due to oxidation, and therefore easily removable [45].
- Mechanical disruption: The microbial cells are mechanically disintegrated by high-pressure homogenization or ultrasonication, thus making PHA granules recuperable [46].
- Osmophilic disruption: The rupture of the cell is caused by the high internal pressure in hypotonic media due to osmotic absorption, which causes the release of the intracellular content [47].
- Biological extraction: This ecological procedure consists of the use of insects, such as the mealworm, that can be fed on lyophilized cells of *Cupriavidus necator*, with intracellular PHB granules. Once the feeding is complete, PHB can be extracted from the fecal pellets of the black soldier fly larvae [48].

The choice of the most suitable recovery method depends on several factors such as the microbial strain, the type of PHA and the required purity grade of the final product. Specifically, the purity of the polymer has a critical importance for biomedical applications. In fact, biological active contaminants, such as endotoxins, can cause undesired immunological responses. For example, the US Food and Drug Administration (FDA) regulations limited the endotoxin content of medical devices to 20 USP endotoxin units per device, and to 2.15 in case of devices associated with the cerebrospinal fluid [49]. So far, different approaches have been suggested, but there is still room for improvement and innovation on this particular aspect. Burniol-Figols et al. evaluated an innovative PHA purification through dilute aqueous ammonia digestion (purity $86 \pm 0.8\%$), and they compared it with reference processes, such as dissolution in chloroform and precipitation in methanol (purity $99 \pm 0.2\%$), or also acid-mediated digestion with H_2SO_4, followed by a treatment with NaOCl and subsequent washing with water and centrifugation (purity $98 \pm 2.6\%$) [50]. Moreover, more environmentally friendly purification processes were proposed like the use of dimethyl carbonate for extraction, followed by a purification step with 1-butanol via reflux. After this purification, the overall purity increased from $91.2 \pm 0.1\%$ to $98.0 \pm 0.1\%$ [51]. Wampfler et al. investigated another possible purification step, particularly experimented for biomedical applications, which implies the filtration through a column filled with activated charcoal (0.5 mL of charcoal per mL of solution to be filtered). The authors stated that endotoxins were almost completely eliminated by this method, removing polymeric impurities with a molecular weight below 10 kDa, as well as the colored impurities [52].

In terms of process development, there are three main steps for industrial PHA production, first the process has to be optimized at laboratory-scale level, and then it is performed in bioreactor and eventually in pilot plant scale with 100–300 L fermenters [53]. After obtaining a globally recognized result at laboratory scale, in the last decades, the industrial PHA market is still gradually increasing, along with the number of independent companies that are investing on PHA production. However, the final result is far from achieved, if we consider, for example, that, in terms of global production capacity, PHA is about 30,000 tons, which is almost ten times less than bio-PE, and almost 20 times less than bio-PET [54]. For successful industrial scale-up PHA production, the influence of oxygen mass transfer and proper agitation are the most important aspects. Therefore, the scale-up strategies need to be based on keeping one of these parameters constant, with respect to the optimized laboratory-scale setup: volumetric oxygen transfer coefficient ($K_L a$), volumetric power consumption (P/V), impeller tip speed of agitator (Vs), and mixing time (t_m) or dissolved oxygen (DO) concentration [55]. To date, worldwide, only a few examples of PHA producers (e.g., Danimer Scientific and Newlight Technologies) have the production capacity to establish collaborations with owners of world-renowned brands in the fields of furniture and food and beverage packaging. This collaboration allows us to boost their economy and lead to a global PHA market growth.

Concurrently, scientific research and technological innovation are engaged for enhancing PHA production efficiency, by optimizing the biosynthesis mechanisms, valorizing cheap and renewable nutrient substrates, and engineering some new bacterial strains or also mixed microbial cultures (MMCs), which do not require sterile conditions and have a wider metabolic potential than single strain [56].

The great structural variety inside the PHA family is reflected in a wide spectrum of physical properties of PHAs, varying from a stiffer behavior, comparable to polystyrene for PHB, to a more flexible behavior with elongation at break values of PHBV similar to those of polypropylene or even low density polyethylene [57,58]. Generally, PHAs are characterized by a low glass transition temperature, between -50 and $0\,^\circ C$, and a melting temperature lower than $200\,^\circ C$ [59]. However, probably the most attractive property of PHAs is their biodegradability, which can occur both in aerobic [60] and anaerobic [61] environments, without developing toxic products. The biodegradation of PHAs evolves in three main stages: (1) biodeterioration, which consists in the colonization of the surface, or the bulk of the material, by microorganisms which modify the physical properties of the polymer; (2) biodepolymerization, which is the conversion of polymers into oligomers and monomers induced by enzymes (i.e., PHA depolymerases), secreted by microorganisms, such as bacteria or fungi, which hydrolyze the ester bond of the PHAs; and (3) assimilation, where these low-molecular-weight molecules are metabolized as carbon and energy sources by microorganisms that convert carbon of PHAs into CO_2, water, and biomass [62,63].

Considering the similarity in mechanical, thermal and barrier properties of PHAs with commodity polymers along with their bio-based origin and biodegradability, this leads to a great interest of PHAs as possible replacements of conventional polymers in different industrial applications [22], such as household or agricultural items manufacturing [64] and packaging [65,66]. However, the higher prices of PHA make them noncompetitive in the current market compared to the fossil-based polymers. In fact, whilst common polyolefins like polyethylene and polypropylene nowadays cost less than 1 €/kg [56], PHAs can range from 2 to 5 €/kg depending on the grade [67]. Their higher prices are mainly due to the cost of carbon sources, substrates and to the low extraction yield at industrial scale [68]. PHAs are largely hydrophobic and soluble in chlorinated hydrocarbons, such as chloroform or dichloromethane. Considering the biomedical applications, the PHA hydrophobic behavior is a suitable property to avoid that the devices undergo a rapid dissolution and a consequent loss of structural properties, once they are implanted in the aqueous body environment. However, it is well-known that wettable scaffolds are conducive to better cellular adhesion, growth and proliferation, due to the ability of maintaining a humid environment and hence promoting fluid exchange between the designed part and the surrounding [69]. In order to tune this hydrophobic behavior, the PHA matrix can be compounded with hydrophilic filler, such as montmorillonite [70], to increase the water affinity of the composites. Two other key properties for PHA medical applications are biocompatibility and biodegradability in physiological environments, which make them suitable for the production of resorbable biomedical devices, which support cellular adhesion, proliferation, and differentiation [49]. A great benefit in biomedicine is the possibility to implant a device that matches the host tissue mechanical property, and hence it decreases stress concentrations at the device–tissue interface. Therefore, the advantage of PHA compared to other polymers clinically used such as poly(lactide-*co*-glycolide) (PLGA), poly(ε-caprolactone) (PCL), poly(glycolic acid) (PGA), or poly(lactic acid) (PLA) is their wide variety of mechanical properties depending on the chemical structure of the monomers. In fact, PLA and PGA have a high Young's modulus (i.e., 3 and 6 GPa respectively) and a limited elongation at break (i.e., around 2%); differently, PCL has an inferior Young's modulus (i.e., 0.35 GPa), but a much higher elongation at break (i.e., 400%). These materials are optimal for specific biomedical applications, according to their inherent properties. Due to the possibility of tailoring the Young's modulus of PHAs, via compounding or synthetic copolymerization, the applicability of this class of biopolymer is potentially much wider and it gives the chance to choose the best grade of copolymer or monomer to mimic the final destination environment [71]. The mechanical properties of human tissue can considerably vary, for example the Young's modulus for granulation tissue is ~0.2 MPa, for fibrous tissue is ~2 MPa, for articular cartilage is 1–20 MPa, for intervertebral disc is 6–50 MPa, for tendon is 1–3 GPa and for mature bone is ~6 GPa [72,73]. Similarly, the Young's modulus for PHA family may range from ~600 MPa for some grade of copolymers such as P(3HB-4HB) to ~3 GPa for PHB. It is important to note that also the Young's modulus of a same copolymer can be tuned by the variation of the molar composition ratio, for example, the P(3HB-4HB) Young's modulus decreases at the increase of 4HB monomer content [74].

Moreover, compared to the abovementioned polymers, PHA has a better interaction with the immune system, due to the unchanged local pH value during its degradation, without toxic or inflammatory reactions [75]. As the other properties, also degradation times for PHAs depend on the chemical structure of the polymer. A previous research study for bioresorbable cardiovascular scaffolds showed that P4HB has a degradation time ranging between two and twelve months. Differently, PGA has an approximate degradation time, starting from six months; PLLA and PCL degradation take longer than two years [76].

3. Overview on the Main Production Techniques for Biomedical Implants Using PHA

Advances in the biomedical field are not limited to their final applications or the materials used, but they may also concern advancements in the processing techniques of the final implants and devices. Considering the thermoplastic behavior and the solubility in organic solvents of PHA, different approaches have been followed for transforming PHA raw material into architectures with various potential biomedical applications. The first PHA biomedical devices were simple systems with no control on the structure development, and they were obtained by traditional methods, such as (1) solvent casting, (2) salt leaching, (3) thermally induced phase separation (TIPS), (4) non-solvent-induced phase separation (NIPS), (5) emulsification, and (6) electrospinning. Here, the main features of these techniques are reported and summarized.

Solvent casting is probably the most common and the simplest technique for polymer film samples production. PHA are dissolved in an organic solvent (e.g., chloroform or dimethyl sulfoxide) at a typical concentration between 2 and 5 wt%; then, the solution is cast into a mold and the solvent is drawn off to obtain a polymer film with a final thickness of about 100 μm [77,78]. An actual problem of this technique is the impossibility of totally controlling the kinetics of the drying process, which could lead to some stress formation into the film structure and to a wrinkled surface.

Salt leaching is a straightforward technique to obtain porous scaffolds, which is a key feature for cell adhesion and proliferation. This process consists in mixing a salt powder, for example, NaCl, with a solution of PHA, and then, after solvent evaporation, leaching out the salt from the structure by soaking the membrane in water [79]. Compared to the solvent cast films, the scaffolds obtained via salt leaching are slightly thicker, varying in a range between 250 and 500 μm [80,81], and with an additional porosity ranging from a few to tens of microns, depending on the size of the salt particles. To avoid using organic solvents, alternatively to the first solvent casting step, a melt molding process is possible. In this case, PHA and salt powders are mixed and poured in a mold, which is first heated above the PHA melting temperature and then cooled down for scaffold solidification. For example, Baek et al. compounded PHBV and hydroxyapatite powder (9:1 *w/w*) with NaCl particles (100–300 μm) at a 1:17 weight ratio and then cast in a mold at 180 °C. The final structure is a porous network with pore sizes ranging from several microns to around 400 μm [82].

Thermally induced phase separation (TIPS) is a common alternative approach used in the fabrication of porous PHA scaffolds. The physical principle on which it is based is the changing of the temperature condition of a polymer solution, in order to induce a separation into two distinct phases. First, PHA is dissolved in an organic solvent and then frozen. Next, the solvent is removed by a sublimation process (e.g., freeze-drying), leaving a final porous structure. As an example, You et al. dissolved PHBH in 1,4-dioxane under vigorous agitation at 65 °C, to promote solubilization. The polymer solution was then frozen at −80 °C and lastly freeze-dried for two days. Vacuum drying was applied to completely remove any possible solvent remaining in the scaffolds. Morphology of the scaffolds showed porous structures with pore sizes of approximately 60–100 μm in diameter and 9.3 ± 1.4% in porosity. Moreover, micropores with 5–10 μm diameters were observed interconnected inside the scaffolds, which may help improve intercellular communication [83,84].

Non-solvent induced phase separation (NIPS) is another technique used to produce films and thin membranes of PHA. In this case, first PHA is dissolved in an organic solvent and then a phase separation is obtained when this solution enters in contact with a non-solvent, and hence PHA precipitate forming a film. This technique can be used with direct injection in local body sites, and in these cases, it is important to use a non-toxic organic solvent (e.g., dimethyl sulfoxide (DMSO)) for dissolution of PHA, and when this solution comes into contact with aqueous body fluid (a non-solvent for PHA), a PHA membrane is formed, and the polymer solution leads to the precipitation of PHA, which consists in film formation. Dai et al. investigated different non harmful organic solvents: *N*-methyl pyrrolidone (NMP), dimethylacetamide (DMAC), 1,4-dioxane (DIOX), dimethyl sulfoxide (DMSO), and 1,4-butanolide (BL) to be used with PHBH, at 15 wt% concentration, as injectable systems in rats at the intra-abdominal position. The results showed that PHBH films with a porous structure were formed and their surface morphologies depended on the different solvent-exchange rate in the phase separation process involving organic solvents and aqueous liquid. PHBH films prepared from NMP, DMAC, and DMSO showed larger porous structures both on the surface and in the cross-section. Those from DIOX and BL had very low porosity on the surfaces [85].

Emulsification is the most prevalent technique to obtain PHA microspheres or nanoparticles, which are further used as drug carriers for pharmacological agents. The derived applications are particularly appropriate for topical therapies at controlled-release rate, to safely achieve the desired therapeutic effects [86]. The oil-in-water emulsion-solvent evaporation method is the standard procedure for PHA nanoparticles fabrication. It consists of mixing an organic phase, PHA polymer dissolved in a solvent, to an aqueous solution with an emulsifier, e.g., poly(vinyl alcohol) (PVA). The organic solvent is then removed by volatilization. Finally, nanoparticles are harvested by centrifugation, washed, and dried. The final dimensions of the nanoparticles are usually between 100 and 200 nm, when ultrasonication is used as mixing step [87–89]; differently, if a homogenizer process is used, the dimensions are slightly higher and they vary into a range between 150 and 300 nm [87].

Electrospinning is a microfiber production method, and, nowadays, it is the most widely used technique for fabrication of fibrous microporous scaffolds, which simulate the structure of the extracellular matrix. Unlike melt-spinning or wet-spinning, electrospinning does not require a thermal or a chemical coagulation step to produce microfibers. A syringe is filled with a PHA solution and then placed in a high-voltage electric field, usually at 20 kV; thereby, the liquid starts to charge electrically. When the voltage is high enough for the electric repulsion to exceed the surface tension of the droplet at the end of the needle, a thin fluid jet erupts in the direction of the collector, which can be a flat metallic plate or a rotating mandrel. During the travel, the solvent evaporates and the jet dries; hence, electrospun microfibers with a mean diameter of about 500 ± 150 nm [90–92] are collected in the form of a microporous film, with a pore size of 1–1.5 μm [92].

Figure 6 summarizes the above-described conventional processing techniques and graphically represent the final shapes and morphologies of different PHA-based medical devices.

From the techniques presented so far, we conclude that the sustainability aspect, coming from the production of a bio-based and biodegradable polymer, is undermined by the technological approaches requesting a high amount of harmful organic solvents. Moreover, all these techniques are only suitable for the manufacturing of devices with a very limited 3D structure and, overall, with a maximum thickness of hundreds of microns, which is an evident drawback for an extensive use for biomedical applications.

With the spreading of **additive manufacturing (AM)** techniques, a new light on the modern research scene has been turned on 3D printing for biomedical applications (e.g., tissue engineering, prosthesis, or drug delivery), due to the possibility of tailoring the final design and the manufacturing of complex structures, eliminating the costs and time needed for the construction of molds [97,98]. Three-dimensional printers are commanded by a sequence of instructions, expressed in a computer numerical control programming language (e.g., g-code), to build a three-dimensional object starting from a computer-aided design (CAD) model. Particularly interesting in biomedical applications is the possibility of customizing and elaborating the starting model, in accordance with the morphological structure of the body in which the device is supposed to be implanted, thus achieving optimal compatibility [99,100]. Moreover, with AM approach is possible to tune the mechanical properties of the final device in order to modify the stiffness of the implant to match that of the original tissue, and hence mitigating the problem of stress concentrations. In fact, varying the structure and the design of the 3D-printed device, it is possible to increase the porosity and thereby to decrease of one order of magnitude the Young's modulus of the implant [73,101,102].

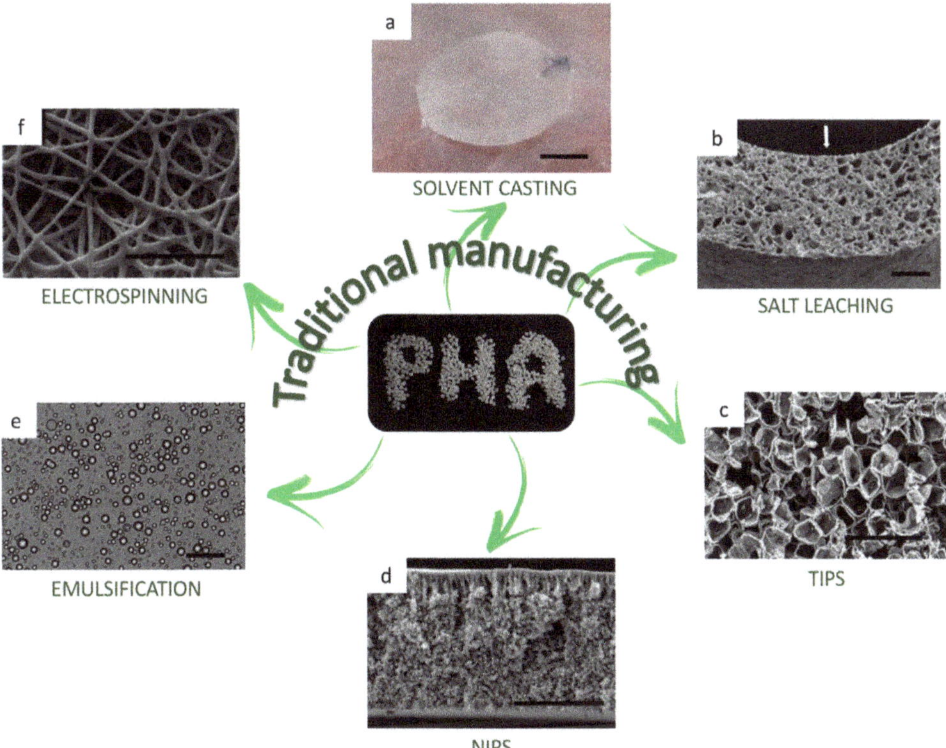

Figure 6. Morphology of PHA scaffolds produced with conventional techniques. (**a**) Visual appearance of a poly(3-hydroxybutyrate-co-3-hydroxyhexanoate), PHBH, film obtained via solvent casting (scale bar = 10 mm; adapted from Reference [93]). (**b**) SEM image of a PHBH conduit cross section with uniform wall porosity obtained via salt leaching; the white arrow indicates the internal side (scale bar = 100 μm; adapted from Reference [81]). (**c**) SEM image of a porous scaffold made of a blend of PHB/PHBH obtained via thermally induced phase separation (TIPS) (scale bar = 500 μm; adapted from Reference [94]). (**d**) SEM image of a PHBV membrane cross-section obtained via non-solvent-induced phase separation (NIPS) (scale bar = 50 μm; adapted from Reference [95]). (**e**) Optical microscopy image of PHBH microspheres prepared via emulsification (scale bar = 50 μm; adapted from Reference [96]). (**f**) SEM image of a porous PHBH film obtained via electrospinning (scale bar = 20 μm; adapted from Reference [92]).

Many different techniques of 3D printing have been invented according to the characteristics of the material processed. For PHA 3D printing, the most applied approach is the one of extrusion-based techniques, in which the biopolymer is either melted or dissolved in a solvent and then extruded through a nozzle and deposited on a printing bed, layer-by-layer. Hereafter, the essential extrusion-based AM techniques used in the production of PHA biomedical applications are discussed and compared to the traditional ones: (1) Direct Ink Writing (DIW), (2) Fused Deposition Modeling (FDM), (3) Selective Laser Sintering (SLS), and (4) Computer Aided Wet-Spinning (CAWS).

Direct Ink Writing (DIW) is an extrusion-based 3D-printing technique in which the material is loaded in the form of an ink with rheological properties that allow flowing through the nozzle, as well as supporting its own weight during assembly. In this technique, unlike FDM, the shape retention does not rely on solidification, but rather on shear thinning behavior of the inks. The material is extruded through a thin nozzle, using a computer-controlled robotic deposition system [103]. The final shape of the CAD model is first sliced into layers of height proportional to the nozzle diameter, and it is achieved layer-by-layer. In the production of PHA biomedical devices, the ink is generally obtained by dissolving the biopolymer in a solvent; however, it is also possible to print directly the biopolymer pellets, using a high-temperature print head and thus exploiting the thermoplastic properties of the material. After printing, a final step of cooling or drying occurs, depending if the material underwent a heating process or not.

Fused Deposition Modeling (FDM) is the most popular AM technique, due to its straightforwardness and its design freedom. It is a layer-by-layer melt-extrusion approach that consists in heating up a continuous filament of a thermoplastic material above its glass transition temperature (T_g), and then deposing the extruded material still hot to ensure the adhesion with the underneath layer, already cooled down and hardened. The result is a fully solidified structure whose final design accuracy is guaranteed by a computer control of movements of both printing platform and 3D-printer extruder head [104]. Although FDM can be considered as the most-used 3D-printing technique in a wide range of applications, with different polymeric materials, its utilization for PHA biomedical devices is still extremely limited. Only four scientific research works were published so far, and they evaluate either the applicability as preliminary investigations [105–107] or the use of this technique for the production of an external medical aid in the form of a finger cast [108].

Selective Laser Sintering (SLS) is another AM technique, and it was the first one investigated for production of PHA-based biomedical devices [109]. This approach uses a high-power laser beam to locally sinter the biopolymeric powder bed. This procedure is repeated layer-by-layer, to form a 3D structure with a predesigned architecture, generated by CAD software and transferred to the 3D printer. Due to a suboptimal definition of the sintering process, pore areas of the printed scaffolds are generally reduced, compared to the initial designs. An important influence over this effect depends on the powder layer thickness (PLT) and the scan spacing (SS). Pereira et al. investigated the effect of the variation of these printing parameters over the morphological structure, and it was demonstrated that the increase of SS reduces the size deviation; for example, with a PLT of 0.18 mm and different SS (0.15, 0.20, and 0.25 mm) pores of 0.60 ± 0.04 mm^2, 0.64 ± 0.04 mm^2, and 0.68 ± 0.05 mm^2 were obtained, respectively. Similarly, the increase of PLT also decreased the reduction of pores with respect to the digital model. Printed scaffolds with SS of 0.15 mm showed pore area values of 0.39 ± 0.07 mm^2, 0.60 ± 0.045 mm^2, and 0.73 ± 0.07 mm^2 for PLT of 0.08, 0.18, and 0.28 mm, respectively [110].

Computer Aided Wet-Spinning (CAWS) can be considered as an evolution of the wet-spinning technique implemented with a computer control. Wet-spinning consists of extruding from a syringe a PHA solution that precipitates and solidifies in a coagulation bath (e.g., ethanol), due to a non-solvent induced phase separation [111]. The novelty introduced by this technique is the computational control layer-by-layer of the syringe movements, affecting the final shape of the 3D-printed object. This technique allows us to obtain structures with high definition, with a fiber diameter of about 100 ± 20 μm [112,113] and a high porosity, above 80% [112,114]. Due to the non-solvent induced phase separation, this particular technique leads to a multi-scale porous structure in which microporosity, inside the single filaments, is added to a designed macroporous structure. This double scale of porosity has a positive effect on cellular interaction and tissue regeneration [115].

Figure 7 displays SEM images of scaffolds 3D printed by different AM techniques, showing the final microstructure of the PHA-based medical devices.

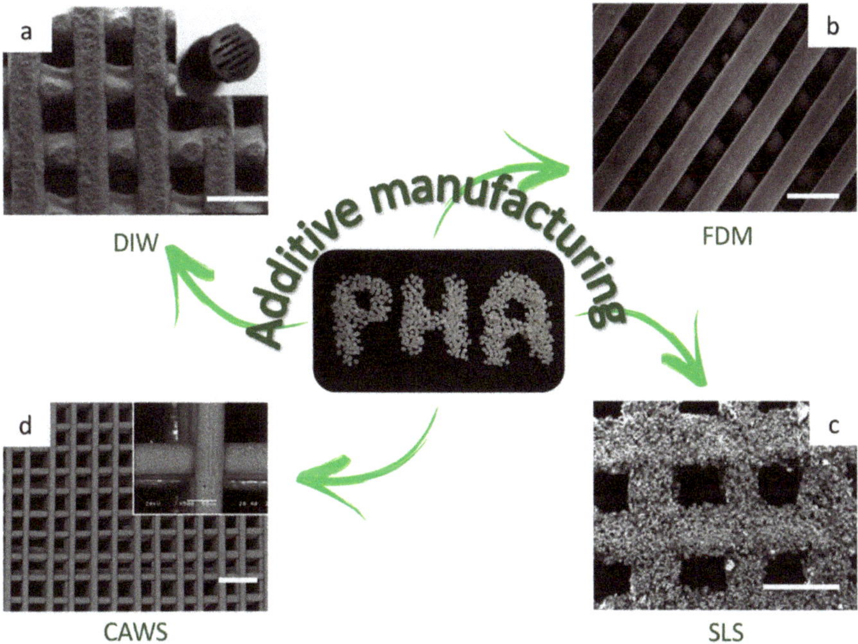

Figure 7. SEM images of PHA scaffolds 3D printed with different AM techniques. (**a**) PHBH scaffolds loaded with anti-tuberculosis drugs 3D printed via Direct Ink Writing (DIW) (scale bar = 500 µm; adapted from Reference [116]). (**b**) PCL/PHBV (50/50) scaffolds 3D printed via Fused Deposition Modeling (FDM) (scale bar = 1 mm; adapted from Reference [107]). (**c**) PHBV scaffolds 3D printed via Selective Laser Sintering (SLS) (scale bar = 500 µm; adapted from Reference [117]). (**d**) Top view of PHBH scaffold 3D printed via Computer Aided Wet-Spinning (CAWS); (insert) detail of the fiber–fiber contact region (scale bar = 500 µm; adapted from Reference [114]).

All presented techniques used with PHAs for biomedical-device production are summarized and compared in Table 1, with an evaluation of the main advantages and disadvantages of each method.

Table 1. Outline and comparison of the traditional and additive manufacturing (AM) techniques used to produce medical devices from PHA.

	Technique	Final Device Shape	Advantage	Disadvantage	Reference
Traditional Techniques	Solvent Casting	film/membrane	+ Easiness and low cost	− Limited to 2D structure − Use of organic solvent − No control on stress formation during drying process	[118]
	Salt Leaching	scaffold	+ Easiness and low cost + Indirect control on pore size	− Small thickness − No customization	[119]

Table 1. *Cont.*

	Technique	Final Device Shape	Advantage	Disadvantage	Reference
	NIPS	film/membrane	+ Easiness and low cost	− Use of organic solvent − No control on final geometry	[120]
	Emulsification	microspheres	+ High surface/volume ratio	− Limited design freedom − No significant 3D development	[121]
	Electrospinning	microporous film	+ Easiness and low cost	− No significant 3D development (thin films) − Dependent on environmental conditions (humidity)	[46,121]
AM techniques	DIW	scaffold	+ Rapidity of processing + Complex geometries + High resolution (low layer height)	− Use of organic solvent − Solvent evaporation (post printing)	[103]
	FDM	scaffold	+ Easiness and low cost + Fast printing speed + Roughness of surface (cell attachment) + Solvent-free process + Complex geometries	− Lower resolution − High temperature processing	[104,105,108]
	SLS	scaffold	+ No need to support material	− Big minimal amount of material − High and not controlled porosity due to not perfect sintering	[122]
	CAWS	scaffold	+ Rapidity of processing + High resolution (low layer height)	− Use of organic solvent	[123]

4. Different Biomedical Applications: From Conventional to Innovative Technologies

In the following sections, for a better and clearer understanding for the reader, we decided to use an iterative structure of the paragraphs, dividing every application according to its final utilization: drug delivery, vessel stenting, bone tissue engineering, and cartilage tissue engineering. Then, for each different medical purpose, initially the traditional fabrication techniques of PHA devices are described, highlighting the most important results obtained. Beyond this, the results achieved with AM techniques are illustrated. Particular importance is given to the advancements that AM techniques introduced in the biomedical field, and to the overcoming of some big limitations, which were encountered with traditional techniques.

4.1. Drug Delivery

Drug delivery was the first biomedical application for PHAs that was investigated [124], and in 1983, Korsatko et al. published the first research work for long term medication dosage [125]. Since then, the use of PHAs as drug carriers met a good success in the biomedical field due to their cytocompatibility and their biodegradation properties in different environments. Particularly for drug carriers, the mechanism of PHA extracellular degradation is important since it is strictly related to the amount and the rate of drug released. The basic idea is to degrade the PHA polymer chains into simpler oligomers or monomers and this can occur via lipase-catalyzed chain scission reactions [126] or via PHA depolymerases enzymatic degradation [127]. Both of them substantially hydrolyze carboxyl-ester bonds in alkanol and alkanoic acid, but they differ according to the substrate preference: lipids for lipases and PHA for depolymerases. However, even lipases showed a degradation activity with PHA polymers [128].

The factors that influence the degradation rate of PHA are different and they can be substantially distinguished between environmental factors and intrinsic PHA properties. Generally, we can state that PHA degrades faster in areas with abundance of bacteria, due to an easy colonization of the biopolymer surface by these microorganisms [129]. However, we have to consider also the PHA chemical structure; for example, if we consider PHA with aromatic side chains, not all microorganisms can decompose them [130]. It was found that an increase in anaerobic conditions [131], temperature [132], and humidity [133] can increment, as well, the degradation rate of the PHA, similar to other biodegradable polymers. On the contrary, an inverse correlation was found between the degradation rate and some properties of the PHA, such as the side chain length [134], the molecular weight, and the degree of crystallinity [75]. Therefore, a useful aspect of this biopolymeric family is the possibility of foreseeing a tunable degradation of the final device, according to the particular application. In Table 2, the main correlations between affecting factors and degradation rate are summarized.

Table 2. Main correlations between the degradation rate and affecting factor of degradation. The ↑ symbol indicates an increase; the symbol ↓ indicates a decrease.

	Factor	Degradation Rate	Reference
Environmental factor	↑ microbial population	↑	[129]
	↑ anaerobic condition	↑	[131]
	↑ temperature	↑	[132]
	↑ humidity	↑	[133]
PHA Properties	↑ side chain length	↓	[134]
	↑ degree of crystallinity	↓	[75]
	↑ molecular weight	↓	[75]

The traditional technique that has undoubtedly met the greatest success is the **emulsification** process, which generates nanoparticles that can be loaded with antimicrobial agents or any other drug. One of the first experiments that used the emulsification/solvent diffusion method is dated back to 2008: Yao et al. realized a drug-delivery system that was composed of PHA nanoparticles, phasin (PhaP), and protein ligands. Varying the protein ligands, these systems were tested both in vitro for macrophages hepatocellular carcinoma and in vivo for liver hepatocellular carcinoma. PHAs were suitable for this application, because, due to their hydrophobicity, they had a good affinity with hydrophobic drugs, such as PhaP bound with ligands, which are able to pull the PhaP–PHA nanoparticles to the targeted cells [89].

Xiong et al. demonstrated, for the first time in 2009, the efficiency of employing PHB and PHBH nanoparticles for intracellular controlled drug release via endocytosis by macrophages, which allow the delivery into the cells without receptor mediation. The intracellular drug release was monitored by the amount of change in cells of the retained lipid-soluble colorant, rhodamine B isothiocyanate (RBITC). Both the PHB and PHBH nanoparticles were prepared at two different average sizes of 160 and 250 nm, with a classic **emulsification** procedure, using dichloromethane as organic solvent. It is noteworthy that the drug-loading efficiency decreases with the increase of the PHA nanoparticles dimensions. This study showed that PHA is a class of biopolymer particularly convenient for this application. In fact, it was proved that PHA uptake by macrophages was not harmful for cell viability; moreover, the use of PHA nanoparticles as carriers extends the drug release time. A control sample of free RBITC, not loaded in nanoparticles, was directly added into the culture medium and absorbed by the macrophages in a week. Differently, the use of PHA nanoparticles led to an intracellular sustained drug release period of at least 20 days, meaning an almost threefold increase in drug release time [87].

More recently, Luo et al. used the **emulsification** technique to produce some PHBH-based polymer micelles loaded with docetaxel (DTX) for melanoma treatment. The PHBH-based system is particularly useful to encapsulate DTX, because it avoids using nonionic surfactants that are currently employed for marketed DTX product and that are reported to cause hemolysis, hypersensitivity reactions, or neuro-toxicity. Interestingly, this micelle formulation shows a drug loading efficiency higher than 90%, it improves DTX solubility in aqueous medium and it reduced hemolysis for better blood compatibility. In vivo tests were run by subcutaneous inoculation of a solid tumor, A375 cells, in a mouse and then applying and comparing a control test with PBS (Phosphate Buffered Saline), a marketed DTX treatment and a DTX-loaded PHBH-based micelle treatment. After a week, the results showed an expected increase of about 450% in final tumor volume for the control group, whereas with commercial DTX and experimental micelle the melanoma underwent a volume reduction of 50% and 80%, respectively. Therefore, the results, shown in Figure 8a, demonstrated not only a better blood compatibility but also a better inhibitory ability of the DTX-loaded PHBH-based micelle, compared to a commercial DTX treatment [135].

Rebia et al. produced a fully natural nanofiber composite via **electrospinning** that can mimic the native extracellular matrix (ECM), and therefore increase the compatibility with the host body. The researchers loaded a PHBH matrix with natural antibacterial reagents (*Centella*, propolis, and hinokitiol) to produce antibacterial wound dressings. The obtained structures have a thickness varying from 50 to 140 µm, and they can withstand only moderate mechanical stresses. The in vitro antibacterial activity was evaluated by using the inhibition zone method both for Gram-positive bacteria, tested with *S. aureus*, as well as for Gram-negative bacteria, tested with *E. coli*. The results with propolis and hinokitiol loading gave promising outcomes (Figure 8b) [91].

Traditional techniques are positively used to fabricate drug nanocarriers, but the biggest limitation is that the obtained devices have a very low versatility in the design structures, which are thin membranes in the case of electrospinning or nanospheres obtained by emulsification. Therefore, the introduction in this application field of AM permitted to obtain complex architectures, extended in all three dimensions, which could operate not only as drug carriers but also as structural support, in the target site.

Duan et al. was one of the first researchers that investigated the 3D printability of PHA for biomedical application. Starting from a micropowder, obtained by double emulsion solvent evaporation method, the researchers decided to further use it, not as a simple drug carrier, but as a powder bed for **SLS** technique. First, a calcium phosphate (Ca-P)/PHBV composite powder loaded with bovine serum albumin (BSA) was prepared, and then the scaffolds (L × W × H = 8 × 8 × 15.5 mm^3) were designed and 3D printed [136].

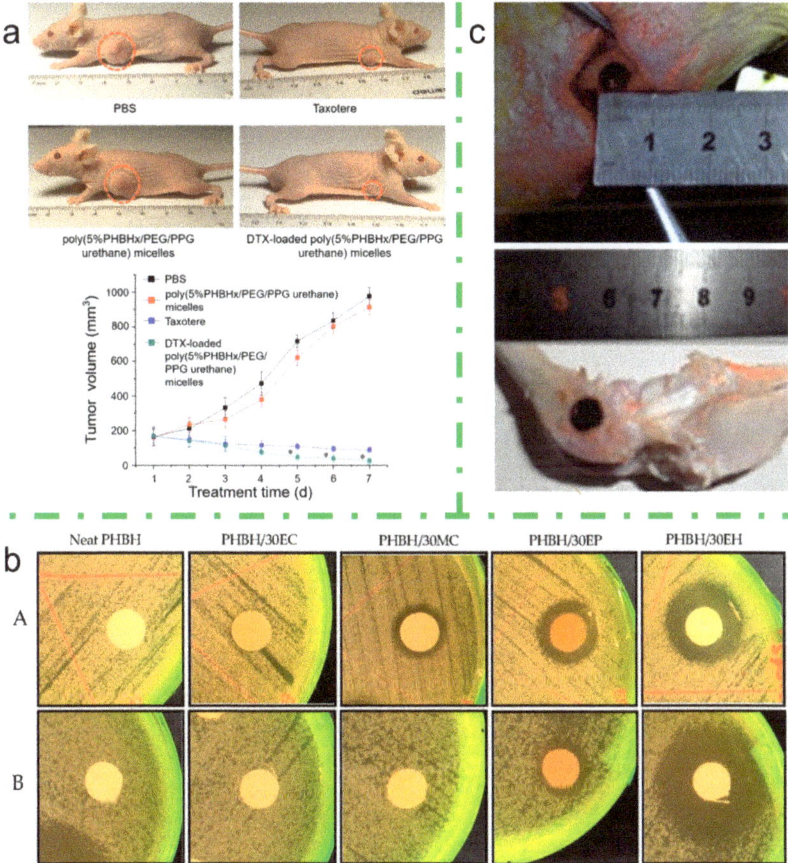

Figure 8. Experimental PHA drug release applications. Panel (**a**) shows in vivo investigation of mice melanoma treatment with PHBH-based polymer micelles loaded with docetaxel (DTX-loaded poly(5%PHBHx/PEG/PPG urethane) and different control groups: Phosphate Buffered Saline (PBS), a commercial docetaxel treatment (Taxotere), and unloaded PHBH-based polymer micelles (poly(5%PHBHx/PEG/PPG urethane). Visual appearance of subcutaneous tumor sizes and tumor volume measurements, within treatment time, are displayed at the top and bottom of the panel, respectively (adapted from Reference [135]). Panel (**b**) represents the inhibition zones of neat PHBH and PHBH composite electrospun nanofibers with centella (30EC) and (30MC), propolis (30EP), and hinokitiol (30EH) on Gram-positive bacteria (*S. aureus*) (A) and Gram-negative bacteria (*E. coli*) (B) (adapted from Reference [91]). Panel (**c**) shows two photographs of a cylindrical scaffold (D × H = 6 × 8 mm^2) 3D printed via DIW and implanted in a rabbit's femur for post-surgical treatment of osteoarticular tuberculosis (adapted from Reference [137]).

Li et al. and Zu et al. suggested an interesting application for a mesoporous bioactive glass (MBG) and PHBV composite, 3D printed via **DIW** starting from a polymer ink dissolved in chloroform and dimethyl sulfoxide. The final goal of this application is meant for post-surgical treatment of osteoarticular tuberculosis, and specifically the 3D-printed scaffolds can be implanted in the surgical defect, combining the osseous regeneration effect with the release of an antituberculotic drug, such as isoniazid or rifampin. The studies investigated in vitro drug release and cellular proliferation, and in vivo surgical procedure was run, implanting the 3D-printed cylindrical scaffolds (D × H = 6 × 8 mm^2) into the femur of different rabbits, represented in Figure 8c. Besides the osteogenetic feature of this material, another attractive property is the slower and controlled release of antituberculotic drug, up to three months, lengthening the healing period and reducing systemic side effects [116,137].

Wu et al. investigated the possibility of 3D printing a clinical device via **FDM**, which could also have an antibacterial activity. They melt-compounded a maleic anhydride grafted PHA (PHA-g-MA) with multi-walled carbon nanotubes (MWCNTs) for the production of a FDM filament, which can be further used to 3D-print different geometries according to the final application. Only a preliminary study of the antimicrobial assay was tested with the inhibition zone method both for Gram-positive bacteria, tested with *S. aureus*, and for Gram-negative bacteria, tested with *E. coli*. Generally, the tested samples demonstrated a higher inhibition zone for *E. coli* rather than *S. aureus*; however, for both class of bacteria, the results showed an increase in antibacterial performance following an increase in MWCNTs content [106].

4.2. Vessel Stenting

One of the most recent developing field of PHA application is the stent vessel production, since biodegradable stents can provide mechanical support while it is needed, for example, for obstructive cardiovascular disease treatments, and then degrade, leaving behind only the healed natural vessel, without any foreign objects in the body.

For vascular application, the most important biological property of PHA to investigate is the hemocompatibility, for example with an erythrocyte contact hemolysis assay. The easiest way to do that was to prepare **solvent cast** films. Qu et al. fabricated samples of PHB, PHBV, and PHBH. Comparing all the films, the best results were obtained with PHBH films, which showed a two-fold reduced hemolytic activity and also a lower number of bound blood platelets, after a 120-min exposure to platelet-rich plasma [138]. Zhang et al. tried to improve other important properties of PHBH, in order to enhance the applicability of this PHA in vascular engineering. Particularly, they blended PHBH and poly(propylene carbonate) (PPC) to obtain a higher flexibility, evidenced by an increase in elongation at break [118].

The former studies were fundamental to characterize and to state the possible use of this class of polyester for vascular engineering applications. However, there was a big technological issue with this traditional technique, because solvent casting is not suitable for the production of final devices with complex and 3D structures, which are meant to be implanted in human blood vessels. Gao et al. suggested the use of **electrospinning** to fabricate two kinds of PHBH vascular grafts, including straight and corrugated structures with 6 mm inner diameters. These devices have been tested mechanically, to undergo radial compression and circumferential tensile stresses, as well as for suture retention strength and radial compliance. Moreover, the biocompatibility was evaluated in vitro with hemolytic and cytotoxicity tests. The results obtained in this study demonstrated good application value in the field of stent vessel engineering, even comparing the final properties of the experimental grafts with those of commercial ones [139]. Electrospinning is a well-known technique for production of microporous films, but the production of devices with an actual 3D structure is time-consuming. For example, in this study, the realization time of a vascular graft with a thickness of 200 μm took 6 h. Figure 9a shows the final macroscopic aspect of such electrospun PHBH vascular grafts.

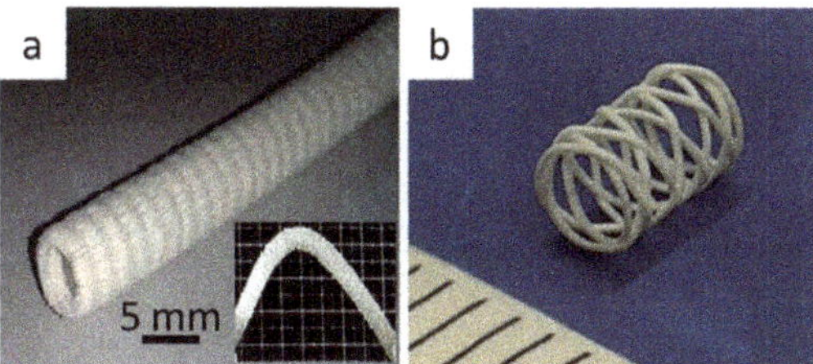

Figure 9. Experimental PHA applications for vessel stenting. (**a**) Macro morphology of corrugated tubular PHBH scaffold obtained via electrospinning (adapted from Reference [139]). (**b**) Representative photograph of a stent 3D printed via CAWS for small-caliber blood vessels (measure unit = 1 mm; adapted from Reference [123]).

Even if the volume of research is still limited, the innovation that AM introduced in this subject of study is noteworthy. Balogová et al. carried out a preliminary study for production of urethra replacement via AM. They prepared a prototype via **DIW** a PLA/PHB tubular structure with the same length and thickness of the aforementioned vascular grafts, which only took 10 min. Compared to the previous research, the use of AM allowed a 36-fold reduction in production time, which is an evident advantage for technological applications. In this first research study, the authors focused on the technological aspect of the production, and they investigated only geometrical and viscoelastic properties of 3D-printed samples, such as the shape retention over time and the deviation from designed sizes. It is possible to state that DIW had sufficient precision to produce tubular samples usable as a replacement for urethra; further mechanical and biological characterizations have to be done to further validate the in vivo implantation [140].

Puppi et al. realized via **CAWS** some PHBH stents for small-caliber blood vessels, and one example is shown in Figure 9b. The developed stents sustained proliferation of human umbilical vein endothelial cells in vitro, and they showed encouraging low levels in terms of thrombogenicity when in contact with human blood. Besides the advance in medical application, this study is also technologically interesting because it widened the field of application of the CAWS technique. It introduced a novel approach that allows the construction of 3D tubular structures by winding the coagulating wet-spun biopolymer fiber around a rotating mandrel with a predefined pattern. The biopolymer solution is extruded through a needle directly above a rotating mandrel immersed in a non-solvent bath of ethanol; the movement of the needle and the mandrel rotational velocity were controlled by an experimental computer-controlled system. The presented technique showed a great versatility in the customization of stent fabrication [141].

4.3. Tissue Engineering

A challenging frontier of modern medicine is the repairing of damaged tissue of the human body, and it is called regenerative medicine. The main goal of this particular application field is to promote and enhance the formation of new viable tissues by biochemical and cellular processes. A key feature is represented by the positive effects of biocompatible materials and the innovations of technologies that can enhance the fabrication of devices able to simulate the original body environment. In order to achieve this, a connection among different disciplines (biomedicine, material science, and engineering) has to be done, and for this reason, a new interdisciplinary research field was created, i.e., tissue engineering. Due to the good cytocompatibility and to the tunable mechanical properties and degradation rate, PHA demonstrated to be suitable for both hard tissues, i.e., bone and cartilage, and soft tissues [142], nerve, tendon, bone marrow, or vascular applications. In tissue engineering the device customization is a great advantage; therefore, we can state that this area is the most promising and with the highest potential for biomedical 3D printing.

4.3.1. Bone Tissue Engineering

One of the first in vitro research studies of biocompatibility of PHBH was conducted by Yang et al., and they demonstrated that bone marrow stromal cells can attach, proliferate, and differentiate into osteoblasts on PHBH films, obtained by **solvent casting** [143]. Wang et al. used the **salt-leaching** technique to obtain porous scaffolds, in order to demonstrate an increased attachment and proliferation of bone marrow cells, as well as an earlier osteogenesis, onto a rough surface. The optimal pore size detected is about 3 µm in diameter. In this study, PHBH scaffolds (Figure 10a) showed better performance for osteoblast proliferation rather than PHB and PLA scaffolds [144]. The same authors investigated also the compounding of PHB and PHBH with hydroxyapatite (HAP), and they found that the mechanical properties (compressive elastic modulus and maximum stress) and the osteoblast response improved for the PHB matrix and decreased for the PHBH blend [145]. More recently, Wu et al. studied how to enhance the cell compatibility of the PHA matrix varying the surface morphology of the solvent cast film by compounding the PHBH with carbon nanotubes (CNTs), which resulted in a higher surface roughness and an electrical conductivity. The proliferation of human mesenchymal stem cells (hMSCs) were demonstrated to be outstanding when nanocomposite films contained 1 wt% CNTs, compared with that on pristine PHBH [77].

Assuming that porosity is an increasing factor of cellular proliferation, Xi et al. investigated the possibility of controlling it and they identified **TIPS** as a straightforward technique that allows the regulation of the scaffold pore diameters by varying the quenching temperature and time. The researchers obtained a series of interconnected highly porous scaffolds with pore sizes ranging from 30 to 150 µm. They demonstrated that the pore diameter decreases with decreasing quenching temperature and consequently also the overall porosity of the scaffold [146].

Figure 10. Experimental PHAs applications for bone tissue engineering. (**a**) SEM images of a porous PHBV scaffold obtained by salt leaching (scale bar = 10 µm; adapted from Reference [144]). (**b**) SEM micrographs of PHBH/silk fibroin (1:1) electrospun films after 14 days of human-umbilical-cord-derived mesenchymal stem cells culture. The white arrow indicates the cells homogeneously distributed on the microporous film (scale bar = 150 µm; adapted from Reference [92]). (**c**) Visual appearance of Ca-P/PHBV scaffolds loaded with BSA and 3D printed via SLS, using different sintering parameters: (A) laser power = 12.5 W and scan spacing = 0.1 mm; (B) laser power = 15 W and scan spacing = 0.1 mm; (C) laser power = 15 W and scan spacing = 0.15 mm (adapted from Reference [136]). (**d**) Visual appearance of PHBH scaffolds 3D printed by CAWS (adapted from Reference [114]). (**e**) Calcium phosphate (Ca–P)/PHBV nanocomposite 3D printed via SLS for the fabrication of a proximal femoral condyle scaffold (scale bar = 1 cm; adapted from Reference [147]).

Ang et al. successfully fabricated **electrospun** films made of PHBH compounded with silk fibroin (SF), and these devices were able to support the human umbilical cord-derived mesenchymal stem cells proliferation and differentiation into the osteogenic lineage. The obtained electrospun films are in the form of a porous matrix with randomly distributed fibers, with an average diameter in the range of 600 and 980 nm. The mean pore diameter of the electrospun films ranged from 1 to 1.5 μm. Silk fibroin demonstrated an enhancing effect on the proliferation and osteogenic differentiation of stem cells, compared to the pristine PHBH. In Figure 10b, the spread of the cells over the electrospun membrane is shown [92].

Even if the techniques described so far have been instrumental in starting to investigate the use of PHA for bone regeneration, traditional scaffolds have some big limitations, such as very little thickness (i.e., hundreds of microns) and no real direct control over porosity, nor over the dislocation and size of the pores. These aspects have been positively overcome with the use of AM, which has widened the range of application. Three-dimensional printing allows us to build geometries with customized and controlled designs, including the internal pattern, and with a development even in height of several centimeters.

SLS was the first AM technique used to fabricate PHA 3D scaffolds. Pereira et al. realized a tetragonal structure squared base (13×13 mm^2) and 26 mm high, with a designed porosity of 1 mm^2 area. PHB scaffolds were 3D printed with different properties, due to the change in the values of the scan spacing (SS) and powder layer thickness (PLT). The results showed that a decrease of the values of PLT or SS involved an increase in the compressive mechanical properties of scaffolds, such as ultimate compressive strength and compressive modulus [110].

Duan et al. studied a system that provided a biomimetic environment for cell attachment, proliferation and differentiation, based on a composite of PHBV compounded with calcium phosphate (Ca-P) nanoparticles, which was proved to be an osteoconductive component. The researchers carried out a study aimed to optimize the **SLS** 3D-printing parameters, i.e., laser power, scan spacing, and layer thickness, according to the final resolution and mechanical properties of a tetragonal porous scaffold (L $\times$ W $\times$ H = $8 \times 8 \times 15.5$ mm^3), of which three examples are shown in Figure 10c [122]. The final nanocomposite revealed to have not only positive mechanical properties but also good cytocompatibility, tested with a human osteoblast-like cell line [117]. To prove the possibility of using the SLS technique for real medical applications, a human proximal femoral condyle model was obtained from computer tomography scans and then 3D printed into a porous scaffold model with a pore size of 2 mm; an image of this medical prosthesis is shown in Figure 10e [147].

In 2013, **DIW** was investigated by Yang et al. for the first time, among all extrusion-based AM approaches, as a possible technique for PHA bone scaffolds production. Yang et al. fabricated composite scaffolds made of PHBH and mesoporous bioactive glass (MBG) through a combination of 3D printing and surface doping. The MBG coating was found to improve surface hydrophilicity and bioactivity, as well as provide a better environment for human mesenchymal stem cells viability, proliferation, and osteogenic differentiation [148]. Based on the promising results of in vitro biological characterization of the nanocomposite, the research was further carried out by Zhao et al., who selected MBG/PHBH composite scaffolds 3D printed via DIW for in vivo evaluation of osteogenic capability. The scaffolds stimulated bone regeneration in rat calvarial defects within eight weeks [149].

Li et al. studied a real application case of interest for a PHBH and MBG drug-loaded scaffold for osteoarticular tuberculosis. After surgery, it is necessary to fill the surgical defect with an implant, which can combine the effects of osseous regeneration and antitubercular drug (e.g., isoniazid and rifampin) local delivery to treat the area affected by the disease and to avoid internal infections. The researchers 3D-printed, via **DIW**, a cylindrical porous scaffold with a height of 8 mm, a diameter of 6 mm, and an area of each pore of 0.25 mm^2. The AM technique was particularly useful in this application, to realize a customized device that could perfectly fit to the size of the hole surgically drilled into the treated bone. The structure was tested both for in vitro compatibility and in vivo implantation in a rabbit femur defect model (Figure 8c). Microtomography evaluations and histology results indicated part degradation of the composite scaffolds and new bone growth in the cavity [116,137].

Mota et al. explored another innovative 3D-printing technique, which was used for the first time with PHA, the **CAWS**. In this study, PHBH 3D-printed scaffolds with different pore sizes and internal architectures were fabricated layer-by-layer, and the processing parameters were investigated for optimization of mechanical compressive properties and biological evaluation. The scaffolds showed a porosity of 79–88%, an extruded filament diameter of 47–76 μm, and a pore size of 123–789 μm; hence, this AM technique allowed the fabrication of scaffolds with a high resolution and a good control over scaffold external shape and internal pattern. The PHBH scaffolds demonstrated also promising results in terms of cell differentiation towards an osteoblast phenotype [114]. Puppi et al. carried on the investigation on PHBH 3D printing for bone scaffold regeneration via CAWS with a pristine PHBH matrix (Figure 10d) [115] and with a PHBH/PCL blend composition [113]. All results showed a promising applicability for in vivo studies and implantations. Recently, they published a work where they used a ternary mixture of PHBH/chloroform/ethanol to prepare the polymeric ink to be used in the 3D printer. With this method, they suggested a more sustainable CAWS process for PHBH scaffolds production, which reduces the employment of halogenated solvent by replacing with ethanol up to 40 $v/v\%$ of the chloroform employed. Besides thus, they evaluated the effect of varying the solvent/non-solvent ratio on structural morphology, such as macro- and microporosity, on tensile properties and on in vitro preosteoblast cells proliferation [112].

4.3.2. Cartilage Tissue Engineering

Differently from bone regeneration, cartilage structure cannot be self-recreated and an excessive wear of this tissue can lead to a cartilage loss and to osteoarthritis problems. Currently, the most common treatments involve only the use of painkillers or surgeries, such as microfracture, osteochondral transfer or autologous chondrocyte implantation. However, these treatments present no actual restoration of cartilage tissue and, in general, an unsatisfactory average long-term result [150]. In the last decade, a new approach for cartilage repair was suggested, and it consists in the use of engineered scaffolds able to support the growth of chondrocytes. However, still further research is required to develop suitable scaffolds, because the neo-generated tissue is often fibrocartilage, which is mechanically inferior and less durable than the one found in healthy articular joints [151]. Since the beginning of the investigation, a particular interest was attributed to PHA as interesting material for the recreation of a favorable environment for the growth of chondrocytes from stem cells. The first works focused on the interaction of chondrocytes with polymer matrices. Deng et al. blended PHBH and PHB and then porous scaffolds were fabricated by the **salt-leaching** method. In order to evaluate the compatibility with this material and the production of extracellular matrix, the chondrocyte cell lines were isolated from rabbit articular cartilage, seeded on the scaffolds and incubated over 28 days [80,152]. Following research explored the best ratios between different component polymers, which could positively combine mechanical properties and biological compatibility. Considering collagen II as a differentiation marker of chondrocytes maturation, blended scaffolds of PHB and PHBH (ratio 1:2) gave the best results, compared with other ratios of PHB/PHBH or even with PLA [153].

The **TIPS** technique was used as another simple approach to fabricate PHB/PHBH porous scaffold upon which human adipose-derived stem cells (hASCs) were seeded to produce neocartilage, subsequent to a chondrogenic differentiation in vitro process. After 14 days of in vitro culture, the differentiated cells grown on the PHB/PHBH scaffold were implanted into the subcutaneous layer nude mice and after 24 weeks, the appearance of a new cartilage-like tissue could be observed [94]. To develop a higher and more homogeneous cell proliferations over the PHBH scaffolds, You et al. experimented a biological coating of the biopolymer scaffolds with PHA granule binding protein (PhaP) fused with RGD peptide (PhaP-RGD coating). Human bone marrow mesenchymal stem cells (hBMSCs) were inoculated in the scaffolds and the findings showed that the proposed PhaP-RGD coating led to a more homogeneous spread of cells, and to a better cell adhesion, proliferation and chondrogenic differentiation [84].

All mentioned research studies provided a strong and valid basis to start investigating the applicability of PHA matrices for cartilage tissue engineering, but a big limitation was represented by the geometrical constraint in the final shapes of the devices obtained by TIPS or salt leaching. Starting from a real case study, Sun et al. analyzed a possible and new route to build and replace a damaged laryngeal cartilage. The noteworthy innovation of this work was the construction of a hollow, semi-flared geometry prepared by a **combination of solvent casting, compression molding** in a polytetrafluorethylene form, **and salt-leaching** methods. The morphology of the implant was shaped according to the anatomy of an adult laryngeal cartilage, as can been seen in Figure 11a. First, chondrocytes were inoculated onto the PHBH scaffold, and after one week of in vitro culture, an in vivo implantation was performed and the results showed that cartilage formed six weeks after the surgery (Figure 11b) [154].

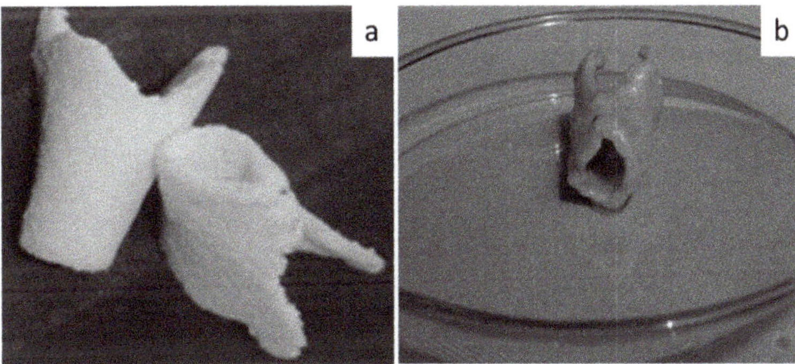

Figure 11. Experimental PHAs applications for cartilage tissue engineering. (**a**) Representation of a PHBH medical device prepared by solvent casting, compression molding, and particulate filtering, with a final hollow semi-flared shape, which intends to mimic the laryngeal cartilage morphology (adapted from Reference [154]). (**b**) Photograph of the laryngeal cartilage PHBH specimen with chondrocytes inoculated, 18 weeks after implantation (adapted from Reference [154]).

The former work had the great advantage to allow the construction of a complex-shaped device; nevertheless, the experimental procedure for the scaffold fabrication is long and expensive, since it involves using a plastic mold, which should be, every time, customized according to the final implant. Moreover, organic solvent and long times of evaporation need to be estimated. With AM, these limitations could be easily overcome, because starting from a different CAD model, the need for the mold would be completely eliminated. Moreover, 3D printing would allow the fabrication of personalized and complex structures, which could encourage cellular growth in preferential directions or which could have architectures that optimize the contact and the stress transmission between bone and cartilage, for example, in the case of articular cartilage. To the authors' knowledge, there is only one recent work dealing with 3D printing of PHA scaffolds for cartilage tissue engineering. De Pascale et al. assessed the properties of collagen I hydrogel 3D scaffolds, strengthened with solvent cast and 3D-printed PHA polymer. The addition of solvent cast and 3D-printed scaffolds increased the mechanical resistance of the structures when compared to the collagen matrix only. Once again, the use of AM technique was an advantage related to traditional techniques, because regarding the compressive stress that the device could undergo, 3D-printed scaffolds showed the highest stiffness compared to the collagen and solvent cast polymer samples [155].

5. Future Perspective

The introduction of PHA and AM in the biomedical field has boosted the advancements of innovative solutions for problems that were so far totally or partially unresolved. The main reasons for this success was certainly due to the high level of customization brought by AM and by the possibility of tailoring the final mechanical properties of 3D-printed materials, in order to mimic the tissue environment. Besides, also the tunable and interesting properties of PHAs played a central role, for example the wide processing and application versatility, the biological origin, the biocompatibility and the biodegradability. Among AM techniques, FDM owns some well-known advantages, namely its simplicity, rapidity, and ecological sustainability; in fact, it does not require the use of any organic solvent. However, FDM used with PHA for biomedical application is still limited; however, according to the abovementioned properties and advantages of PHA and FDM, we believe that its use will be increasingly investigated and the number of 3D-printed devices by FDM will grow significantly in the next years.

In the field of PHA 3D-printed medical devices, the most promising results were obtained with non-toxic and safely resorbable scaffolds containing living cells that were used for hard tissue regeneration, bone and cartilage particularly. However, no studies were carried on the production of more complex-shaped devices like prosthesis or surgical implants, because these are generally 3D printed with synthetic biopolymer, such as PCL.

The production of synthetic polymer requires the use of chemical solvents, different catalysts (e.g., metal-based, organic, or even enzymatic systems) and also reaction conditions that are particularly energy consuming [156]. If compared to a bacterial synthesis of PHAs, it is quite evident the inconvenience in terms of ecological sustainability. As an indication of possible future developments, in Figure 12 a PHBH clavicle plate 3D printed by FDM is shown, which could be used to treat a broken fracture. Especially due to the resorbability, to the biocompatibility and to the osteogenesis induction of PHAs, this class of material allows us to think of a future medicine, where all components are bio-based, perfectly compatible with human body and devices can be harmlessly reabsorbed by our organism, when they are not needed anymore.

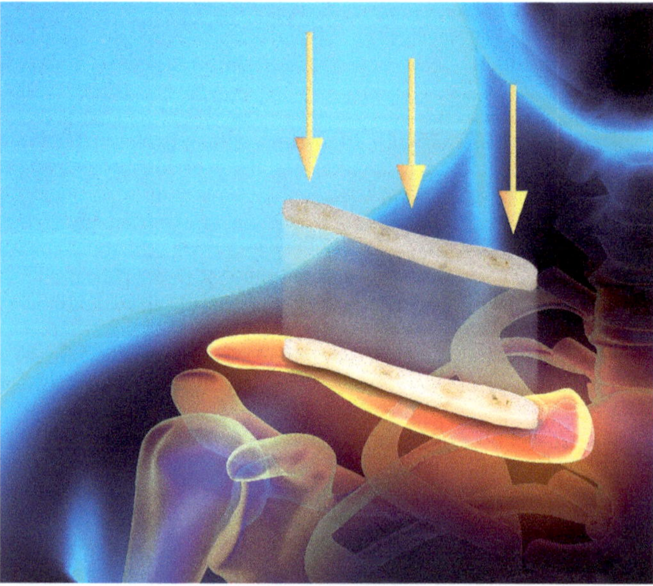

Figure 12. Graphical representation of a PHBH clavicle plate 3D printed by FDM and its final surgical application for bone regeneration (adapted from istock.com/yodiyim, accessed on 21 September 2020).

In conclusion, we can foresee a quick and important development in this research field and we think that the next frontier and challenge in biomedical application of PHA could be the 3D printing by FDM of entire prosthesis, or complex surgical implants, which can replace the materials used until now, and which will notably improve the biomedical knowledge and technological state-of-the-art.

Author Contributions: All authors have contributed to the conceptualization and the writing of this manuscript. All authors have read and agreed to the published version of the manuscript.

Funding: This research received no external funding.

Conflicts of Interest: The authors declare no conflict of interest.

References

1. ASTM D6866-20. *Standard Test Methods for Determining the Biobased Content of Solid, Liquid, and Gaseous Samples Using Radiocarbon Analysis*; ASTM Standard: West Conshohocken, PA, USA, 2020.
2. ISO 14855-1:2012. *Determination of the Ultimate Aerobic Biodegradability of Plastic Materials under Controlled Composting Conditions—Method by Analysis of Evolved Carbon Dioxide—Part 1: General Method*; ISO Standard: Geneva, Switzerland, 2012.
3. EN 13432:2000. *Packaging—Requirements for Packaging Recoverable through Composting and Biodegradation—Test Scheme and Evaluation Criteria for the Final Acceptance of Packaging*; EN Standard: Brussels, Belgium, 2000.
4. Tarrahi, R.; Fathi, Z.; Seydibeyoğlu, M.Ö.; Doustkhah, E.; Khataee, A. Polyhydroxyalkanoates (PHA): From production to nanoarchitecture. *Int. J. Biol. Macromol.* **2020**, *146*, 596–619. [CrossRef] [PubMed]
5. Velu, R.; Calais, T.; Jayakumar, A.; Raspall, F. A comprehensive review on bio-nanomaterials for medical implants and feasibility studies on fabrication of such implants by additive manufacturing technique. *Materials* **2020**, *13*, 92. [CrossRef]
6. Li, Y.; Feng, Z.; Hao, L.; Huang, L.; Xin, C.; Wang, Y.; Bilotti, E.; Essa, K.; Zhang, H.; Li, Z.; et al. A review on functionally graded materials and structures via additive manufacturing: From multi-scale design to versatile functional properties. *Adv. Mater. Technol.* **2020**, *5*, 1900981. [CrossRef]
7. Li, P.; Wang, X.; Su, M.; Zou, X.; Duan, L.; Zhang, H. Characteristics of plastic pollution in the environment: A review. *Bull. Environ. Contam. Toxicol.* **2020**. [CrossRef] [PubMed]
8. George, A.; Sanjay, M.R.; Srisuk, R.; Parameswaranpillai, J.; Siengchin, S. A comprehensive review on chemical properties and applications of biopolymers and their composites. *Int. J. Biol. Macromol.* **2020**, *154*, 329–338. [CrossRef]
9. Ray, S.; Kalia, V.C. Microbial cometabolism and polyhydroxyalkanoate Co-polymers. *Indian J. Microbiol.* **2017**, *57*, 39–47. [CrossRef] [PubMed]
10. Kumar, M.; Rathour, R.; Singh, R.; Sun, Y.; Pandey, A.; Gnansounou, E.; Andrew Lin, K.Y.; Tsang, D.C.W.; Thakur, I.S. Bacterial polyhydroxyalkanoates: Opportunities, challenges, and prospects. *J. Clean. Prod.* **2020**, *263*, 121500. [CrossRef]
11. Snell, K.D.; Singh, V.; Brumbley, S.M. Production of novel biopolymers in plants: Recent technological advances and future prospects. *Curr. Opin. Biotechnol.* **2015**, *32*, 68–75. [CrossRef]
12. Zulfiqar, A.R.; Sharjeel, A.; Ibrahim, M.B. Polyhydroxyalkanoates: Characteristics, production, recent developments and applications. *Int. Biodeterior. Biodegrad.* **2018**, *126*, 45–56.
13. Lemoigne, M. Produit de deshydratation et de polymerisation de l'acide β=oxybutyrique. *Bull. Soc. Chim. Biol.* **1926**, *8*, 770–782.
14. Licciardello, G.; Catara, A.F.; Catara, V. Production of polyhydroxyalkanoates and extracellular products using Pseudomonas corrugata and P. mediterranea: A review. *Bioengineering* **2019**, *6*, 105. [CrossRef] [PubMed]
15. Nomura, C.T.; Tanaka, T.; Gan, Z.; Kuwabara, K.; Abe, H.; Takase, K.; Taguchi, K.; Doi, Y. Effective enhancement of short-chain-length—Medium-chain-length polyhydroxyalkanoate copolymer production by coexpression of genetically engineered 3-ketoacyl-acyl-carrier-protein synthase III (fabH) and polyhydroxyalkanoate synthesis genes. *Biomacromolecules* **2004**, *5*, 1457–1464. [CrossRef]
16. Koller, M. Chemical and biochemical engineering approaches in manufacturing polyhydroxyalkanoate (PHA) biopolyesters of tailored structure with focus on the diversity of building blocks. *Chem. Biochem. Eng. Q.* **2018**, *32*, 413–438. [CrossRef]
17. Puppi, D.; Pecorini, G.; Chiellini, F. Biomedical processing of polyhydroxyalkanoates. *Bioengineering* **2019**, *6*, 108. [CrossRef] [PubMed]
18. Wang, Y.; Guo-Qiang, C. Polyhydroxyalkanoates: Sustainability, production, and industrialization. In *Sustainable Polymers from Biomass*; Tang, C., Ryu, C.Y., Eds.; Wiley-VCH Verlag GmbH & Co. KGaA: Weinheim, Germany, 2017; pp. 11–33.
19. Luzi, F.; Torre, L.; Kenny, J.M.; Puglia, D. Bio- and fossil-based polymeric blends and nanocomposites for packaging: Structure-property relationship. *Materials* **2019**, *12*, 471. [CrossRef] [PubMed]
20. Urtuvia, V.; Maturana, N.; Peña, C.; Díaz-Barrera, A. Accumulation of poly(3-hydroxybutyrate-co-3-hydroxyvalerate) by Azotobacter vinelandii with different 3HV fraction in shake flasks and bioreactor. *Bioprocess. Biosyst. Eng.* **2020**, *43*, 1469–1478. [CrossRef]
21. Willson, A.; Takashi, K. Effects of glass fibers on mechanical and thermal properties of Poly(3-hydroxybutyrate-co-3-hydroxyhexanoate). *Polym. Compos.* **2018**, *39*, 491–503.

22. Muneer, F.; Rasul, I.; Azeem, F.; Siddique, M.H.; Zubair, M.; Nadeem, H. Microbial Polyhydroxyalkanoates (PHAs): Efficient replacement of synthetic polymers. *J. Polym. Environ.* **2020**, *28*, 2301–2323. [CrossRef]
23. Aljuraifani, A.A.; Berekaa, M.M.; Ghazwani, A.A. Bacterial biopolymer (polyhydroxyalkanoate) production from low-cost sustainable sources. *Microbiologyopen* **2019**, *8*, 1–7. [CrossRef] [PubMed]
24. Kunasundari, B.; Sudesh, K. Isolation and recovery of microbial polyhydroxyalkanoates. *Express Polym. Lett.* **2011**, *5*, 620–634. [CrossRef]
25. Higuchi-Takeuchi, M.; Morisaki, K.; Toyooka, K.; Numata, K. Synthesis of high-molecular-weight polyhydroxyalkanoates by marine photosynthetic purple bacteria. *PLoS ONE* **2016**, *11*, 1–17. [CrossRef]
26. Reddy, C.S.K.; Ghai, R.; Kalia, V.C. Polyhydroxyalkanoates: An overview. *Bioresour. Technol.* **2003**, *87*, 137–146. [CrossRef]
27. Winnacker, M. Polyhydroxyalkanoates: Recent advances in their synthesis and applications. *Eur. J. Lipid Sci. Technol.* **2019**, *121*, 1–9. [CrossRef]
28. Elmowafy, E.; Abdal-Hay, A.; Skouras, A.; Tiboni, M.; Casettari, L.; Guarino, V. Polyhydroxyalkanoate (PHA): Applications in drug delivery and tissue engineering. *Expert Rev. Med. Devices* **2019**, *16*, 467–482. [CrossRef] [PubMed]
29. Luo, Z.; Wu, Y.L.; Li, Z.; Loh, X.J. Recent progress in polyhydroxyalkanoates-based copolymers for biomedical applications. *Biotechnol. J.* **2019**, *14*, 1–16. [CrossRef]
30. Hazer, B.; Steinbüchel, A. Increased diversification of polyhydroxyalkanoates by modification reactions for industrial and medical applications. *Appl. Microbiol. Biotechnol.* **2007**, *74*, 1–12. [CrossRef] [PubMed]
31. Rai, R.; Keshavarz, T.; Roether, J.A.; Boccaccini, A.R.; Roy, I. Medium chain length polyhydroxyalkanoates, promising new biomedical materials for the future. *Mater. Sci. Eng. R Rep.* **2011**, *72*, 29–47. [CrossRef]
32. Obruca, S.; Benesova, P.; Marsalek, L.; Marova, I. Use of lignocellulosic materials for PHA production. *Chem. Biochem. Eng. Q.* **2015**, *29*, 135–144. [CrossRef]
33. Aneesh, B.P.; Arjun, J.K.; Kavitha, T.; Harikrishnan, K. Production of Short chain length polyhydroxyalkanoates by Bacillus megaterium PHB29 from starch feed stock. *Int. J. Curr. Microbiol. Appl. Sci.* **2016**, *5*, 816–823. [CrossRef]
34. Arikawa, H.; Matsumoto, K.; Fujiki, T. Polyhydroxyalkanoate production from sucrose by Cupriavidus necator strains harboring csc genes from Escherichia coli W. *Appl. Microbiol. Biotechnol.* **2017**, *101*, 7497–7507. [CrossRef]
35. Povolo, S.; Romanelli, M.G.; Fontana, F.; Basaglia, M.; Casella, S. Production of Polyhydroxyalkanoates from Fatty Wastes. *J. Polym. Environ.* **2012**, *20*, 944–949. [CrossRef]
36. Walsh, M.; O'Connor, K.; Babu, R.; Woods, T.; Kenny, S. Plant oils and products of their hydrolysis as substrates for polyhydroxyalkanoate synthesis. *Chem. Biochem. Eng. Q.* **2015**, *29*, 123–133. [CrossRef]
37. Jiang, G.; Hill, D.J.; Kowalczuk, M.; Johnston, B.; Adamus, G.; Irorere, V.; Radecka, I. Carbon sources for polyhydroxyalkanoates and an integrated biorefinery. *Int. J. Mol. Sci.* **2016**, *17*, 1157. [CrossRef]
38. Cui, Y.W.; Shi, Y.P.; Gong, X.Y. Effects of C/N in the substrate on the simultaneous production of polyhydroxyalkanoates and extracellular polymeric substances by Haloferax mediterranei via kinetic model analysis. *RSC Adv.* **2017**, *7*, 18953–18961. [CrossRef]
39. Raza, Z.A.; Tariq, M.R.; Majeed, M.I.; Banat, I.M. Recent developments in bioreactor scale production of bacterial polyhydroxyalkanoates. *Bioprocess. Biosyst. Eng.* **2019**, *42*, 901–919. [CrossRef] [PubMed]
40. Sreekanth, M.S.; Vijayendra, S.V.N.; Joshi, G.J.; Shamala, T.R. Effect of carbon and nitrogen sources on simultaneous production of α-amylase and green food packaging polymer by Bacillus sp. CFR 67. *J. Food Sci. Technol.* **2013**, *50*, 404–408. [CrossRef]
41. Aramvash, A.; Moazzeni Zavareh, F.; Gholami Banadkuki, N. Comparison of different solvents for extraction of polyhydroxybutyrate from Cupriavidus necator. *Eng. Life Sci.* **2018**, *18*, 20–28. [CrossRef] [PubMed]
42. Ramsay, J.A.; Berger, E.; Voyer, R.; Chavarie, C.; Ramsay, B.A. Extraction of PHB using chlorinated solvents. *Biotechnol. Tech.* **1994**, *8*, 589–594. [CrossRef]
43. Jiang, G.; Johnston, B.; Townrow, D.E.; Radecka, I.; Koller, M.; Chaber, P.; Adamus, G.; Kowalczuk, M. Biomass extraction using non-chlorinated solvents for biocompatibility improvement of polyhydroxyalkanoates. *Polymers* **2018**, *10*, 731. [CrossRef]
44. Neves, A.; Müller, J. Use of enzymes in extraction of polyhydroxyalkanoates produced by Cupriavidus necator. *Biotechnol. Prog.* **2012**, *28*, 1575–1580. [CrossRef]
45. Madkour, M.H.; Heinrich, D.; Alghamdi, M.A.; Shabbaj, I.I.; Steinbüchel, A. PHA recovery from biomass. *Biomacromolecules* **2013**, *14*, 2963–2972. [CrossRef]
46. Sanhueza, C.; Acevedo, F.; Rocha, S.; Villegas, P.; Seeger, M.; Navia, R. Polyhydroxyalkanoates as biomaterial for electrospun scaffolds. *Int. J. Biol. Macromol.* **2019**, *124*, 102–110. [CrossRef]
47. Koller, M.; Niebelschütz, H.; Braunegg, G. Strategies for recovery and purification of poly[(R)-3-hydroxyalkanoates] (PHA) biopolyesters from surrounding biomass. *Eng. Life Sci.* **2013**, *13*, 549–562. [CrossRef]
48. Haddadi, M.H.; Asadolahi, R.; Negahdari, B. The bioextraction of bioplastics with focus on polyhydroxybutyrate: A review. *Int. J. Environ. Sci. Technol.* **2019**, *16*, 3935–3948. [CrossRef]
49. Singh, A.K.; Srivastava, J.K.; Chandel, A.K.; Sharma, L.; Mallick, N.; Singh, S.P. Biomedical applications of microbially engineered polyhydroxyalkanoates: An insight into recent advances, bottlenecks, and solutions. *Appl. Microbiol. Biotechnol.* **2019**, *103*, 2007–2032. [CrossRef]
50. Burniol-Figols, A.; Skiadas, I.V.; Daugaard, A.E.; Gavala, H.N. Polyhydroxyalkanoate (PHA) purification through dilute aqueous ammonia digestion at elevated temperatures. *J. Chem. Technol. Biotechnol.* **2020**, *95*, 1519–1532. [CrossRef]

51. De Souza Reis, G.A.; Michels, M.H.A.; Fajardo, G.L.; Lamot, I.; de Best, J.H. Optimization of green extraction and purification of PHA produced by mixed microbial cultures from sludge. *Water* **2020**, *12*, 1185. [CrossRef]
52. Wampfler, B.; Ramsauer, T.; Rezzonico, S.; Hischier, R.; Köhling, R.; Thöny-Meyer, L.; Zinn, M. Isolation and purification of medium Chain length Poly(3-hydroxyalkanoates) (mcl-PHA) for medical applications using nonchlorinated solvents. *Biomacromolecules* **2010**, *11*, 2716–2723. [CrossRef]
53. Gahlawat, G.; Kumari, P.; Bhagat, N.R. Technological advances in the production of polyhydroxyalkanoate biopolymers. *Curr. Sustain. Energy Rep.* **2020**, *7*, 73–83. [CrossRef]
54. RameshKumar, S.; Shaiju, P.; O'Connor, K.E.; Babu, P.R. Bio-based and biodegradable polymers—State-of-the-art, challenges and emerging trends. *Curr. Opin. Green Sustain. Chem.* **2020**, *21*, 75–81. [CrossRef]
55. Zhou, Y.; Han, L.R.; He, H.W.; Sang, B.; Yu, D.L.; Feng, J.T.; Zhang, X. Effects of agitation, aeration and temperature on production of a novel glycoprotein gp-1 by streptomyces kanasenisi zx01 and scale-up based on volumetric oxygen transfer coefficient. *Molecules* **2018**, *23*, 125. [CrossRef]
56. Sabapathy, P.C.; Devaraj, S.; Meixner, K.; Anburajan, P.; Kathirvel, P.; Ravikumar, Y.; Zabed, H.M.; Qi, X. Recent developments in Polyhydroxyalkanoates (PHAs) production—A review. *Bioresour. Technol.* **2020**, *306*, 123132. [CrossRef]
57. Rastogi, V.K.; Samyn, P. Bio-based coatings for paper applications. *Coatings* **2015**, *5*, 887–930. [CrossRef]
58. Anjum, A.; Zuber, M.; Zia, K.M.; Noreen, A.; Anjum, M.N.; Tabasum, S. Microbial production of polyhydroxyalkanoates (PHAs) and its copolymers: A review of recent advancements. *Int. J. Biol. Macromol.* **2016**, *89*, 161–174. [CrossRef] [PubMed]
59. Grigore, M.E.; Grigorescu, R.M.; Iancu, L.; Ion, R.M.; Zaharia, C.; Andrei, E.R. Methods of synthesis, properties and biomedical applications of polyhydroxyalkanoates: A review. *J. Biomater. Sci. Polym. Ed.* **2019**, *30*, 695–712. [CrossRef] [PubMed]
60. Johnston, B.; Radecka, I.; Hill, D.; Chiellini, E.; Ilieva, V.I.; Sikorska, W.; Musioł, M.; Zięba, M.; Marek, A.A.; Keddie, D.; et al. The microbial production of Polyhydroxyalkanoates from Waste polystyrene fragments attained using oxidative degradation. *Polymers* **2018**, *10*, 957. [CrossRef]
61. Gómez, E.F.; Michel, F.C. Biodegradability of conventional and bio-based plastics and natural fiber composites during composting, anaerobic digestion and long-term soil incubation. *Polym. Degrad. Stab.* **2013**, *98*, 2583–2591. [CrossRef]
62. Emadian, S.M.; Onay, T.T.; Demirel, B. Biodegradation of bioplastics in natural environments. *Waste Manag.* **2017**, *59*, 526–536. [CrossRef] [PubMed]
63. Gebauer, B.; Jendrossek, D. Assay of poly(3-hydroxybutyrate) depolymerase activity and product determination. *Appl. Environ. Microbiol.* **2006**, *72*, 6094–6100. [CrossRef]
64. Kalia, V.C. *Biotechnological Applications of Polyhydroxyalkanoates*; Springer Nature: Singapore, 2019; ISBN 9789811337598.
65. Khosravi-Darani, K.; Bucci, D.Z. Application of poly(hydroxyalkanoate) in food packaging: Improvements by nanotechnology. *Chem. Biochem. Eng. Q.* **2015**, *29*, 275–285. [CrossRef]
66. Mangaraj, S.; Yadav, A.; Bal, L.M.; Dash, S.K.; Mahanti, N.K. Application of biodegradable polymers in food packaging industry: A comprehensive review. *J. Packag. Technol. Res.* **2019**, *3*, 77–96. [CrossRef]
67. Gholami, A.; Mohkam, M.; Rasoul-Amini, S.; Ghasemi, Y. industrial production of polyhydroxyalkanoates by bacteria: Opportunities and challenges. *Minerva Biotecnol.* **2016**, *28*, 59–74. [CrossRef]
68. Leong, Y.K.; Show, P.L.; Lan, J.C.W.; Loh, H.S.; Lam, H.L.; Ling, T.C. Economic and environmental analysis of PHAs production process. *Clean Technol. Environ. Policy* **2017**, *19*, 1941–1953. [CrossRef]
69. Montalbano, G.; Fiorilli, S.; Caneschi, A.; Vitale-Brovarone, C. Type I collagen and strontium containing mesoporous glass particles as hybrid material for 3D printing of bone-like materials. *Materials* **2018**, *11*, 700. [CrossRef] [PubMed]
70. Follain, N.; Crétois, R.; Lebrun, L.; Marais, S. Water sorption behaviour of two series of PHA/montmorillonite films and determination of the mean water cluster size. *Phys. Chem. Chem. Phys.* **2016**, *18*, 20345–20356. [CrossRef]
71. Ang, H.Y.; Huang, Y.Y.; Lim, S.T.; Wong, P.; Joner, M.; Foin, N. Mechanical behavior of polymer-based vs. metallic based bioresorbable stents. *J. Thorac. Dis.* **2017**, *9*, S923–S934. [CrossRef] [PubMed]
72. Byrne, D.P.; Lacroix, D.; Planell, J.A.; Kelly, D.J.; Prendergast, P.J. Simulation of tissue differentiation in a scaffold as a function of porosity, Young's modulus and dissolution rate: Application of mechanobiological models in tissue engineering. *Biomaterials* **2007**, *28*, 5544–5554. [CrossRef]
73. Abar, B.; Alonso-Calleja, A.; Kelly, A.; Kelly, C.; Gall, K.; West, J.L. 3D printing of high-strength, porous, elastomeric structures to promote tissue integration of implants. *J. Biomed. Mater. Res. Part A* **2021**, *109*, 54–63. [CrossRef] [PubMed]
74. Singh, M.; Kumar, P.; Ray, S.; Kalia, V.C. Challenges and opportunities for customizing polyhydroxyalkanoates. *Indian J. Microbiol.* **2015**, *55*, 235–249. [CrossRef] [PubMed]
75. Koller, M. Biodegradable and biocompatible polyhydroxy-alkanoates (PHA): Auspicious microbial macromolecules for pharmaceutical and therapeutic applications. *Molecules* **2018**, *23*, 362. [CrossRef] [PubMed]
76. Generali, M.; Dijkman, P.E.; Hoerstrup, S.P. Bioresorbable scaffolds for cardiovascular tissue engineering. *EMJ Interv. Cardiol.* **2014**, *1*, 91–99.
77. Wu, L.P.; You, M.; Wang, D.; Peng, G.; Wang, Z.; Chen, G.Q. Fabrication of carbon nanotube (CNT)/poly(3-hydroxybutyrate-co-3-hydroxyhexanoate) (PHBHHx) nanocomposite films for human mesenchymal stem cell (hMSC) differentiation. *Polym. Chem.* **2013**, *4*, 4490–4498. [CrossRef]

78. Yu, B.Y.; Chen, P.Y.; Sun, Y.M.; Lee, Y.T.; Young, T.H. Response of human mesenchymal stem cells (hMSCs) to the topographic variation of poly(3-Hydroxybutyrate-co-3-Hydroxyhexanoate) (PHBHHx) films. *J. Biomater. Sci. Polym. Ed.* **2012**, *23*, 1–26. [CrossRef] [PubMed]

79. Cho, Y.S.; Kim, B.S.; You, H.K.; Cho, Y.S. A novel technique for scaffold fabrication: SLUP (salt leaching using powder). *Curr. Appl. Phys.* **2014**, *14*, 371–377. [CrossRef]

80. Deng, Y.; Lin, X.-S.; Zheng, Z.; Deng, J.-G.; Chen, J.-C.; Ma, H.; Chen, G.-Q. Poly(hydroxybutyrate-co-hydroxyhexanoate) promoted production of extracellular matrix of articular cartilage chondrocytes in vitro. *Biomaterials* **2003**, *24*, 4273–4281. [CrossRef]

81. Bian, Y.-Z.; Wang, Y.; Aibaidoula, G.; Chen, G.-Q.; Wu, Q. Evaluation of poly(3-hydroxybutyrate-co-3-hydroxyhexanoate) conduits for peripheral nerve regeneration. *Biomaterials* **2009**, *30*, 217–225. [CrossRef] [PubMed]

82. Baek, J.Y.; Xing, Z.C.; Kwak, G.; Yoon, K.B.; Park, S.Y.; Park, L.S.; Kang, I.K. Fabrication and characterization of collagen-immobilized porous PHBV/HA nanocomposite scaffolds for bone tissue engineering. *J. Nanomater.* **2012**, *2012*, 171804. [CrossRef]

83. Degli Esposti, M.; Chiellini, F.; Bondioli, F.; Morselli, D.; Fabbri, P. Highly porous PHB-based bioactive scaffolds for bone tissue engineering by in situ synthesis of hydroxyapatite. *Mater. Sci. Eng. C* **2019**, *100*, 286–296. [CrossRef]

84. You, M.; Peng, G.; Li, J.; Ma, P.; Wang, Z.; Shu, W.; Peng, S.; Chen, G.-Q. Chondrogenic differentiation of human bone marrow mesenchymal stem cells on polyhydroxyalkanoate (PHA) scaffolds coated with PHA granule binding protein PhaP fused with RGD peptide. *Biomaterials* **2011**, *32*, 2305–2313. [CrossRef]

85. Dai, Z.-W.; Zou, X.-H.; Chen, G.-Q. Poly(3-hydroxybutyrate-co-3-hydroxyhexanoate) as an injectable implant system for prevention of post-surgical tissue adhesion. *Biomaterials* **2009**, *30*, 3075–3083. [CrossRef]

86. Barouti, G.; Jaffredo, C.G.; Guillaume, S.M. Advances in drug delivery systems based on synthetic poly(hydroxybutyrate) (co)polymers. *Prog. Polym. Sci.* **2017**, *73*, 1–31. [CrossRef]

87. Xiong, Y.C.; Yao, Y.C.; Zhan, X.Y.; Chen, G.Q. Application of polyhydroxyalkanoates nanoparticles as intracellular sustained drug-release vectors. *J. Biomater. Sci. Polym. Ed.* **2010**, *21*, 127–140. [CrossRef]

88. Wu, L.P.; Wang, D.; Parhamifar, L.; Hall, A.; Chen, G.Q.; Moghimi, S.M. Poly(3-hydroxybutyrate-co-R-3-hydroxyhexanoate) nanoparticles with polyethylenimine coat as simple, safe, and versatile vehicles for cell targeting: Population characteristics, cell uptake, and intracellular trafficking. *Adv. Healthc. Mater.* **2014**, *3*, 817–824. [CrossRef]

89. Yao, Y.-C.; Zhan, X.-Y.; Zhang, J.; Zou, X.-H.; Wang, Z.-H.; Xiong, Y.-C.; Chen, J.; Chen, G.-Q. A specific drug targeting system based on polyhydroxyalkanoate granule binding protein PhaP fused with targeted cell ligands. *Biomaterials* **2008**, *29*, 4823–4830. [CrossRef]

90. Rebia, R.A.; Rozet, S.; Tamada, Y.; Tanaka, T. Biodegradable PHBH/PVA blend nanofibers: Fabrication, characterization, in vitro degradation, and in vitro biocompatibility. *Polym. Degrad. Stab.* **2018**, *154*, 124–136. [CrossRef]

91. Rebia, R.A.; Sadon, N.S.B.; Tanaka, T. Natural antibacterial reagents (Centella, propolis, and hinokitiol) loaded into poly[(R)-3-hydroxybutyrate-co-(R)-3-hydroxyhexanoate] composite nanofibers for biomedical applications. *Nanomaterials* **2019**, *9*, 1665. [CrossRef] [PubMed]

92. Ang, S.L.; Shaharuddin, B.; Chuah, J.A.; Sudesh, K. Electrospun poly(3-hydroxybutyrate-co-3-hydroxyhexanoate)/silk fibroin film is a promising scaffold for bone tissue engineering. *Int. J. Biol. Macromol.* **2020**, *145*, 173–188. [CrossRef]

93. Qu, X.-H.; Wu, Q.; Zhang, K.-Y.; Chen, G.Q. In vivo studies of poly(3-hydroxybutyrate-co-3-hydroxyhexanoate) based polymers: Biodegradation and tissue reactions. *Biomaterials* **2006**, *27*, 3540–3548. [CrossRef]

94. Ye, C.; Hu, P.; Ma, M.-X.; Xiang, Y.; Liu, R.-G.; Shang, X.-W. PHB/PHBHHx scaffolds and human adipose-derived stem cells for cartilage tissue engineering. *Biomaterials* **2009**, *30*, 4401–4406. [CrossRef]

95. Tomietto, P.; Carré, M.; Loulergue, P.; Paugam, L.; Audic, J.L. Polyhydroxyalkanoate (PHA) based microfiltration membranes: Tailoring the structure by the non-solvent induced phase separation (NIPS) process. *Polymer* **2020**, *204*, 122813. [CrossRef]

96. Zhang, S.-L.; Zheng, D.-J.; Fan, W.-Z.; Wei, D.-X.; Peng, S.-W.; Tang, M.-M.; Chen, G.-Q.; Wei, C.-J. Transient embolization with microspheres of polyhydroxyalkanoate renders efficient adenoviral transduction of pancreatic capillary in vivo. *J. Gene Med.* **2012**, *14*, 530–539. [CrossRef] [PubMed]

97. Zhu, W.; Ma, X.; Gou, M.; Mei, D.; Zhang, K.; Chen, S. 3D printing of functional biomaterials for tissue engineering. *Curr. Opin. Biotechnol.* **2016**, *40*, 103–112. [CrossRef] [PubMed]

98. Szymczyk-Ziółkowska, P.; Łabowska, M.B.; Detyna, J.; Michalak, I.; Gruber, P. A review of fabrication polymer scaffolds for biomedical applications using additive manufacturing techniques. *Biocybern. Biomed. Eng.* **2020**, *40*, 624–638. [CrossRef]

99. Loterie, D.; Delrot, P.; Moser, C. High-resolution tomographic volumetric additive manufacturing. *Nat. Commun.* **2020**, *11*, 1–6. [CrossRef] [PubMed]

100. Wu, D.; Spanou, A.; Diez-Escudero, A.; Persson, C. 3D-printed PLA/HA composite structures as synthetic trabecular bone: A feasibility study using fused deposition modeling. *J. Mech. Behav. Biomed. Mater.* **2020**, *103*, 103608. [CrossRef]

101. Chuan, Y.L.; Hoque, M.E.; Pashby, I. Prediction of patient-specific tissue engineering scaffolds for optimal design. *Int. J. Model. Optim.* **2013**, *3*, 468–470. [CrossRef]

102. Meng, Z.; He, J.; Cai, Z.; Wang, F.; Zhang, J.; Wang, L.; Ling, R.; Li, D. Design and additive manufacturing of flexible polycaprolactone scaffolds with highly-tunable mechanical properties for soft tissue engineering. *Mater. Des.* **2020**, *189*, 1–9. [CrossRef]

103. Li, L.; Lin, Q.; Tang, M.; Duncan, A.J.E.; Ke, C. Advanced polymer designs for direct-ink-write 3D printing. *Chem. A Eur. J.* **2019**, *25*, 10768–10781. [CrossRef] [PubMed]

104. Vyavahare, S.; Teraiya, S.; Panghal, D.; Kumar, S. Fused deposition modelling: A review. *Rapid Prototyp. J.* **2020**, *26*, 176–201. [CrossRef]

105. Gonzalez Ausejo, J.; Rydz, J.; Musioł, M.; Sikorska, W.; Sobota, M.; Włodarczyk, J.; Adamus, G.; Janeczek, H.; Kwiecień, I.; Hercog, A.; et al. A comparative study of three-dimensional printing directions: The degradation and toxicological profile of a PLA/PHA blend. *Polym. Degrad. Stab.* **2018**, *152*, 191–207. [CrossRef]

106. Wu, C.S.; Liao, H.T. Interface design of environmentally friendly carbon nanotube-filled polyester composites: Fabrication, characterisation, functionality and application. *Express Polym. Lett.* **2017**, *11*, 187–198. [CrossRef]

107. Kosorn, W.; Sakulsumbat, M.; Uppanan, P.; Kaewkong, P.; Chantaweroad, S.; Jitsaard, J.; Sitthiseripratip, K.; Janvikul, W. PCL/PHBV blended three dimensional scaffolds fabricated by fused deposition modeling and responses of chondrocytes to the scaffolds. *J. Biomed. Mater. Res. Part B Appl. Biomater.* **2017**, *105*, 1141–1150. [CrossRef] [PubMed]

108. Giubilini, A.; Siqueira, G.; Clemens, F.J.; Sciancalepore, C.; Messori, M.; Nystrom, G.; Bondioli, F. 3D printing nanocellulose-poly(3-hydroxybutyrate-co-3-hydroxyhexanoate) biodegradable composites by fused deposition modeling. *ACS Sustain. Chem. Eng.* **2020**, *8*, 10292–10302. [CrossRef]

109. Mazzoli, A. Selective laser sintering in biomedical engineering. *Med. Biol. Eng. Comput.* **2013**, *51*, 245–256. [CrossRef]

110. Pereira, T.F.; Silva, M.A.C.; Oliveira, M.F.; Maia, I.A.; Silva, J.V.L.; Costa, M.F.; Thiré, R.M.S.M. Effect of process parameters on the properties of selective laser sintered Poly(3-hydroxybutyrate) scaffolds for bone tissue engineering. *Virtual Phys. Prototyp.* **2012**, *7*, 275–285. [CrossRef]

111. Puppi, D.; Chiellini, F. Computer-Aided Wet-Spinning. In *Computer-Aided Tissue Engineering: Methods and Protocols*; Rainer, A., Moroni, L., Eds.; Springer: New York, NY, USA, 2021; pp. 101–110. ISBN 978-1-0716-0611-7.

112. Puppi, D.; Braccini, S.; Ranaudo, A.; Chiellini, F. Poly(3-hydroxybutyrate-co-3-hydroxyexanoate) scaffolds with tunable macro- and microstructural features by additive manufacturing. *J. Biotechnol.* **2020**, *308*, 96–107. [CrossRef] [PubMed]

113. Puppi, D.; Morelli, A.; Chiellini, F. Additive manufacturing of poly(3-hydroxybutyrate-co-3-hydroxyhexanoate)/poly(ε-caprolactone) blend scaffolds for tissue engineering. *Bioengineering* **2017**, *4*, 49. [CrossRef]

114. Mota, C.; Wang, S.Y.; Puppi, D.; Gazzarri, M.; Migone, C.; Chiellini, F.; Chen, G.Q.; Chiellini, E. Additive manufacturing of poly[(R)-3-hydroxybutyrate-co-(R)-3-hydroxyhexanoate] scaffolds for engineered bone development. *J. Tissue Eng. Regen. Med.* **2017**, *11*, 175–186. [CrossRef]

115. Puppi, D.; Pirosa, A.; Morelli, A.; Chiellini, F. Design, fabrication and characterization of tailored poly[(R)-3-hydroxybutyrate-co-(R)-3-hydroxyexanoate] scaffolds by computer-aided wet-spinning. *Rapid Prototyp. J.* **2018**, *24*, 1–8. [CrossRef]

116. Min, Z.; Kun, L.; Yufang, Z.; Jianhua, Z.; Xiaojian, Y. 3D-printed hierarchical scaffold for localized isoniazid/rifampin drug delivery and osteoarticular tuberculosis therapy. *Acta Biomater.* **2015**, *16*, 145–155. [CrossRef]

117. Duan, B.; Wang, M.; Zhou, W.Y.; Cheung, W.L.; Li, Z.Y.; Lu, W.W. Three-dimensional nanocomposite scaffolds fabricated via selective laser sintering for bone tissue engineering. *Acta Biomater.* **2010**, *6*, 4495–4505. [CrossRef] [PubMed]

118. Zhang, L.; Zheng, Z.; Xi, J.; Gao, Y.; Ao, Q.; Gong, Y.; Zhao, N.; Zhang, X. Improved mechanical property and biocompatibility of poly(3-hydroxybutyrate-co-3-hydroxyhexanoate) for blood vessel tissue engineering by blending with poly(propylene carbonate). *Eur. Polym. J.* **2007**, *43*, 2975–2986. [CrossRef]

119. Tessmar, J.; Holland, T.; Mikos, A. Salt leaching for polymer scaffolds: Laboratory-scale manufacture of cell carriers. In *Scaffolding in Tissue Engineering*; Ma, P.X., Elisseeff, J., Eds.; Taylor & Francis Group: Boca Raton, FL, USA, 2006; pp. 111–124.

120. Kim, J.F.; Hoon, K.; Lee, Y.M. Thermally induced phase separation and electrospinning methods for emerging membrane applications: A review. *Am. Inst. Chem. Eng. J.* **2016**, *62*, 461–490. [CrossRef]

121. Shatrohan Lal, R.K. Synthesis of organic nanoparticles and their applications in drug delivery and food nanotechnology: A review. *J. Nanomater. Mol. Nanotechnol.* **2014**, *3*, 150. [CrossRef]

122. Duan, B.; Cheung, W.L.; Wang, M. Optimized fabrication of Ca-P/PHBV nanocomposite scaffolds via selective laser sintering for bone tissue engineering. *Biofabrication* **2011**, *3*, 015001. [CrossRef] [PubMed]

123. Puppi, D.; Chiellini, F. Wet-spinning of biomedical polymers: From single-fibre production to additive manufacturing of three-dimensional scaffolds. *Polym. Int.* **2017**, *66*, 1690–1696. [CrossRef]

124. Misra, S.K.; Valappil, S.P.; Roy, I.; Boccaccini, A.R. Polyhydroxyalkanoate (PHA)-inorganic phase composites for tissue engineering applications. *Biomacromolecules* **2006**, *7*, 2249–2258. [CrossRef]

125. Korsatko, W.; Wabnegg, B.; Tillian, H.; Braunegg, G.; Lafferty, R.M. Poly-D-(—)-3-hydroxybutyric acid-a biodegradable carrier for long term medication dosage. II: The biodegradation in animal organism and in vitro in vivo correlation of pharmaceuticals from parenteral matrix retard tablets. *Pharm. Ind.* **1983**, *45*, 1004–1007.

126. Kanmani, P.; Kumaresan, K.; Aravind, J.; Karthikeyan, S.; Balan, R. Enzymatic degradation of polyhydroxyalkanoate using lipase from Bacillus subtilis. *Int. J. Environ. Sci. Technol.* **2016**, *13*, 1541–1552. [CrossRef]

127. Blevins, H.M.; Blue, M.D.; Cobbs, B.D.; Ricotilli, T.A.; Kyler, S.L.; Shuey, C.T.; Thompson, W.D.; Baron, S.F. Characterization of an Extracellular Polyhydroxyalkanoate Depolymerase from Streptomyces sp. SFB5A. *J. Bioremediation Biodegrad.* **2018**, *9*, 452. [CrossRef]

128. Sharma, P.K.; Mohanan, N.; Sidhu, R.; Levin, D.B. Colonization and degradation of polyhydroxyalkanoates by lipase-producing bacteria. *Can. J. Microbiol.* **2019**, *65*, 461–475. [CrossRef]

129. Boyandin, A.N.; Prudnikova, S.V.; Karpov, V.A.; Ivonin, V.N.; Dỗ, N.L.; Nguyễn, T.H.; Lê, T.M.H.; Filichev, N.L.; Levin, A.L.; Filipenko, M.L.; et al. Microbial degradation of polyhydroxyalkanoates in tropical soils. *Int. Biodeterior. Biodegrad.* **2013**, *83*, 77–84. [CrossRef]

130. Ishii-Hyakutake, M.; Mizuno, S.; Tsuge, T. Biosynthesis and characteristics of aromatic polyhydroxyalkanoates. *Polymers* **2018**, *10*, 1267. [CrossRef] [PubMed]

131. Bátori, V.; Åkesson, D.; Zamani, A.; Taherzadeh, M.J.; Sárvári Horváth, I. Anaerobic degradation of bioplastics: A review. *Waste Manag.* **2018**, *80*, 406–413. [CrossRef] [PubMed]

132. Dilkes-Hoffman, L.S.; Lant, P.A.; Laycock, B.; Pratt, S. The rate of biodegradation of PHA bioplastics in the marine environment: A meta-study. *Mar. Pollut. Bull.* **2019**, *142*, 15–24. [CrossRef]

133. Kolstad, J.J.; Vink, E.T.H.; De Wilde, B.; Debeer, L. Assessment of anaerobic degradation of Ingeo™ polylactides under accelerated landfill conditions. *Polym. Degrad. Stab.* **2012**, *97*, 1131–1141. [CrossRef]

134. Kanesawa, Y.; Tanahashi, N.; Doi, Y.; Saito, T. Enzymatic degradation of microbial poly(3-hydroxyalkanoates). *Polym. Degrad. Stab.* **1994**, *45*, 179–185. [CrossRef]

135. Luo, Z.; Jiang, L.; Ding, C.; Hu, B.; Loh, X.J.; Li, Z.; Wu, Y.L. Surfactant Free Delivery of Docetaxel by Poly[(R)-3-hydroxybutyrate-(R)-3-hydroxyhexanoate]-Based Polymeric Micelles for Effective Melanoma Treatments. *Adv. Healthc. Mater.* **2018**, *7*, 1–12. [CrossRef] [PubMed]

136. Duan, B.; Wang, M. Encapsulation and release of biomolecules from Ca-P/PHBV nanocomposite microspheres and three-dimensional scaffolds fabricated by selective laser sintering. *Polym. Degrad. Stab.* **2010**, *95*, 1655–1664. [CrossRef]

137. Li, K.; Zhu, M.; Xu, P.; Xi, Y.; Cheng, Z.; Zhu, Y.; Ye, X. Three-dimensionally plotted MBG/PHBHHx composite scaffold for antitubercular drug delivery and tissue regeneration. *J. Mater. Sci. Mater. Med.* **2015**, *26*, 102. [CrossRef]

138. Qu, X.H.; Wu, Q.; Chen, G.Q. In vitro study on hemocompatibility and cytocompatibility of poly(3-hydroxybutyrate-co -3-hydroxyhexanoate). *J. Biomater. Sci. Polym. Ed.* **2006**, *17*, 1107–1121. [CrossRef]

139. Gao, J.; Guo, H.; Tian, S.; Qiao, Y.; Han, J.; Li, Y.; Wang, L. Preparation and mechanical performance of small-diameter PHBHHx vascular graft by electrospinning. *Int. J. Polym. Mater. Polym. Biomater.* **2019**, *68*, 575–581. [CrossRef]

140. Findrik Balogová, A.; Hudák, R.; Tóth, T.; Schnitzer, M.; Feranc, J.; Bakoš, D.; Živčák, J. Determination of geometrical and viscoelastic properties of PLA/PHB samples made by additive manufacturing for urethral substitution. *J. Biotechnol.* **2018**, *284*, 123–130. [CrossRef] [PubMed]

141. Puppi, D.; Pirosa, A.; Lupi, G.; Erba, P.A.; Giachi, G.; Chiellini, F. Design and fabrication of novel polymeric biodegradable stents for small caliber blood vessels by computer-aided wet-spinning. *Biomed. Mater.* **2017**, *12*, 035011. [CrossRef] [PubMed]

142. Rathbone, S.; Furrer, P.; Lübben, J.; Zinn, M.; Cartmell, S. Biocompatibility of polyhydroxyalkanoate as a potential material for ligament and tendon scaffold material. *J. Biomed. Mater. Res. Part. A* **2010**, *93*, 1391–1403. [CrossRef]

143. Yang, M.; Zhu, S.; Chen, Y.; Chang, Z.; Chen, G.; Gong, Y.; Zhao, N.; Zhang, X. Studies on bone marrow stromal cells affinity of poly (3-hydroxybutyrate-co-3-hydroxyhexanoate). *Biomaterials* **2004**, *25*, 1365–1373. [CrossRef]

144. Wang, Y.-W.; Wu, Q.; Chen, G.-Q. Attachment, proliferation and differentiation of osteoblasts on random biopolyester poly(3-hydroxybutyrate-co-3-hydroxyhexanoate) scaffolds. *Biomaterials* **2004**, *25*, 669–675. [CrossRef]

145. Wang, Y.-W.; Wu, Q.; Chen, J.; Chen, G.-Q. Evaluation of three-dimensional scaffolds made of blends of hydroxyapatite and poly(3-hydroxybutyrate-co-3-hydroxyhexanoate) for bone reconstruction. *Biomaterials* **2005**, *26*, 899–904. [CrossRef]

146. Xi, J.; Li, J.; Zhu, L.; Gong, Y.; Zhao, N.; Zhang, X. Effects of Quenching Temperature and Time on Pore Diameter of Poly(3-hydroxybutyrate-co-3-hydroxyhexanoate) Porous Scaffolds and MC3T3-E1 Osteoblast Response to the Scaffolds. *Tsinghua Sci. Technol.* **2007**, *12*, 366–371. [CrossRef]

147. Duan, B.; Wang, M. Customized Ca-P/PHBV nanocomposite scaffolds for bone tissue engineering: Design, fabrication, surface modification and sustained release of growth factor. *J. R. Soc. Interface* **2010**, *7*, S615–S629. [CrossRef]

148. Yang, S.; Wang, J.; Tang, L.; Ao, H.; Tan, H.; Tang, T.; Liu, C. Mesoporous bioactive glass doped-poly (3-hydroxybutyrate-co-3-hydroxyhexanoate) composite scaffolds with 3-dimensionally hierarchical pore networks for bone regeneration. *Colloids Surf. B Biointerfaces* **2014**, *116*, 72–80. [CrossRef]

149. Zhao, S.; Zhu, M.; Zhang, J.; Zhang, Y.; Liu, Z.; Zhu, Y.; Zhang, C. Three dimensionally printed mesoporous bioactive glass and poly(3-hydroxybutyrate-co-3-hydroxyhexanoate) composite scaffolds for bone regeneration. *J. Mater. Chem. B* **2014**, *2*, 6106–6118. [CrossRef] [PubMed]

150. Gao, G.; Cui, X. Three-dimensional bioprinting in tissue engineering and regenerative medicine. *Biotechnol. Lett.* **2016**, *38*, 203–211. [CrossRef] [PubMed]

151. Turnbull, G.; Clarke, J.; Picard, F.; Zhang, W.; Riches, P.; Li, B.; Shu, W. 3D biofabrication for soft tissue and cartilage engineering. *Med. Eng. Phys.* **2020**, *82*, 13–39. [CrossRef] [PubMed]

152. Deng, Y.; Zhao, K.; Zhang, X.; Hu, P.; Chen, G.-Q. Study on the three-dimensional proliferation of rabbit articular cartilage-derived chondrocytes on polyhydroxyalkanoate scaffolds. *Biomaterials* **2002**, *23*, 4049–4056. [CrossRef]

153. Zheng, Z.; Deng, Y.; Lin, X.S.; Zhang, L.X.; Chen, G.Q. Induced production of rabbit articular cartilage-derived chondrocyte collagen II on polyhydroxyalkanoate blends. *J. Biomater. Sci. Polym. Ed.* **2003**, *14*, 615–624. [CrossRef]

154. Sun, A.; Meng, Q.; Li, W.; Liu, S.; Chen, W. Construction of tissue-engineered laryngeal cartilage with a hollow, semi-flared shape using poly(3-hydroxybutyrate-co-3-hydroxyhexanoate) as a scaffold. *Exp. Ther. Med.* **2015**, *9*, 1482–1488. [CrossRef]
155. De Pascale, C.; Marcello, E.; Getting, S.J.; Roy, I.; Locke, I.C. Populated collagen hydrogel and polyhydroxyalkanoate composites: Novel matrices for cartilage repair and regeneration? *Osteoarthr. Cartil.* **2019**, *27*, S432–S433. [CrossRef]
156. Labet, M.; Thielemans, W. Synthesis of polycaprolactone: A review. *Chem. Soc. Rev.* **2009**, *38*, 3484–3504. [CrossRef] [PubMed]

 *bioengineering*

Review

Additive Biomanufacturing with Collagen Inks

Weng Wan Chan [1,†], David Chen Loong Yeo [1,†], Vernice Tan [1], Satnam Singh [1], Deepak Choudhury [1,*] and May Win Naing [1,2,*]

1   Biomanufacturing Technology, Bioprocessing Technology Institute (BTI), Agency for Science, Technology and Research (A*STAR), Singapore City 138668, Singapore; chan_weng_wan@bti.a-star.edu.sg (W.W.C.); david_yeo@bti.a-star.edu.sg (D.C.L.Y.); vtan026@e.ntu.edu.sg (V.T.); Satnam_Singh@bti.a-star.edu.sg (S.S.)
2   Singapore Institute of Manufacturing Technology (SIMTech), Agency for Science, Technology and Research (A*STAR), 2 Fusionopolis Way, #08-04, Innovis, Singapore City 138634, Singapore
*   Correspondence: Deepak_choudhury@bti.a-star.edu.sg (D.C.); winnaingm@simtech.a-star.edu.sg (M.W.N.)
†   Co-first authors.

Received: 15 May 2020; Accepted: 25 June 2020; Published: 1 July 2020

Abstract: Collagen is a natural polymer found abundantly in the extracellular matrix (ECM). It is easily extracted from a variety of sources and exhibits excellent biological properties such as biocompatibility and weak antigenicity. Additionally, different processes allow control of physical and chemical properties such as mechanical stiffness, viscosity and biodegradability. Moreover, various additive biomanufacturing technology has enabled layer-by-layer construction of complex structures to support biological function. Additive biomanufacturing has expanded the use of collagen biomaterial in various regenerative medicine and disease modelling application (e.g., skin, bone and cornea). Currently, regulatory hurdles in translating collagen biomaterials still remain. Additive biomanufacturing may help to overcome such hurdles commercializing collagen biomaterials and fulfill its potential for biomedicine.

Keywords: collagen; ECM; extracellular matrix; bioinks; biomanufacturing

---

1. Introduction

Collagen is by far the most prevalent extracellular matrix (ECM) molecule found in adult mammals with an estimated 30% of protein mass of multicellular organisms [1]. Although the collagen molecule has 29 subtypes (variants) [2,3], approximately 90% of collagen consists of variants types I, II, III [4]. Collagen extracellular matrix can be found throughout the body in both soft and hard connective tissues including bones, skin, tendon, cartilage, cornea, lung, liver etc. [5].

Its fundamental structural unit is a 300 nm protein consisting of 3 braided α-subunits of 1050 amino acids in length. Each strand comprises the repeating amino acid motif: Gly-Pro-X (X is any amino acid). These strands form hydrogen bonds between the NH bond of a glycine and a carbonyl (C=O) group from an adjacent strand that holds the structure together and form their characteristic triple helix structure [4,6]. Collagen is a hierarchical biomaterial that is self-assembled into fibrils (containing numerous structural units) of ~1 cm length and ~500 nm in diameter (using type 1 Collagen as the archetype). Fascinatingly, the individual tropocollagen monomers are unstable at body temperature and favour random coil conformations. However, collagen fibrillogenesis gives rise to triple helix macromolecular structures with favourable mechanical strength in 3-dimensions, with resistance to enzymatic degradation [6]. Through the introduction of energy (e.g., heat energy from the surroundings), the H-bonds maintaining the orderly collagen structure are separated, causing the individual strands of the triple-helix to separate, resulting in a disorganized, denatured state known as gelatin (please see Figure 1 for more information).

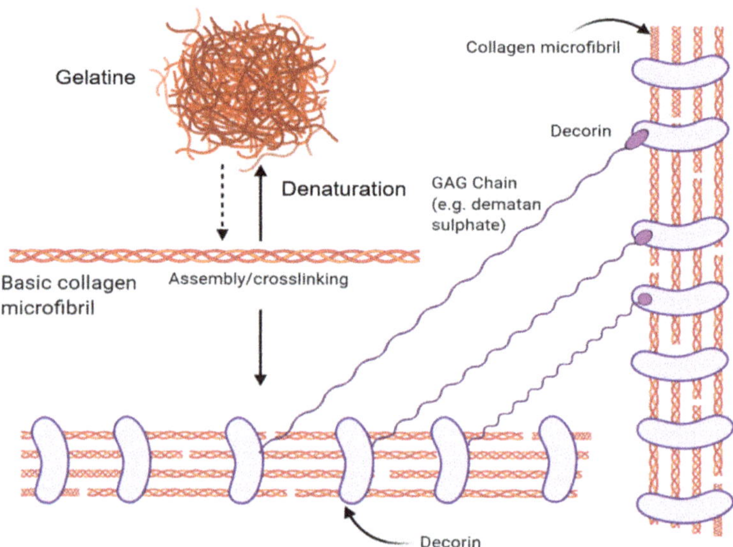

Figure 1. The structural forms of collagen and their native interactions. The basic collagen unit is a triple-helix microfiber that denatures into gelatine or can be assembled into collagen fibrils. Decorin proteins wrap around collagen fibrils in their native context and bind with glycosaminoglycan chains such as dermatan sulphate. Created with BioRender.com.

The Gly-Pro-X amino acid arrangement is critical to the collagen molecule as seen from disease-causing mutations that lead to osteogenesis imperfecta or "brittle bone" disease. A single misplacement of glycine due to the mutation results in unstable helices [4]. In their native microenvironment, collagen molecules interact with other biological molecules. Negatively-charged Glycosaminoglycans (linear polysaccharides) sequester growth factors within the ECM [7]. These have been used to generate bio-active collagen scaffolds for cell growth [8]. Furthermore, Collagen interacts with Elastin fibers to provide recoil to the ECM, as well as fibronectin to mediate cell attachment and function [1]. Collagen molecules can also interact with reducing sugars in the body which result in its glycation. Glycation molecules result in the formation of advanced glycation end products (AGEs) which gives rise to the loss of soft tissue biomechanical properties and is associated with various diseases such as atherosclerosis, osteoporosis, diabetes and renal failure [9].

Collagen biomaterials have been utilised for decades to enhance cell culture/function [10]. A number of collagen or collagen-derivative based protocols and commercial culture products have been used extensively ranging from cell culture surfaces to hydrogels [10]. These include culture well inserts [11,12] (MilliCell®, Transwell®), sponge/gels (Matrigel™, Extracel™) and microcarriers (GEM™). While matrigel is derived from Engelbreth–Holm–Swarm (EHS) tumor and found to contain collagen IV, laminin and heparin sulfate, GEM ™ microcarriers coat an alginate core with gelatin to aid cell attachment.

Beyond cell culture reagents, collagen biomaterials have been used for tissue engineering applications including: bone, tendon, cardiovascular therapies and disease models [13], cornea [5], skin, skeletal muscle, artery [14] etc. One usage with great popularity is using collagen scaffolds as dermal regeneration templates for severe wounds and other trauma such as burns. To date, a number of scaffolds/templates containing collagen ingredients are commercially available including: Helistat (Integra ®), Instat (Johnson & Johnson), SkinTemp (BioCor), Helitene (Integra ®), Fibracol (J&J), Biobrane (UDL Laboratories), and Chronicure (Derma Sciences)–not an exhaustive list, which is currently presented in fibre, powder, composite forms etc. [15]. Collagen biomaterials as dermal templates have seen the greatest number of commercial translations to date. Recently, novel applications in sustainable cellular agriculture using collagen biomaterials include making artificial leather and bio-artificial muscle [16].

Despite plentiful collagen biomaterial applications developed, collagen has several limitations that curtail its widespread usage: generally poor mechanical properties (vascular tissue engineering applications), thrombogenicity, contamination, source and batch variability [13]. These limitations leave many collagen biomaterial applications in the earlier technology development stages, hindering technology translation.

The emerging field of biomaterials printing - bioprinting, provides the means to create structures from collagen biomaterials, additives and cells in a reproducible and scalable way [17,18]. Adapted from methods first used to manufacture inorganic materials [19], bioprinting is an additive manufacturing approach to produce living tissue and organ analogs for regenerative medicine, tissue engineering, pharmacokinetic and disease/developmental modelling [20]. By patterning various combinations of biomaterials and cells, a goal is to reproduce complex biological architecture to recreate the anatomy in reproducible ways [21,22]. Thus, bioprinting potentially mitigates concerns of product variability by increasing process reproducibility. Moreover, increasing production throughput with bioprinting circumvents bottlenecks in production capacity, making collagen biomaterial products more cost-effective.

This article focuses on the bioprinting of collagen biomaterials/bioinks for (mostly) therapeutic purposes. Bioinks differ from biomaterials in that cells are introduced with the materials and printed, even in situ [23]. On the other hand, biomaterial scaffolds are printed alone before cellular components are added. We discuss how collagen biomaterials are isolated from different sources, processed and analysed post-processing. Thereafter, we discuss various printing methods for collagen biomaterials ranging from manually-casted production (the simplest and lowest throughput) to stereo-/digital light printing (additive manufacturing suited for producing complex shapes). The article concludes with a discussion about translational regulatory, cost and strategy issues using bioprinted collagen biomaterials/bioinks for regeneration and therapy applications.

2. Processing Parameters

Each step in the processing of collagen for additive manufacturing alters the properties and structure of collagen. Depending on the sources of collagen, extraction steps and crosslinking methods (chemical, physical), the resultant properties will differ. The effects of these processes as well as methods for analyzing collagen biomaterials will be discussed.

2.1. Sources of Collagen

For additive biomanufacturing, fibril-forming sub-types of collagen (type I, type II, type III, type V, type XI, type XXIV and type XXVII) are preferred because they contribute to the mechanical integrity of the ECM [15,24]. Fibrillar collagen is formed from the assembly of collagen molecules because of the intermolecular bonds between the individual strands to create the signature triple-helix collagen molecule (see the introduction section). These fibrils further assemble into fibre-bundles with tensile strength in tendons and skin [3] or into orthogonal transparent layers (e.g., cornea) [25].

Fibrillar collagen can be extracted from various sources. As animal skin/tendons and cartilaginous tissues are abundant in type I and type II collagen respectively, these tissues are sources of fibrillar collagen extraction [26]. Cells cultured in vitro are used to synthesize collagen as well [27,28]. Cells such as fibroblast and chondrocytes which specialize in type I and type II collagen production respectively can be cultured and the synthesised collagen harvested from media or cell layers. Recombinant collagen production is using genetically engineered microorganisms, plants or animals such as bacteria, yeast, transgenic corn and silkworms [29,30]. Synthetic peptides mimicking collagen trimeric structure have also been investigated to produce collagen-like peptides [31,32]. Collagen from cells grown in vitro, recombinant protein production as well as peptide synthesis have very low yield and are not as cost-effective as collagen extraction from animal tissues. Hence, most commercial collagen extraction relies on animal sources. While there are variations in collagen between different animal species and tissue sources, variation of collagen exists as well, within the same species due to the nature of collagen. As the collagen molecules in animals form mature crosslinks over time, the age, gender, activity and physical state of the animals play a significant role in forming these crosslinks [2]. The variability of collagen between batches of extraction affects fibrillation and self-assembly properties, and in turn the final collagen biomaterial product.

2.2. Collagen Extraction

Collagen extraction depends on its solubility in the chosen solvent and composition of collagen types in the tissue sources [26]. Collagen extraction can be broken down into 3 stages: Pre-treatment, extraction and purification. During the pre-treatment step, non-collagen proteins are removed to increase the yield of the collagen extraction process. Depending on the tissue source, removal of the non-collagen proteins (lipids, calcium, etc.) is achieved using alkali solutions, neutral saline solutions, alcohol solutions or a combination of solution [33]. Following pre-treatment of tissues, collagen is then extracted via acid-solubilisation or enzymatic-digestion.

In the extraction of collagen by acid-solubilisation, the pre-treated tissue is added into a dilute acidic solution, typically acetic acid, to disrupt weaker hydrogen bonds between collagen molecules [26]. This allows tissue swelling and acid-soluble collagen (ASC) from the loosened structure to dissolve in dilute acid [34]. However, dilute acid does not disrupt the triple helix structure of collagen due to the strong intermolecular forces between the polypeptide strands [35]. The extracted collagen still retains its telopeptide region and is known as telocollagen.

In the extraction of collagen by enzymatic-digestion, pre-treated tissue is added into a proteolytic enzyme solution, typically pepsin which cleaves non-helical telopeptide at the ends of the collagen microfibrils. Selective cleaving of the telopeptide region results in the destabilisation of the fibril structure and increases collagen dissolution [34]. The triple helix structure of collagen is unaffected due to the selective pepsin enzyme digestion. The extracted collagen molecule does not retain its telopeptide regions and is known as atelocollagen.

While clinical use of collagen use both telocollagen as well as atelocollagen in dermal substitute product showed no collagen induced adverse immunogenic response, the removal the telopeptide regions is suspected to play a role in the immunogenicity and antigenicity of collagen [36]. This is because the immune response in the body targets the antigenic determinant are found in mostly the telopeptides of collagen [37]. However, the antigenic determinants which arise from the helical structure and the amino acid sequence of the collagen also contribute to the immunogenicity and antigenicity of collagen [37]. Additionally, antigenic determinates for immune responses in the body depends on the species as well [36].

These extraction methods are not exclusive and can be performed together. Enzymatic-digestion can be done on acid insoluble collagen to obtain higher yields [26]. The extracted collagen is then filtered to remove impurities and purified through repeated salt precipitation, centrifugation and dissolution in acetic acid. Alternatively, the filtered extract undergoes dialysis for purification before freezing and freeze-drying.

2.2.1. Various Forms–Native, Gelatin (Disordered), Collagen Peptides

Depending on extraction methods used, the molecular weight, α-chain composition, and molecular structure are affected, in turn resulting in a change to the properties of the collagen (e.g., solubility, viscosity, etc.) From the extraction process, collagen can further be processed into denatured forms. Using thermal energy, acids, enzymes or a combination of methods, the intramolecular bonds between the α-chains are broken. As a result, the native helix structure transforms into a random coiled structure known as gelatin. Gelatin is formed as a result of the hydrolytic cleavage of collagen into individual protein strands [34]. Further processing of gelatin into smaller peptide chains is achieved through proteolytic enzymes resulting in hydrolysed collagen. Hydrolysed collagens molecular weight is significantly smaller (3–6 kDA) compared to their native structure (~300 kDa) [38]. As a result, hydrolysed collagen is much less viscous and more soluble than its native counterpart. For this review, we will only limit our discussion to collagen-based inks for bio-additive manufacturing. While collagen is favoured for its excellent biocompatibility, it exhibits poor mechanical properties [34]. This limitation can be overcome by crosslinking collagen molecules which will be discussed later (Section 2.3 Methods of Collagen Crosslinking).

2.2.2. Collagen Biocomposites

To enhance/modify the biological and mechanical properties of collagen, a mixture of synthetic or natural polymers are used. Blending of collagen together with synthetic polymers gives the final product enhanced mechanical and biological properties. The use of biocompatible synthetic polymers such as poly(lactic acid) (PLA) and polycaprolactone (PCL) allow products with excellent mechanical properties [39,40]. These composites have both the beneficial biological properties of collagen and the mechanical stiffness of synthetic polymers. Blending collagen with natural polymers (such as hyaluronic acid [41], alginate [42], glycosaminoglycans (GAGs) [43], growth factors [44], etc. [15]) for biomaterials that better mimic the native ECM environment or elicit a desired cellular response. Other than natural and synthetic polymers, inorganic compounds such as hydroxyapatite [45] or Tricalcium Phosphate (TCP) [46] can be incorporated to elicit desired cellular responses as well.

2.3. Methods of Collagen Crosslinking

Additional crosslinking of collagen molecules can be used to enhance the mechanical properties of collagen to provide structural integrity for additive bio-manufacturing such as for muscle tissue (8–20 kPa), cartilage tissue (20–30 kPa) and bone tissue (2–30 GPa) [45,47]. These "artificial" crosslinking bonds can be generated using chemical agents or physical treatment. Increasing concentration of chemical and treatment times generally increase collagen crosslinking. However, when using chemical agents for crosslinker, residual unreacted chemicals and/or chemical byproducts are often left behind [15]. This needs to be managed by washing to minimize cytotoxicity.

2.3.1. Chemical Crosslinking

A commonly used aldehyde for collagen crosslinking is glutaraldehyde (GA). As a dialdehyde, the crosslinker reacts with available amide groups on the collagen chains via Schiff base reactions resulting in covalent imide linkages [48]. These covalent linkages stabilise the intramolecular and intermolecular collagen structure. However, unreacted GA is cytotoxic as it crosslinks cellular proteins which disrupt cellular functions, causing cytotoxicity. GA is used in varying concentrations (0.0025–2.5% wt/v) and treatment times (20 min to 24 h) [49–55]. Increasing concentration and treatment times lead to increased collagen crosslinking.

Carbodiimides can also be used for collagen crosslinking such as 1-ethyl-3-(3-dimethylaminopropyl) carbodiimide (EDC). EDC crosslinks the amino and carboxyl groups collagen in a 2-step process: EDC first activates the carboxyl groups of collagen, the activated group then forms an amide linkage with primary amines in collagen [56]. This results in zero-length crosslinking where covalent bond formed is directly between the amino and carboxyl groups without addition of EDC. Crosslinking stabilises the intramolecular and intermolecular collage structure, improving overall mechanical stiffness of collagen as well as the bending stiffness of collagen fibrils. Typically, the use of EDC is accompanied with N-hydroxysuccinimide (NHS), which allows a higher conversion of crosslinks due to amine-reactive intermediates stabilizing [57]. EDC or EDC together with NHS are used in varying concentrations (0.01–2.5% wt/v) and treatment times (2 h to 48 h) [46,51,54–56,58–62]. Increasing concentration and treatment times lead to increased crosslinking of collagen.

Hexamethylene di-isocyanate (HDI), an isocyanate is also used for crosslinking as HDI reacts with available amide groups on the collagen in a nucleophilic addition reaction [63]. The resultant reaction forms a urea linkage to stabilize the intramolecular and intermolecular collage structure [64]. HDI was used in varying concentrations (1.5–5%) and treatment times (5 h–overnight) [55,63,65]. Increasing concentration and treatment times lead to increased crosslinking of collagen.

Plant extracts such as tannic acid and genipin have been explored as sustainable crosslinking agents as well. Tannic acid (TA) is a polyphenol extracted from plants which stabilises the intermolecular bonds of collagen via hydrogen bonds and hydrophobic interactions between TA and collagen molecules [66]. Tannic acids of varying concentrations (0.1% to 6% wt/v) and treatment times(10 min to 120 h) [66–68]. Increasing concentration and treatment times lead to increased collagen crosslinking. Genipin is an Iridoid glycoside compound extracted from plants able to crosslink the free primary amines in protein [69]. This allows genipin crosslink primary amides in collagen, stabilizing the intramolecular and intermolecular collagen structure. Genipin is used in varying concentrations (0.00025% to 0.6%) and treatment times (1 h to 48 h) [69–72].

2.3.2. Physical Crosslinking

The use of chemical crosslinkers inevitably faces issues with cytotoxicity. Physical methods such as dehydrothermal (DHT) treatment and ultraviolet (UV) irradiation are used to create covalent bonds between intermolecular collagen structures.

Dehydrothermal treatment is a thermal treatment process that subjects collagen to high temperatures (>90 °C) for several hours or days (12 h to 5 days) under vacuum [40,51,54,73,74]. As a result, condensation reactions occur: between the free amino and hydroxyl groups of collagen (esterification); or between the carboxyl and free amino groups (amide linkage formation) [73]. These ester and amide bonds stabilise intramolecular and intermolecular collagen bonds. Despite the low water content of the collagen in vacuum, due to high temperatures, hydrolysis of the peptide bonds occurs resulting in the collagen triple-helix structure denaturing [73]. Though the mechanical properties of collagen improve with longer treatment times and higher temperature, collagen denaturing increases as well.

UV crosslinking involves irradiating collagen (15 min to 240 min) [74]. The mechanism of crosslinking is a result of free radical formation from peptide bond scissions. UV irradiation forms aromatic radicals which in turn attack the peptide bonds in collagen. These radicals then interact and crosslink, which stabilises intramolecular and intermolecular collagen structure. The effectiveness of UV irradiation depends on the sample preparation, irradiation dose and time of exposure [75]. While UV irradiation improves mechanical properties, it also denatures collagen triple-helix structures [75].

Gamma irradiation crosslinking is similar to UV crosslinking where the collagen structure is irradiated for a period of time (250 min to 1250 min) depending on the desired irradiation dosage. Gamma irradiation "radio-lyzes" water, creating radicals. The effectiveness of crosslinking depends on irradiation dose and exposure time. Compared to UV irradiation, the higher energy of gamma irradiation is able to deeper penetrate thicker collagen structures. However, its downside is denaturing collagen's triple-helix structure. Furthermore, gamma irradiation is often used for sterilization, making it unsuitable to crosslink cell-laden bioinks [76].

While cytotoxic compounds are not formed using physical crosslinking methods, they generally lead to collagen denaturation. Furthermore, physical crosslinking methods are less effective in improving mechanical properties of collagen compared to chemical methods [70].

2.4. Collagen Analytical Methods

Understanding the structural, morphological, and chemical composition of collagen is critical since additive bio-manufacturing processes may give rise to significant changes. Understanding the structural, morphological and chemical composition allows better design and processing of the collagen raw material to meet the needs of the final product [34,77].

2.4.1. Structural Analysis

Differential Scanning Calorimetry (DSC) can determine collagen thermal stability. DSC compares and measures heat flow differences between a specimen and control when heat is supplied. Using this information, the denaturation temperature of collagen can be determined due to endothermic processes observed during collagen denaturation [78]. Using DSC, the denaturation temperature of soluble fish collagen was determined to be 10 °C lower than soluble porcine collagen [79].

Sodium Dodecyl Sulfate Polyacrylamide Gel Electrophoresis (SDS-Page) is used to visualise molecular size distribution of collagen protein fragments. SDS-Page uses an electric field to drive charged proteins through gel. Larger fragments move slower, while smaller fragments move quicker through the gel. Following separation by size, the fragments are stained with Coomassie blue or silver to obtain protein bands. Comparing these with controls of known molecular weight, can determine protein fragment weight. By comparing the banding pattern of known type I collagen chains (α1(I): 97 kDa; α2(I): 95 kDa), SDS-PAGE was used to determine the molecular weight of type V collagen chains through the relationship between relative molecular weight and migration rate [80].

Circular Dichroism (CD) is an absorption spectroscopy method to determine the presence of secondary and tertiary collagen structures. CD measures the differences in absorption of left circularly polarised light and right circularly polarised light. Due to the nature of the peptide bonds and structures, it results in characteristic absorption spectrums. From this information, the secondary and tertiary protein structures such as α-helices (negative bands at 222 nm and 208 nm; positive band at 193 nm), β-pleated sheets (negative band at 218 nm; positive band at 195 nm), triple helical conformation (negative band at 195 nm; positive band at 220 nm) can be determined respectively [81,82].

Raman spectroscopy is a label-free and non-destructive method used to determine the bonds and protein structures present in collagen. Raman spectroscopy measures inelastic light scattering of a sample from incident light generated by a laser source. The bonds and protein structures result in distinct shifts in wavelength of scattered light and hence distinct spectrum peaks such as Amide I band (1655 cm^{-1}), Amide III band (1268 cm^{-1}), α-helix shoulder (1630 cm^{-1}) and β-pleated sheet peak (1675 cm^{-1}). From this information, the relative quantities of bonds and protein structures can be determined for collagen [83].

FTIR is a spectroscopy method to determine the bonds and protein structures present in collagen. FTIR measures absorbance or emission of infrared radiation from a sample after irradiation from an infrared source. The bonds and protein structures result in distinct infrared spectrum peaks. Typical peaks of type Collagen are: Amide A (3299 cm^{-1}), (N–H) stretching; Amide B (2919 cm^{-1}), (CH$_3$) asymmetric stretching; amide I (1628 cm^{-1}), (C=O) stretching; amide II (1540 cm^{-1}), (N–H) bending & (C–N) stretching; amide III (1234 cm^{-1}), (–CONH$_2$) stretching. From this information we can determine the presence of bonds and protein structures and their relative quantities in collagen [84]. Additionally, the ratio peak intensity of 1 between the amide III peak and 1450 cm^{-1} is indicative of the helix structure of collagen [84].

2.4.2. Morphological Analysis

Scanning Electron Microscopy (SEM) uses focused beams of electrons to image surface topography of collagen samples. SEM measures the energies of elastic and inelastic-scattered electrons incident upon the sample to recreate surface topography. Typically, SEM can examine porosity of collagen sponges as well as assembled collagen fibre structures. SEM was used to study pore morphology of collagen sponge, collagen-I fibrin gel, collagen 2D nanofibers (oriented and random) [85].

Confocal microscopy can be used for structural visualization too. Confocal microscopy sections images for each focal plane using a laser source before compilation into a 3D image volume of high resolution. There are two modes of image acquisition: fluorescence [86] and reflectance [87]. Fluorescence image acquisition uses fluorescent dyes or autofluorescent properties of collagen to generate image contrast, while reflectance image acquisition relies on differences in refractive indexes. Collagen fibril diameters and pore sizes have been studied using both modes of acquisition [86,87].

Transmission electron microscopy (TEM) is a microscope used to visualise banded collagen fibril structures. TEM generates an image by transmitting an electron beam through a thin specimen on a copper grid, the image is then magnified and projected onto a stage. The regular array of gaps and overlaps in collagen microfibrils result in differences in packing density along the assembled collagen fibre. This leads to the banded structure of the collagen fibrils (64–67 nm). Cryo-TEM was used to analyse fibrillar collagen from mineralized and non-mineralized tissue [88].

Atomic force microscopy (AFM) also visualises banded collagen fibril structures. AFM generates an image by measuring deflection of a cantilever probe across collagen fibres. This information is then rendered into a topographic images. AFM is able to detect differences in packing densities that arise from the array of gaps and overlaps in collagen microfibres [89].

2.4.3. Chemical Assays

Hydroxyproline is a colorimetric assay for quantifying hydroxyproline in collagen. Due to the hydroxyproline amino acid composition being approximately constant across the different types of collagen 11.3% (type I) and 15% (type III), it can indicate the amount of collagen within a sample [90].

Sircol assay is a colormetic assay to quantify collagen, binding to the $[Gly-X-Y]_n$ helical structure in collagen. Collagen content can be obtained by comparing it to standard curves for calibration [91].

2,4,6-Trinitrobenzne sulfronic acid (TNBS) assay is a colorimetric assay used to quantify free primary amines found in collagen. The amount of free primary amino groups can be obtained by comparing it to known quantities. The amount of TNBS can determine the degree of collagen methacrylation [92].

Ninhydrin assay is a colorimetric assay to quantify free primary amino groups. The dye binds to primary amines found in collagen. It was used to determine the change of free amino groups on collagen nanofibers following pre-treatment of L-lysine [93].

Western blot is a method used to identify the type of collagen following SDS-page analysis. Using monoclonal antibodies specific to the collagen types and visualisation through immunofluorescent staining, the type of collagen can be identified. Western blot was used to confirm Collagen VI chains from cell extracts and culture media [94].

Mass spectroscopy identifies proteins from gaseous ions generated from the protein fragments. These are sorted using an electric field according to mass-to-charge ratio. The relative quantities of ions are recorded. By comparing profiles of protein fragments with a database, the proteins can be identified. Mass spectroscopy was able to identify crosslinked pyridinoline and deoxypyridinoline amino acid in hydrolysed collagen [95].

3. Collagen-Based Ink Printing Applications

The application of collagen-based ink in both non-additive and additive manufacturing requires understanding of collagen processing, as well as the various printing methods. In this section, the principles behind the printing methods and their applications are examined.

3.1. Non-Additive Manufacturing

Non-additive manufacturing methods, casting and electrospinning of collagen-based inks and their applications are discussed. Casting involves pouring a liquid material into a mold of desired shape before solidifying and removal. Typically for collagen-based biomaterials, highly porous 3D structures (sponges) are obtained via the freeze-drying process while thin-films are obtained via air drying [34]. Freeze drying is a complex process where ice crystals in the frozen mold are removed by sublimation under vacuum. Pore size and direction of the sponge can be controlled during freeze-drying [96,97].

Collagen sponges are used extensively in wound healing and tissue engineering as scaffolds for bone [98], skin and soft tissues [99]. The porous nature of collagen sponges allow cell migration as well as nutrient diffusion into the scaffold while providing a substrate for growth. The collagen sponge can be loaded with drugs, growth factors and bio-additives to enhance scaffold bioactivity [50,60,98–100].

Collagen-glycosaminoglycan scaffolds have been successfully used to regenerate skin from full thickness burns [50]. Additionally, by varying the glycosaminoglycan concentration and pore size, peripheral nerve tissue was successfully regenerated too [101]. Loading TGF-β1 into a collagen sponge allowed controlled release of growth factors, enhancing bone regeneration of a rabbit skull defect [44].

When collagen is laid out to dry, a thin-film of collagen is obtained via evaporation. As water and solvents evaporate, fibres and molecules are brought closer together due to surface tension of the solvent giving rise to a thin-film layer upon drying [34]. Thin collagen films are typically used in cornea treatment owing to their optically transparent nature and biological properties [61]. However, collagen films are not limited to ocular tissue engineering, micropatterns can also be designed onto the film as part of the casting process to influence osteoblast cell orientation [54]. By stacking the collagen film layer by layer, the resulting biomatrix encouraged neo-tissue formation in a hernia repair model [71]. The films can also be wrapped into tubes for nerve grafting applications [49]. While functioning as a barrier membrane, collagen films can also be loaded with drugs, growth factors and bio-additives to enhance bioactivity. Additionally, collagen film degeneration and its mechanical properties can be controlled by varying crosslinking to control the release of its contents via degradation [102,103]. Collagen films are also suitable as edible food packaging [104].

Electrospinning

Electrospinning consists of loading a desired biomaterial and a volatile solvent into a syringe. By applying a voltage to the needle tip, an electric field forms between the needle tip and the collector. Once, the electrostatic forces of repulsion are greater than the surface tension of the extruded liquid, a taylor cone is formed and the charged liquid is ejected onto the collector. The volatile solvent evaporates, resulting in fine nano/microscale fibres. These fibres are then deposited onto the metallic collector. By varying the extrusion rate, voltage of charged material, needle gauge and distance between the needle and collector the fibre diameters can be controlled [105].

Processing materials via electrospinning is appealing due to the ability to produce fibre meshes with diameters similar to the native fibrillar network present in the extracellular matrix (20 nm to 40 μm) [106]. Electrospinning can be performed using pure collagen or synthetic polymer additives such as PLLA or PCL to increase mechanical stiffness. Various electrospinning set-up can be used to produce different scaffolds for a variety of applications. A co-electrospinning system containing 2 mixtures of collagen and synthetic polymers was used to produce a scaffold with different regions to mimic muscle-tendon junction properties [107]. Using multi-layered electrospinning, an arterial structure was fabricated using a PCL, elastin and collagen layer was able to achieve significant improvement in mechanical properties and designed to mimic native arterial tissue [108]. A combination of electrospinning and electrospraying technology was used to produce 3D constructs which improved cell infiltration and controlled release of bio-additives [109].

However, the solvents used in electrospinning can significantly denature collagen. Typical fluoroalcohols used in electrospinning such as 1,1,1,3,3,3-Hexafluoro-2-propanol (HFP) cause a loss of collagen's triple helical structure [106]. Fortunately, solvents have been designed to minimize collagen denaturation when electrospun using "less harsh" solvents such as acetic acid/DMSO and PBS/ethanol [110].

3.2. Additive Biomanufacturing

In this section, four additive bio-manufacturing technologies will be discussed: extrusion bioprinting, inkjet bioprinting, laser-assisted bioprinting and stereolithographic/digital light processing bioprinting. The main advantage of additive bio-manufacturing is to produce complex shapes with internal structures at high resolution and accuracy without molds or shaping tools required by non-additive methods. Moreover, additive bio-manufacturing is amenable to printing with cell-laden inks (bio-inks) [24].

While all additive biomanufacturing processes create structures via layer-by-layer deposition of biomaterials, not all collagen-based inks can be printed using the following methods. As such, flexible printing method such as extrusion printing have a larger number of applications and variation of printing formulations, whereas more restrictive printing methods such as inkjet, laser-assisted, and stereolithography printing have fewer applications.

3.2.1. Extrusion

In extrusion bioprinting, biomaterial inks are loaded into a syringe and printed as filaments onto a stage via a mechanical or pneumatic dispensing system. Precise deposition of material is controlled by a dispensing stage along the x, y, and z axis. This method of bioprinting accommodates a large range of ink viscosities (30–60×10^7 mPa·s) [18]. Through multiple print heads, multiple materials and formulations can be printed together. However, this printing method is limited by the print resolution (100 μm), which is determined by the nozzle diameter [111]. Furthermore, printed cells experience high shear stresses when extruded under high pressure and small nozzles, resulting in lower cell viability.

Due to the nature of extrusion bioprinting, the viscosity of the collagen-based ink plays an important role in the printing process. The tendency of collagen to self-assemble into fibrillar structures at neutral pH when incubated at 37 °C allows collagen to form stable structures after printing [2]. Pure collagen was formulated to be self-supporting by either increasing the concentration or neutralising pH prior to extrusion. Following the extrusion process, scaffolds self-assembled in a neutral buffer to support self-assembly. This process produced tissue spheroid scaffolds as well as printing cell-laden inks into pre-set extrusion designs [47,112].

Combining collagen with other polymers, it is possible to design self-supporting structures by incorporating polymers rather than solely relying on pure collagen. An example was the use of cell-laden collagen/gelatin/alginate ink, by taking advantage of a two-step process involving thermal crosslinking with gelatin at low temperatures followed by crosslinking alginate in calcium solution [113]. The construct was printed at low temperature for gelatin to thermally crosslink and support the structure. Thereafter, it was immersed in calcium solution for ionic crosslinking of alginate to fix its shape. Gelatin and alginate was removed via diffusion and sodium citrate respectively, leaving behind a cell-laden collagen structure. A similar approach was used in cell-laden collagen/alginate ink where coaxial extrusion of collagen-alginate inks with calcium solution allowed the printed ink to be self-supporting [114]. In another, Pluronic F-127/Collagen ink was used to modify the gelation of the printed collagen ink, allowing it to be self-supported and be removed via diffusion in media [115]. A process unique to extrusion bioprinting known as freeform reversible embedding of suspended hydrogels (FRESH), non-self-supporting collagen ink formulations can print complex collagen scaffolds which are then self-assembled and collected from the hydrogel suspension [116].

Following the extrusion printing process, additional crosslinking of collagen (mentioned in earlier sections) can tune the mechanical properties of the collagen scaffold as desired [41,46,58,59,67,68,70,117]. Additionally, the extrusion printing process was able to generate collagen-composite scaffolds loaded with bio-additives such as silk fibroin, β-TCP, HA via-freeze-drying process for bone tissue regeneration [46,59]. Extrusion bioprinting can be combined with inkjet bioprinting for a one-step process to produce cell-laden 3D skin tissue (Figure 2A) [118].

3.2.2. Inkjet Printing

In inkjet bioprinting, biomaterials in a liquid state are loaded into a cartridge and deposited onto a substrate via droplets. The propulsion of droplets is achieved through pulses of pressure generated via thermal, acoustic or piezoelectric elements. The precise deposition of material is controlled by the dispensing system along the x, y-axis and print platform along the z-axis [18]. Through multiple print heads and cartridges, different material formulations can be combined. Additionally, as a nozzle-less systems, cell viability via inkjet bioprinting is higher compared to extrusion bioprinting. However, there is a material viscosity limit (10 mPa·s) for the inks printed due to the limited force generated to propel droplets onto the substrate [119]. Due to low-viscosity inks used in the system, additional processing steps are required to form 3D structures.

The viscosity limit of inkjet bioprinting restricts bioink formulations and bioink cell concentration. However, the self-assembly of collagen after printing allows it to be printed at low viscosity and crosslinked to produce cornea-like structures loaded with corneal stromal keratocytes (Figure 2B) [120]. Collagen ink blended with agarose in cell-laden printing gave rise to mesenchymal stem cells (MSCs) with a spread morphology, resulting in osteogenic differentiation [121]. Inkjet bioprinting was also used to generate collagen ink patterns onto which smooth muscle cells as well as neuronal cells were cultured, resulting in complex cellular patterns [122,123]. Additionally, inkjet bioprinting was applied to create in vitro cancer model microtissue arrays for drug testing and studying tumor progression [124]. Moreover, by controlling the thickness of the collagen gels printed via inkjet printing and seeding cells between the layers of the 3D construct, cell aggregates have been shown to fuse together, demonstrating potential for organ printing [125].

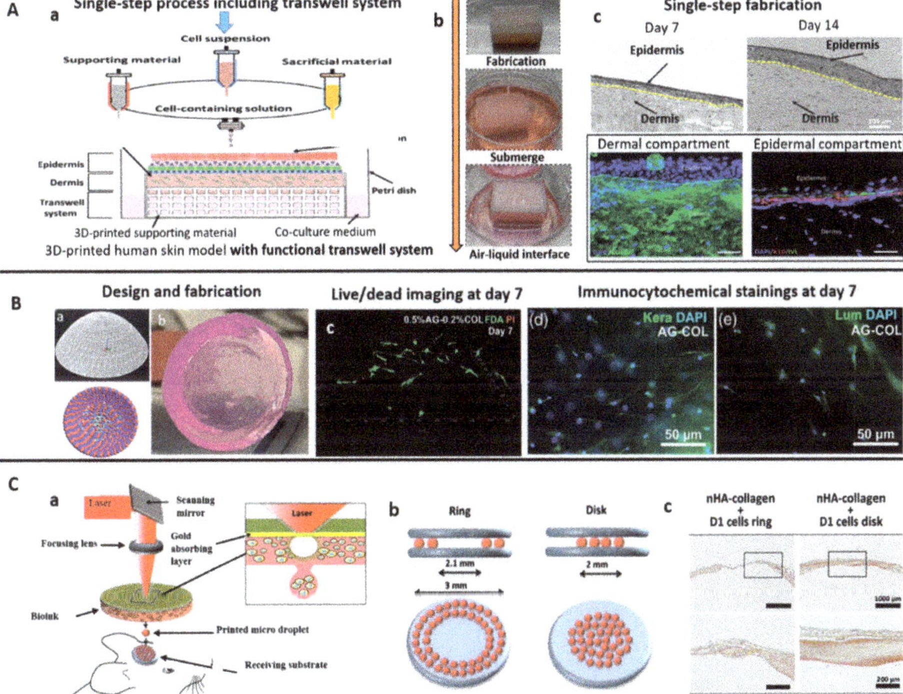

Figure 2. Bioprinting of collagen-based inks for tissue engineering. (**A**) (a,b) Hybrid system (extrusion-based and inkjet-based dispensing modules) used for bioprinting of collagen bioink for developing human skin models, (c) bioprinted model showed good structural features and respective dermis (Col) and epidermis (K10) biomarkers [118]; (**B**) (a,b) Drop-on-demand (DoD) bioprinting was used for bioprinting collagen bioink to develop functional biomimetic 3D corneal model, (c) 3D view of human CSK 7 days after bioprinting stained with live/dead staining, most of cells found viable, (d) Smooth muscle actin immunocytochemical stainings of CSK-loaded agarose-collagen blends 7 days after bioprinting, observed positive keratocan (Kera) and lumican (Lum) expression [120]; (**C**) (a) Laser-assisted bioprinting was explored for in-situ bioprinting of collagen-based bioinks for bone regeneration applications, (b) two different printed designs: a ring and a disk, and (c) disk printed geometry showed homogeneous regeneration throughout the defect, in contrast with the ring geometry, where regeneration is mainly observed at the periphery [126].

3.2.3. Laser-Assisted Printing

In laser-assisted bioprinting (LAB), a layer of biomaterial is deposited onto a substrate via laser-induced forward transfer. A pulsed laser beam is focused on to a donor substrate coated with a laser-energy absorbing layer and a biomaterial layer. Energy absorbed by the donor would drive the biomaterial from the donor substrate onto the receiving substrate. The precise deposition of biomaterial is achieved by the movement of the donor substrate in the x, y axis and the receiving substrate in the x,y and z axis [18]. By coating the donor film with different materials and focusing the laser beam on different locations of the donor substrate for deposition, a heterogenous 3D structure can be obtained. This method of bioprinting, like extrusion bioprinting, also allows a large range of ink viscosities (1–300 mPa·s) [127]. It has the highest print resolution (10 µm) amongst additive bio-manufacturing methods and allows for a high concentration of cell loading [128]. However, preparation of a homogenous donor substrate for each cell type and biomaterials is time-consuming and may be difficult with multiple cells and material formulations.

Collagen-based inks are a suitable donor substrate due to cell biocompatibility and their potential for self-assembly and crosslinking. Laser-assisted bioprinting has been used to recreate skin substitutes [129,130] and corneal stroma-like tissue [131]. Additionally, in vivo bone regeneration was achieved by in situ printing of mesenchymal stromal cells using LAB (Figure 2C) [126].

3.2.4. Stereolithography Printing

In stereolithography/digital light process (SLA/DLP) bioprinting, the ink is crosslinked by photopolymerisation. A reservoir of photo-sensitive ink is exposed to a predefined light pattern and crosslinked layer by layer onto a platform to produce a 3D structure [18]. The use of light patterns allow for high print resolution (50 µm) and accuracy [132]. Similar to nozzle free systems such as inkjet and laser-assisted bioprinting, SLA/DLP systems do not face clogging issues during printing. SLA accommodates inks with greater viscosities (<5 Pa) [133]. However, its restriction is the requirement for photopolymerisation crosslinking since not all materials are compatible for printing. Furthermore, photo-curing agents can be cytotoxic if residual components remain after printing [132]. Unlike previous methods, SLA/DLP bioprinting is unable to incorporate multiple ink formulations.

While collagen can be crosslinked by UV irradiation, on its own, it cannot crosslink sufficiently fast for viable bioprinting. This necessitates functionalisation of collagen molecules. Typically, free amine groups in collagen are replaced with methacrylate groups which can participate in free radical polymerisation (methacrylation). Additionally, this functionalised collagen retains the ability to self-assemble into fibrillar structures upon neutralisation. Modified collagen has shown successful 3D photopatterning of hydrogels loaded with human mesenchymal stem cells [134].

To aid the reader, Table 1 has been provided to summarise applications of additive bioprinting methods for collagen biomaterials/biocomposites and bioinks (cell-laden).

Table 1. Applications of additive bioprinting methods for collagen-based inks.

Bioprinting Method	Collagen-Based Ink Formulation	Outcome	Ref.
Extrusion	Methacrylated type I collagen; Sodium alginate	Fabrication of structures that resembles native human corneal stroma with cell-laden bioink via extrusion bioprinting.	[116]
Extrusion	Collagen Type I; Alginic acid sodium salt from brown algae; $CaCl_2$ solution	Core-sheath coaxial extrusion of alginate/collagen bioink with $CaCl_2$ allows creation of scaffolds with low collagen centration despite its low viscosity.	[114]
Extrusion	Rat tail type I collagen; Gelatin (type A); Sodium alginate	Extrusion bioprinting of collagen scaffold via gelatin/alginate system with controllable degradation time based on amount of sodium citrate during incubation.	[113]
Extrusion	Type I collagen was extracted from tendons obtained from rat tails	Identified storage modulus as the best predictor of collagen bioink printability during deposition.	[117]
Extrusion	PureCol Purified Bovine Collagen Solution; Soldium alginate (low viscosity)	Fabrication of interwoven hard (PLLA) and soft (bioink) scaffolds which support cell attachment and proliferation using a modified desktop 3D printer.	[135]
Extrusion	Methacrylated COL I; Heprasil; Photoinitiator	Successful bioprinting of liver model. Printed primary hepatocytes retained function over 2 weeks exhibiting appropriate response to toxic drugs.	[41]
Extrusion	Lyophilized Atelo-collagen, Matrixen-PSP	Pre-set extrusion bioprinting technique is able to create heterogeneous, multicellular and multi-material structures which perform better than traditional bioprinting.	[112]
Extrusion	Collagen Type I extracted from rat tails; Pluronic® F127	Fabrication of 3D constructs without chemical or photocrosslinking before and after printing via thermally-controlled extrusion.	[115]
Extrusion	Lyophilized sterile collagen, Viscoll	Formation of scaffolds which support spatial arrangement of tissue spheroids as well as support cell adhesion and proliferation.	[47]
Extrusion	Type-I collagen, Matrixen-PSP; Tannic acid	Fabrication of 3D porous structures which support cell migration and proliferation for long periods of culture. Determined optimal tannic acid crosslinking.	[67]
Extrusion	Collagen Type I; Sodium Alginate	Improved mechanical strength and bioactivity via the addition of collagen. Higher cartilage gene markers expressed, preservation of chondrocyte phenotype.	[42]
Extrusion	Type-I collagen, Matrixen-PSP	Established a crosslinking process using tannic acid. High printed preosteoblast viability and well-defined pore size and strut dimensions for bone regeneration.	[68]
Extrusion	Type-I collagen, Matrixen-PSP; Decellularised extracellular matrix (dECM); Silk Fibroin(SF)	Hybrid collagen/dECM/SF scaffold with enhanced cellular activity and mechanical properties. Enhanced cell differentiation, mechanical properties, amenable for hard tissue regeneration.	[59]
Extrusion	Atelocollagen Type I powder	Novel self-assembly induced 3D printing to produce macro/nano porous collagen scaffolds with reasonable mechanical properties, excellent biocompatibility and mimicking native ECM.	[58]
Extrusion	Type-I collagen, Matrixen-PSP; Polycaprolactone (PCL); Hydroxyapatite (HA)/β-tricalcium-phosphate (TCP); Platelet-rich plasma(PRP)	Fabrication of collagen/PCL biocomposites loaded with bio-additives via 3D extrusion printing. Collagen/PCL biocomposites allow controlled release of HA/TCP bio-additives, which promote osteogenesis. PRP biocomposites demonstrate increased mineralisation.	[46]

might involve developing technology in a modular manner. For example, development could begin with a minimally-modified biomaterial using the 510 (K) pathway to initiate revenue generation, before developing combination products suited for the PMA route. The likelihood of obtaining approval for the 2nd product with more complex features could be enhanced by the original (basic) product, because of its regulatory predicate [137].

A further consideration concerns differences between the EU and USA in regulating 3D bioprinted tissue engineering products. Whereas they may be considered biologics in USA, they are regulated as combined advanced therapy medicinal products (ATMPs) in EU. In general, the authors found that existing frameworks fail to address aspects of computer-aided 3D-bioprinting for additive manufacturing of customised tissue products [142]. They concluded, early and regular dialogue with regulatory authorities may alleviate these bottlenecks in manufacturing and quality development [142].

5. Concluding Remarks

As the most ubiquitous extracellular matrix material, collagen is an obvious candidate biomaterial with great promise for regenerative medicine. Collagen is a natural polymer with high biocompatibility, biodegradability and weak antigenicity [13]. Other benefits include: its evolutionary conservation [143]—suggesting it can be derived from many sources including (but not limited to) common commercial sources: rat tail, porcine tendon, bovine skin, fish skin etc. Thus, several xenogeneic acellular matrices have already obtained clinical approval [143]. Collagen is also extracted relatively easily, increasing the ease of availability. However, issues of ethical derivation and sustainability of collagen have arisen, which makes transgenic sources an attractive proposition [29]. Collagen is also a highly versatile biomaterial, denaturing into gelatin (and other derivatives), increasing crosslinking degree through chemical and physical means—rendering control over physical properties such as: mechanical stiffness, pore size and biodegradability. Its versatility extends to formulating biocomposites with inorganic and natural polymers to provide appropriate mechanical stiffness (e.g., PCL), gelation properties (e.g., alginates) etc. to develop suitable collagen bioinks and biomaterials for therapy.

Producing collagen-derived therapeutic and testing products with additive bioprinting methods provides significant benefits over non-additive production. Additive bioprinting exquisitely controls ink deposition, facilitating spatial patterning (mimicking the heterogeneity of skin dermis) [141], reproducibility, customisation, higher throughput, cost-effectiveness etc. [19]. On the other hand, non-additive methods like manual casting may limit product complexity and reproducibility, while electrospinning is limited in throughput and product complexity. These attractive attributes of additive bioprinting may significantly lower barriers to utilising collagen-based products in regenerative therapy and disease modelling etc. With increased process reproducibility, the inter-batch variability during manufacturing is likely to decrease, resulting in smaller tolerances reflected in its device master file (cGMP requirement). Therefore, strategic considerations of regulatory and cost issues in the application of additive bioprinting will help to ensure collagen biomaterials fulfil their tremendous potential in biomedicine and bioscience.

Author Contributions: Conceptualization, D.C.L.Y., and W.W.C.; software, D.C.; writing—original draft preparation, D.C.L.Y., W.W.C., V.T.; writing—review and editing, D.C.L.Y., W.W.C., D.C. and S.S.; supervision, D.C.L.Y., D.C.; project administration, D.C.L.Y., D.C.; funding acquisition, M.W.N. All authors have read and agreed to the published version of the manuscript. Please turn to the CRediT taxonomy for the term explanation.

Funding: This research was funded by SCIENCE AND ENGIEERING RESEARCH COUNCIL (SERC), A*STAR grant number A18A8b0059.

Conflicts of Interest: The authors declare no conflict of interest.

References

1.   Frantz, C.; Stewart, K.M.; Weaver, V.M. The extracellular matrix at a glance. *J. Cell Sci.* **2010**, *123*, 4195–4200. [CrossRef] [PubMed]

2. Sorushanova, A.; Delgado, L.M.; Wu, Z.; Shologu, N.; Kshirsagar, A.; Raghunath, R.; Mullen, A.M.; Bayon, Y.; Pandit, A.; Raghunath, M.; et al. The Collagen Suprafamily: From biosynthesis to advanced biomaterial development. *Adv. Mater.* **2019**, *31*, 1801651. [CrossRef] [PubMed]

3. Sorushanova, A.; Coentro, J.Q.; Pandit, A.; Zeugolis, D.I.; Raghunath, M. Collagen: Materials Analysis and Implant Uses. In *Comprehensive Biomaterials II*; Elsevier Ltd.: Amsterdam, The Netherlands, 2017; pp. 332–350, ISBN 9780081006924.

4. Lodish, H.; Berk, A.; Zipursky, S.L.; Matsudaira, P.; Baltimore, D.; Darnell, J.E. Collagen: The Fibrous Proteins of the Matrix. In *Molecular Cell Biology*, 4th ed.; W. H. Freeman and Company: New York, NY, USA, 2000.

5. Cen, L.; Liu, W.; Cui, L.; Zhang, W.; Cao, Y. Collagen tissue engineering—Development of novel biomaterials and applications. *Pediatr. Res.* **2008**, *63*, 492–496. [CrossRef] [PubMed]

6. Shoulders, M.D.; Raines, R.T. Collagen structure and stability. *Annu. Rev. Biochem.* **2009**, *78*, 929–958. [CrossRef]

7. Hortensius, R.A.; Harley, B.A. The use of bioinspired alterations in the glycosaminoglycan content of collagen-GAG scaffolds to regulate cell activity. *Biomaterials* **2013**, *34*, 7645–7652. [CrossRef]

8. Caliari, S.R.; Ramirez, M.A.; Harley, B.A. The development of collagen-GAG scaffold-membrane composites for tendon tissue engineering. *Biomaterials* **2011**, *32*, 8990–8998. [CrossRef]

9. Fournet, M.; Bonté, F.; Desmoulière, A. Glycation Damage: A Possible hub for major pathophysiological disorders and aging. *Aging Dis.* **2018**, *9*, 880–900. [CrossRef]

10. Justice, B.A.; Badr, N.A.; Felder, R.A. 3D cell culture opens new dimensions in cell-based assays. *Drug Discov. Today* **2009**, *14*, 102–107. [CrossRef]

11. Grobstein, C. Morphogenetic interaction between embryonic mouse tissues separated by a membrane filter. *Nature* **1953**, *172*, 869–871. [CrossRef]

12. Steele, R.E.; Preston, A.S.; Johnson, J.P.; Handler, J.S. Porous-bottom dishes for culture of polarized cells. *Am. J. Physiol. Physiol.* **1986**, *251*, C136–C139. [CrossRef]

13. Copes, F.; Pien, N.; Van Vlierberghe, S.; Boccafoschi, F.; Mantovani, D. Collagen-based tissue engineering strategies for vascular medicine. *Front. Bioeng. Biotechnol.* **2019**, *7*, 166. [CrossRef] [PubMed]

14. Hinderer, S.; Layland, S.L.; Schenke-Layland, K. ECM and ECM-like materials—Biomaterials for applications in regenerative medicine and cancer therapy. *Adv. Drug Deliv. Rev.* **2016**, *97*, 260–269. [CrossRef]

15. Chattopadhyay, S.; Raines, R.T. Collagen-based biomaterials for wound healing. *Biopolymers* **2014**, *101*, 821–833. [CrossRef]

16. Stephens, N.; Di Silvio, L.; Dunsford, I.; Ellis, M.; Glencross, A.; Sexton, A. Bringing cultured meat to market: Technical, socio-political, and regulatory challenges in cellular agriculture. *Trends Food Sci.Technol.* **2018**, *78*, 155–166. [CrossRef] [PubMed]

17. Singh, S.; Choudhury, D.; Yu, F.; Mironov, V.; Naing, M.W. In situ bioprinting—Bioprinting from benchside to bedside? *Acta Biomater.* **2020**, *101*, 14–25. [CrossRef] [PubMed]

18. Choudhury, D.; Anand, S.; Naing, M.W. The arrival of commercial bioprinters—Towards 3d bioprinting revolution! *Int. J. Bioprinting.* **2018**, *4*, 139. [CrossRef]

19. Murphy, S.V.; Atala, A. 3D bioprinting of tissues and organs. *Nat. Biotechnol.* **2014**, *32*, 773–785. [CrossRef]

20. Leberfinger, A.N.; Ravnic, D.J.; Dhawan, A.; Ozbolat, I.T. Concise review: Bioprinting of stem cells for transplantable tissue fabrication. *Stem Cells Transl. Med.* **2017**, *6*, 1940–1948. [CrossRef]

21. Lee, A.; Hudson, A.R.; Shiwarski, D.J.; Tashman, J.W.; Hinton, T.J.; Yerneni, S.; Bliley, J.M.; Campbell, P.G.; Feinberg, A.W. 3D bioprinting of collagen to rebuild components of the human heart. *Science* **2019**, *365*, 482–487. [CrossRef]

22. Hinton, T.J.; Jallerat, Q.; Palchesko, R.N.; Park, J.H.; Grodzicki, M.S.; Shue, H.-J.; Ramadan, M.H.; Hudson, A.R.; Feinberg, A.W. Three-dimensional printing of complex biological structures by freeform reversible embedding of suspended hydrogels. *Sci. Adv.* **2015**, *1*. [CrossRef]

23. Choudhury, D.; Tun, H.W.; Wang, T.; Naing, M.W. Organ-derived decellularized extracellular matrix: A Game Changer for Bioink Manufacturing? *Trends Biotechnol.* **2018**, *36*, 787–805. [CrossRef] [PubMed]

24. Marques, C.F.; Diogo, G.S.; Pina, S.; Oliveira, J.M.; Silva, T.H.; Reis, R.L. Collagen-based bioinks for hard tissue engineering applications: A comprehensive review. *J. Mater. Sci Mater. Med.* **2019**, *30*, 32. [CrossRef] [PubMed]

25. Holmes, D.F.; Gilpin, C.J.; Baldock, C.; Ziese, U.; Koster, A.J.; Kadler, K.E. Corneal collagen fibril structure in three dimensions: Structural insights into fibril assembly, mechanical properties, and tissue organization. *PNAS* **2001**, *98*, 7307–7312. [CrossRef]

26. Schmidt, M.M.; Dornelles, R.C.P.; Mello, R.O.; Kubota, E.H.; Mazutti, M.A.; Kempka, A.P.; Demiate, I.M. Collagen extraction process. *Int. Food Res. J.* **2015**, *23*, 913–922.

27. Sawicki, L.A.; Choe, L.H.; Wiley, K.L.; Lee, K.H.; Kloxin, A.M. Isolation and Identification of Proteins Secreted by Cells Cultured within Synthetic Hydrogel-Based Matrices. *ACS Biomater. Sci. Eng.* **2018**, *4*, 836–845. [CrossRef] [PubMed]

28. Kleinman, H.K. Isolation of laminin-1 and type IV collagen from the EHS sarcoma. *J. Tissue Cult. Methods* **1994**, *16*, 231–233. [CrossRef]

29. Wang, T.; Lew, J.; Premkumar, J.; Poh, C.L.; Win Naing, M. Production of recombinant collagen: State of the art and challenges. *Eng. Biol.* **2017**, *1*, 18–23. [CrossRef]

30. Tomita, M.; Munetsuna, H.; Sato, T.; Adachi, T.; Hino, R.; Hayashi, M.; Shimizu, K.; Nakamura, N.; Tamura, T.; Yoshizato, K. Transgenic silkworms produce recombinant human type III procollagen in cocoons. *Nat. Biotechnol.* **2003**, *21*, 52–56. [CrossRef]

31. Rele, S.; Song, Y.; Apkarian, R.P.; Qu, Z.; Conticello, V.P.; Chaikof, E.L. D-periodic collagen-mimetic microfibers. *J. Am. Chem. Soc.* **2007**, *129*, 14780–14787. [CrossRef]

32. O'Leary, L.E.R.; Fallas, J.A.; Bakota, E.L.; Kang, M.K.; Hartgerink, J.D. Multi-hierarchical self-assembly of a collagen mimetic peptide from triple helix to nanofibre and hydrogel. *Nat. Chem.* **2011**, *3*, 821–828. [CrossRef]

33. Li, Z.-R.; Wang, B.; Chi, C.; Zhang, Q.-H.; Gong, Y.; Tang, J.-J.; Luo, H.; Ding, G. Isolation and characterization of acid soluble collagens and pepsin soluble collagens from the skin and bone of Spanish mackerel (Scomberomorous niphonius). *Food Hydrocoll.* **2013**, *31*, 103–113. [CrossRef]

34. Meyer, M. Processing of collagen based biomaterials and the resulting materials properties. *Biomed. Eng. Online* **2019**, *18*, 24. [CrossRef] [PubMed]

35. Davison, P.F.; Cannon, D.J.; Andersson, L.P. The effects of acetic acid on collagen cross-links. *Connect. Tissue Res.* **1972**, *1*, 205–216. [CrossRef]

36. Lynn, A.K.; Yannas, I.V.; Bonfield, W. Antigenicity and immunogenicity of collagen. *J. Biomed. Mater. Res. Part B Appl. Biomater.* **2004**, *71B*, 343–354. [CrossRef]

37. Parenteau-Bareil, R.; Gauvin, R.; Berthod, F. Collagen-based biomaterials for tissue engineering applications. *Materials* **2010**, *3*, 1863–1887. [CrossRef]

38. Leon-Lopez, A.; Morales-Penaloza, A.; Martinez-Juarez, V.M.; Vargas-Torres, A.; Zeugolis, D.I.; Aguirre-Alvarez, G. Hydrolyzed collagen-sources and applications. *Molecules* **2019**, *24*, 4031. [CrossRef]

39. Law, J.X.; Liau, L.L.; Saim, A.; Yang, Y.; Idrus, R. Electrospun collagen nanofibers and their applications in skin tissue engineering. *Tissue Eng. Regen. Med.* **2017**, *14*, 699–718. [CrossRef]

40. Hiraoka, Y.; Kimura, Y.; Ueda, H.; Tabata, Y. Fabrication and biocompatibility of collagen sponge reinforced with poly (glycolic acid) fiber. *Tissue Eng.* **2004**, *9*, 1101–1112. [CrossRef]

41. Mazzocchi, A.; Devarasetty, M.; Huntwork, R.; Soker, S.; Skardal, A. Optimization of collagen type I-hyaluronan hybrid bioink for 3D bioprinted liver microenvironments. *Biofabrication* **2018**, *11*, 15003. [CrossRef]

42. Yang, X.; Lu, Z.; Wu, H.; Li, W.; Zheng, L.; Zhao, J. Collagen-alginate as bioink for three-dimensional (3D) cell printing based cartilage tissue engineering. *Mater. Sci. Eng. C* **2018**, *83*, 195–201. [CrossRef]

43. Hortensius, R.A.; Harley, B.A.C. Collagen-GAG Materials. In *Comprehensive Biomaterials II*; Elsevier: Amsterdam, The Netherlands, 2017; pp. 351–380. ISBN 9780081006924.

44. Ueda, H.; Hong, L.; Yamamoto, M.; Shigeno, K.; Inoue, M.; Toba, T.; Yoshitani, M.; Nakamura, T.; Tabata, Y.; Shimizu, Y. Use of collagen sponge incorporating transforming growth factor-beta1 to promote bone repair in skull defects in rabbits. *Biomaterials* **2002**, *23*, 1003–1010. [CrossRef]

45. Wahl, D.A.; Czernuszka, J.T. Collagen-hydroxyapatite composites for hard tissue repair. *Eur. Cell Mater.* **2006**, *11*, 43–56. [CrossRef] [PubMed]

46. Kim, W.; Jang, C.H.; Kim, G. Optimally designed collagen/polycaprolactone biocomposites supplemented with controlled release of HA/TCP/rhBMP-2 and HA/TCP/PRP for hard tissue regeneration. *Mater. Sci. Eng. C* **2017**, *78*, 763–772. [CrossRef] [PubMed]

47. Osidak, E.O.; Karalkin, P.A.; Osidak, M.S.; Parfenov, V.A.; Sivogrivov, D.E.; Pereira, F.D.A.S.; Gryadunova, A.A.; Koudan, E.V.; Khesuani, Y.D.; Kasyanov, V.A.; et al. Viscoll collagen solution as a novel bioink for direct 3D bioprinting. *J. Mater. Sci. Mater. Med.* **2019**, *30*, 31. [CrossRef]

48. Olde Damink, L.H.H.; Dijkstra, P.J.; Van Luyn, M.J.A.; Van Wachem, P.B.; Nieuwenhuis, P.; Feijen, J. Glutaraldehyde as a crosslinking agent for collagen-based biomaterials. *J. Mater. Sci. Mater. Med.* **1995**, *6*, 460–472. [CrossRef]

49. Ahmed, R.M.; Venkateshwarlu, U.; Jayakumar, R. Multilayered peptide incorporated collagen tubules for peripheral nerve repair. *Biomaterials* **2004**, *25*, 2585–2594. [CrossRef]

50. Yannas, I.V. Tissue regeneration by Use of Glycosaminoglycan Copolymer. *Clin. Mater.* **1992**, *9*, 179–187. [CrossRef]

51. Haugh, M.G.; Murphy, C.M.; McKiernan, R.C.; Altenbuchner, C.; O'Brien, F.J. Crosslinking and mechanical properties significantly influence cell attachment, proliferation, and migration within collagen glycosaminoglycan scaffolds. *Tissue Eng. Part A* **2010**, *17*, 1201–1208. [CrossRef]

52. Cote, M.F.; Sirois, E.; Doillon, C.J. In vitro contraction rate of collagen in sponge-shape matrices. *J. Biomater. Sci. Polym. Ed.* **1992**, *3*, 301–313. [CrossRef]

53. White, M.J.; Kohno, I.; Rubin, A.L.; Stenzel, K.H.; Miyata, T. Collagen films—Effect of cross-linking on physical and biological properties. *Biomater. Med. Devices. Artif. Organs* **1973**, *1*, 703–715. [CrossRef]

54. Ber, S.; Torun Kose, G.; Hasirci, V. Bone tissue engineering on patterned collagen films: An in vitro study. *Biomaterials* **2005**, *26*, 1977–1986. [CrossRef] [PubMed]

55. Vasudev, S.C.; Chandy, T.; Sharma, C.P.; Mohanty, M.; Umasankar, P.R. Effects of double cross-linking technique on the enzymatic degradation and calcification of bovine pericardia. *J. Biomater. Appl.* **2000**, *14*, 273–295. [CrossRef] [PubMed]

56. Damink, L.H.O.; Dijkstra, P.J.; van Luyn, M.J.A.; van Wachem, P.B.; Nieuwenhuis, P.; Feijen, J. Cross-linking of dermal sheep collagen using a water-soluble carbodiimide. *Biomaterials* **1996**, *17*, 765–773. [CrossRef]

57. Sehgal, D.; Vijay, I.K. A Method for the High Efficiency of water-soluble carbodiimide-mediated amidation. *Anal. Biochem.* **1994**, *218*, 87–91. [CrossRef] [PubMed]

58. Shin, K.-H.; Kim, J.-W.; Koh, Y.-H.; Kim, H.-E. Novel self-assembly-induced 3D plotting for macro/nano-porous collagen scaffolds comprised of nanofibrous collagen filaments. *Mater. Lett.* **2015**, *143*, 265–268. [CrossRef]

59. Lee, H.; Yang, G.H.; Kim, M.; Lee, J.; Huh, J.; Kim, G. Fabrication of micro/nanoporous collagen/dECM/silk-fibroin biocomposite scaffolds using a low temperature 3D printing process for bone tissue regeneration. *Mater. Sci. Eng. C* **2018**, *84*, 140–147. [CrossRef]

60. Wu, J.M.; Xu, Y.Y.; Li, Z.H.; Yuan, X.Y.; Wang, P.F.; Zhang, X.Z.; Liu, Y.Q.; Guan, J.; Guo, Y.; Li, R.X.; et al. Heparin-functionalized collagen matrices with controlled release of basic fibroblast growth factor. *J. Mater. Sci. Mater. Med.* **2011**, *22*, 107–114. [CrossRef]

61. Tanaka, Y.; Baba, K.; Duncan, T.J.; Kubota, A.; Asahi, T.; Quantock, A.J.; Yamato, M.; Okano, T.; Nishida, K. Transparent, tough collagen laminates prepared by oriented flow casting, multi-cyclic vitrification and chemical cross-linking. *Biomaterials* **2011**, *32*, 3358–3366. [CrossRef]

62. Liu, Y.; Ren, L.; Yao, H.; Wang, Y. Collagen films with suitable physical properties and biocompatibility for corneal tissue engineering prepared by ion leaching technique. *Mater. Lett.* **2012**, *87*, 1–4. [CrossRef]

63. Damink, L.H.O.; Dijkstra, P.J.; van Luyn, M.J.A.; van Wachem, P.B.; Nieuwenhuis, P.; Feijen, J. Crosslinking of dermal sheep collagen using hexamethylene diisocyanate. *J. Mater. Sci. Mater. Med.* **1995**, *6*, 429–434. [CrossRef]

64. Paul, R.G.; Bailey, A.J. Chemical stabilisation of collagen as a biomimetic. *Sci. World J.* **2003**, *3*, 138–155. [CrossRef] [PubMed]

65. Zeugolis, D.I.; Paul, G.R.; Attenburrow, G. Cross-linking of extruded collagen fibers—A biomimetic three-dimensional scaffold for tissue engineering applications. *J. Biomed. Mater. Res. Part A* **2009**, *89A*, 895–908. [CrossRef]

66. Heijmen, F.H.; du Pont, J.S.; Middelkoop, E.; Kreis, R.W.; Hoekstra, M.J. Cross-linking of dermal sheep collagen with tannic acid. *Biomaterials* **1997**, *18*, 749–754. [CrossRef]

67. Yeo, M.G.; Kim, G.H. A cell-printing approach for obtaining {hASC}-laden scaffolds by using a collagen/polyphenol bioink. *Biofabrication* **2017**, *9*, 25004. [CrossRef] [PubMed]

68. Lee, J.; Yeo, M.; Kim, W.; Koo, Y.; Kim, G.H. Development of a tannic acid cross-linking process for obtaining 3D porous cell-laden collagen structure. *Int. J. Biol. Macromol.* **2018**, *110*, 497–503. [CrossRef]

69. Macaya, D.; Ng, K.K.; Spector, M. Injectable. *Adv. Funct. Mater.* **2011**, *21*, 4788–4797. [CrossRef]

70. Kim, Y.B.; Lee, H.; Kim, G.H. Strategy to achieve highly porous/biocompatible macroscale cell blocks, using a collagen/genipin-bioink and an optimal 3d printing process. *ACS Appl. Mater. Interfaces* **2016**, *8*, 32230–32240. [CrossRef] [PubMed]

71. Kumar, V.A.; Caves, J.M.; Haller, C.A.; Dai, E.; Li, L.; Grainger, S.; Chaikof, E.L. Collagen-based substrates with tunable strength for soft tissue engineering. *Biomater. Sci.* **2013**, *1*, 1193–1202. [CrossRef]

72. HW, S.; RN, H.; LL, H.; CC, T.; CT, C. Feasibility study of a natural crosslinking reagent for biological tissue fixation. *J. Biomed Mater Res.* **1998**, *42*, 560–567. [CrossRef]

73. Haugh, M.G.; Jaasma, M.J.; O'Brien, F.J. The effect of dehydrothermal treatment on the mechanical and structural properties of collagen-GAG scaffolds. *J. Biomed. Mater. Res. Part A* **2009**, *89A*, 363–369. [CrossRef]

74. Weadock, K.S.; Miller, E.J.; Bellincampi, L.D.; Zawadsky, J.P.; Dunn, M.G. Physical crosslinking of collagen fibers—Comparison of ultraviolet irradiation and dehydrothermal treatment. *J. Biomed. Mater. Res.* **1995**, *29*, 1373–1379. [CrossRef] [PubMed]

75. Davidenko, N.; Bax, D.V.; Schuster, C.F.; Farndale, R.W.; Hamaia, S.W.; Best, S.M.; Cameron, R.E. Optimisation of UV irradiation as a binding site conserving method for crosslinking collagen-based scaffolds. *J. Mater. Sci. Mater. Med.* **2016**, *27*, 14. [CrossRef]

76. Zhang, X.; Xu, L.; Huang, X.; Wei, S.; Zhai, M. Structural study and preliminary biological evaluation on the collagen hydrogel crosslinked by γ-irradiation. *J. Biomed. Mater. Res. A* **2012**, *100*, 2960–2969. [CrossRef]

77. Abraham, L.C.; Zuena, E.; Perez-Ramirez, B.; Kaplan, D.L. Guide to collagen characterization for biomaterial studies. *J. Biomed Mater Res B Appl Biomater* **2008**, *87*, 264–285. [CrossRef] [PubMed]

78. Schroepfer, M.; Meyer, M. DSC investigation of bovine hide collagen at varying degrees of crosslinking and humidities. *Int. J. Biol. Macromol.* **2017**, *103*, 120–128. [CrossRef] [PubMed]

79. Nomura, Y.; Sakai, H.; Ishii, Y.; Shirai, K. Preparation and some properties of type i collagen from fish scales. *Biosci. Biotechnol. Biochem.* **1996**, *60*, 2092–2094. [CrossRef] [PubMed]

80. Wu, J.; Li, Z.; Yuan, X.; Wang, P.; Liu, Y.; Wang, H. Extraction and isolation of type I, III and V collagens and their SDS-PAGE analyses. *Trans. Tianjin Univ.* **2011**, *17*, 111. [CrossRef]

81. Drzewiecki, K.E.; Grisham, D.R.; Parmar, A.S.; Nanda, V.; Shreiber, D.I. Circular dichroism spectroscopy of collagen fibrillogenesis: A new use for an old technique. *Biophys. J.* **2016**, *111*, 2377–2386. [CrossRef]

82. Greenfield, N.J. Using circular dichroism spectra to estimate protein secondary structure. *Nat. Protoc.* **2006**, *1*, 2876–2890. [CrossRef]

83. Martinez, M.G.; Bullock, A.J.; MacNeil, S.; Rehman, I.U. Characterisation of structural changes in collagen with Raman spectroscopy. *Appl. Spectrosc. Rev.* **2019**, *54*, 509–542. [CrossRef]

84. Riaz, T.; Zeeshan, R.; Zarif, F.; Ilyas, K.; Muhammad, N.; Safi, S.Z.; Rahim, A.; Rizvi, S.A.A.; Rehman, I.U. FTIR analysis of natural and synthetic collagen. *Appl. Spectrosc. Rev.* **2018**, *53*, 703–746. [CrossRef]

85. Beier, J.P.; Klumpp, D.; Rudisile, M.; Dersch, R.; Wendorff, J.H.; Bleiziffer, O.; Arkudas, A.; Polykandriotis, E.; Horch, R.E.; Kneser, U. Collagen matrices from sponge to nano: New perspectives for tissue engineering of skeletal muscle. *BMC Biotechnol.* **2009**, *9*, 34. [CrossRef] [PubMed]

86. Lucchese, A.; Pilolli, G.P.; Petruzzi, M.; Crincoli, V.; Scivetti, M.; Favia, G. Analysis of collagen distribution in human crown dentin by confocal laser scanning microscopy. *Ultrastruct. Pathol.* **2008**, *32*, 107–111. [CrossRef] [PubMed]

87. Dewavrin, J.-Y.; Hamzavi, N.; Shim, V.P.W.; Raghunath, M. Tuning the architecture of three-dimensional collagen hydrogels by physiological macromolecular crowding. *Acta Biomater.* **2014**, *10*, 4351–4359. [CrossRef]

88. Quan, B.D.; Sone, E.D. Cryo-TEM analysis of collagen fibrillar structure. *Methods Enzym.* **2013**, *532*, 189–205. [CrossRef]

89. Chernoff, E.A.G.; Chernoff, D.A. Atomic force microscope images of collagen fibers. *J. Vac. Sci. Technol. A* **1992**, *10*, 596–599. [CrossRef]

90. Cissell, D.D.; Link, J.M.; Hu, J.C.; Athanasiou, K.A. A Modified hydroxyproline assay based on hydrochloric acid in ehrlich's solution accurately measures tissue collagen content. *Tissue Eng. Part C. Methods* **2017**, *23*, 243–250. [CrossRef]

91. Lareu, R.R.; Zeugolis, D.I.; Abu-Rub, M.; Pandit, A.; Raghunath, M. Essential modification of the Sircol Collagen Assay for the accurate quantification of collagen content in complex protein solutions. *Acta Biomater.* **2010**, *6*, 3146–3151. [CrossRef]

92. Liang, H.; Russell, S.J.; Wood, D.J.; Tronci, G. A hydroxamic acid–methacrylated collagen conjugate for the modulation of inflammation-related MMP upregulation. *J. Mater. Chem. B* **2018**, *6*, 3703–3715. [CrossRef]

93. Lai, J.-Y.; Wang, P.-R.; Luo, L.-J.; Chen, S.-T. Stabilization of collagen nanofibers with L-lysine improves the ability of carbodiimide cross-linked amniotic membranes to preserve limbal epithelial progenitor cells. *Int. J. Nanomed.* **2014**, *9*, 5117–5130. [CrossRef]

94. Castagnaro, S.; Chrisam, M.; Cescon, M.; Braghetta, P.; Grumati, P.; Bonaldo, P. Extracellular collagen vi has prosurvival and autophagy instructive properties in mouse fibroblasts. *Front. Physiol.* **2018**, *9*, 1129. [CrossRef] [PubMed]

95. Van Huizen, N.A.; Ijzermans, J.N.M.; Burgers, P.C.; Luider, T.M. Collagen analysis with mass spectrometry. *Mass Spectrom. Rev.* **2019**. [CrossRef] [PubMed]

96. O'Brien, F.J.; Harley, B.A.; Yannas, I.V.; Gibson, L. Influence of freezing rate on pore structure in freeze-dried collagen-GAG scaffolds. *Biomaterials* **2004**, *25*, 1077–1086. [CrossRef]

97. Caliari, S.R.; Weisgerber, D.W.; Ramirez, M.A.; Kelkhoff, D.O.; Harley, B.A.C. The influence of collagen-glycosaminoglycan scaffold relative density and microstructural anisotropy on tenocyte bioactivity and transcriptomic stability. *J. Mech. Behav. Biomed. Mater.* **2012**, *11*, 27–40. [CrossRef] [PubMed]

98. John, A.; Hong, L.; Ikada, Y.; Tabata, Y. A trial to prepare biodegradable collagen-hydroxyapatite composites for bone repair. *J. Biomater. Sci. Polym. Ed.* **2001**, *12*, 689–705. [CrossRef]

99. Yannas, I.V.; Tzeranis, D.S.; So, P.T.C. Regeneration of injured skin and peripheral nerves requires control of wound contraction, not scar formation. *Wound Repair Regen.* **2017**, *25*, 177–191. [CrossRef]

100. Yannas, I.V.; Tzeranis, D.S.; Harley, B.A.; So, P.T. Biologically active collagen-based scaffolds: Advances in processing and characterization. *Philos. Trans. R. Soc. A Math. Phys. Eng. Sci.* **2010**, *368*, 2123–2139. [CrossRef]

101. Chamberlain, L.J.; Yannas, I.V.; Hsu, H.-P.; Strichartz, G.; Spector, M. Collagen-GAG substrate enhances the quality of nerve regeneration through collagen tubes up to level of autograft. *Exp. Neurol.* **1998**, *154*, 315–329. [CrossRef]

102. Cascone, M.G.; Sim, B.; Sandra, D. Blends of synthetic and natural polymers as drug delivery systems for growth hormone. *Biomaterials* **1995**, *16*, 569–574. [CrossRef]

103. Maeda, M.; Kadota, K.; Kajihara, M.; Sano, A.; Fujioka, K. Sustained release of human growth hormone (hGH) from collagen film and evaluation of effect on wound healing in db/db mice. *J. Control. Release* **2001**, *77*, 261–272. [CrossRef]

104. Wang, Z.; Hu, S.; Wang, H. Scale-up preparation and characterization of collagen/sodium alginate blend films. *J. Food Qual.* **2017**, *2017*, 4954259. [CrossRef]

105. Fuller, K.; Pandit, A.; Zeugolis, D.I. The multifaceted potential of electro-spinning in regenerative medicine. *Pharm. Nanotechnol.* **2014**, *2*, 23–34. [CrossRef]

106. Zeugolis, D.I.; Khew, S.T.; Yew, E.S.Y.; Ekaputra, A.K.; Tong, Y.W.; Yung, L.-Y.L.; Hutmacher, D.W.; Sheppard, C.; Raghunath, M. Electro-spinning of pure collagen nano-fibres—Just an expensive way to make gelatin? *Biomaterials* **2008**, *29*, 2293–2305. [CrossRef]

107. Ladd, M.R.; Lee, S.J.; Stitzel, J.D.; Atala, A.; Yoo, J.J. Co-electrospun dual scaffolding system with potential for muscle-tendon junction tissue engineering. *Biomaterials* **2011**, *32*, 1549–1559. [CrossRef] [PubMed]

108. McClure, M.J.; Sell, S.A.; Simpson, D.G.; Walpoth, B.H.; Bowlin, G.L. A three-layered electrospun matrix to mimic native arterial architecture using polycaprolactone, elastin, and collagen: A preliminary study. *Acta Biomater.* **2010**, *6*, 2422–2433. [CrossRef] [PubMed]

109. Ekaputra, A.K.; Prestwich, G.D.; Cool, S.M.; Hutmacher, D.W. Combining electrospun scaffolds with electrosprayed hydrogels leads to three-dimensional cellularization of hybrid constructs. *Biomacromolecules* **2008**, *9*, 2097–2103. [CrossRef]

110. Dong, B.; Arnoult, O.; Smith, M.E.; Wnek, G.E. Electrospinning of collagen nanofiber scaffolds from benign solvents. *Macromol. Rapid Commun.* **2009**, *30*, 539–542. [CrossRef]

111. Ozbolat, I.T.; Yu, Y. Bioprinting toward organ fabrication: Challenges and future trends. *IEEE Trans. Biomed. Eng.* **2013**, *60*, 691–699. [CrossRef]

112. Kang, D.; Ahn, G.; Kim, D.; Kang, H.W.; Yun, S.; Yun, W.S.; Shim, J.H.; Jin, S. Pre-set extrusion bioprinting for multiscale heterogeneous tissue structure fabrication. *Biofabrication* **2018**, *10*, 35008. [CrossRef]

113. Wu, Z.; Su, X.; Xu, Y.; Kong, B.; Sun, W.; Mi, S. Bioprinting three-dimensional cell-laden tissue constructs with controllable degradation. *Sci. Rep.* **2016**, *6*, 24474. [CrossRef]

114. Zhu, K.; Chen, N.; Liu, X.; Mu, X.; Zhang, W.; Wang, C.; Zhang, Y.S. A General strategy for extrusion bioprinting of bio-macromolecular bioinks through alginate-templated dual-stage crosslinking. *Macromol. Biosci.* **2018**, *18*, e1800127. [CrossRef] [PubMed]

115. Moncal, K.K.; Ozbolat, V.; Datta, P.; Heo, D.N.; Ozbolat, I.T. Thermally-controlled extrusion-based bioprinting of collagen. *J. Mater. Sci. Mater. Med.* **2019**, *30*, 55. [CrossRef]

116. Isaacson, A.; Swioklo, S.; Connon, C.J. 3D bioprinting of a corneal stroma equivalent. *Exp. Eye Res.* **2018**, *173*, 188–193. [CrossRef] [PubMed]

117. Diamantides, N.; Wang, L.; Pruiksma, T.; Siemiatkoski, J.; Dugopolski, C.; Shortkroff, S.; Kennedy, S.; Bonassar, L.J. Correlating rheological properties and printability of collagen bioinks: The effects of riboflavin photocrosslinking and pH. *Biofabrication* **2017**, *9*, 34102. [CrossRef]

118. Kim, B.S.; Lee, J.S.; Gao, G.; Cho, D.W. Direct 3D cell-printing of human skin with functional transwell system. *Biofabrication* **2017**, *9*, 25034. [CrossRef] [PubMed]

119. Kim, J.D.; Choi, J.S.; Kim, B.S.; Chan Choi, Y.; Cho, Y.W. Piezoelectric inkjet printing of polymers: Stem cell patterning on polymer substrates. *Polymer* **2010**, *51*, 2147–2154. [CrossRef]

120. Duarte Campos, D.F.; Rohde, M.; Ross, M.; Anvari, P.; Blaeser, A.; Vogt, M.; Panfil, C.; Yam, G.H.; Mehta, J.S.; Fischer, H.; et al. Corneal bioprinting utilizing collagen-based bioinks and primary human keratocytes. *J. Biomed. Mater. Res. Part A* **2019**, *107*, 1945–1953. [CrossRef]

121. Duarte Campos, D.F.; Blaeser, A.; Buellesbach, K.; Sen, K.S.; Xun, W.; Tillmann, W.; Fischer, H. Bioprinting organotypic hydrogels with improved mesenchymal stem cell remodeling and mineralization properties for bone tissue engineering. *Adv. Healthc. Mater.* **2016**, *5*, 1336–1345. [CrossRef]

122. Roth, E.A.; Xu, T.; Das, M.; Gregory, C.; Hickman, J.J.; Boland, T. Inkjet printing for high-throughput cell patterning. *Biomaterials* **2004**, *25*, 3707–3715. [CrossRef]

123. Sanjana, N.E.; Fuller, S.B. A fast flexible ink-jet printing method for patterning dissociated neurons in culture. *J. Neurosci. Methods* **2004**, *136*, 151–163. [CrossRef] [PubMed]

124. Park, T.M.; Kang, D.; Jang, I.; Yun, W.S.; Shim, J.H.; Jeong, Y.H.; Kwak, J.Y.; Yoon, S.; Jin, S. Fabrication of in vitro cancer microtissue array on fibroblast-layered nanofibrous membrane by inkjet printing. *Int. J. Mol. Sci.* **2017**, *18*, 2348. [CrossRef]

125. Boland, T.; Mironov, V.; Gutowska, A.; Roth, E.A.; Markwald, R.R. Cell and organ printing 2: Fusion of cell aggregates in three-dimensional gels. *Anat. Rec. Part A Discov. Mol. Cell. Evol. Biol.* **2003**, *272A*, 497–502. [CrossRef]

126. Keriquel, V.; Oliveira, H.; Remy, M.; Ziane, S.; Delmond, S.; Rousseau, B.; Rey, S.; Catros, S.; Amedee, J.; Guillemot, F.; et al. In situ printing of mesenchymal stromal cells, by laser-assisted bioprinting, for in vivo bone regeneration applications. *Sci. Rep.* **2017**, *7*, 1778. [CrossRef]

127. Guillotin, B.; Guillemot, F. Cell patterning technologies for organotypic tissue fabrication. *Trends Biotechnol.* **2011**, *29*, 183–190. [CrossRef] [PubMed]

128. Guillotin, B.; Catros, S.; Guillemot, F. Laser assisted bio-printing (LAB) of Cells and Bio-materials Based on Laser Induced Forward Transfer (LIFT). In *Laser Technology in Biomimetics*; Biological and Medical Physics, Biomedical Engineering; Springer: Berlin/Heidelberg, Germany, 2013; pp. 193–209, ISBN1 978-3-642-41340-7, ISBN2 978-3-642-41341-4.

129. Koch, L.; Deiwick, A.; Schlie, S.; Michael, S.; Gruene, M.; Coger, V.; Zychlinski, D.; Schambach, A.; Reimers, K.; Vogt, P.M.; et al. Skin tissue generation by laser cell printing. *Biotechnol. Bioeng.* **2012**, *109*, 1855–1863. [CrossRef] [PubMed]

130. Michael, S.; Sorg, H.; Peck, C.T.; Koch, L.; Deiwick, A.; Chichkov, B.; Vogt, P.M.; Reimers, K. Tissue engineered skin substitutes created by laser-assisted bioprinting form skin-like structures in the dorsal skin fold chamber in mice. *PLoS ONE* **2013**, *8*, e57741. [CrossRef] [PubMed]

131. Sorkio, A.; Koch, L.; Koivusalo, L.; Deiwick, A.; Miettinen, S.; Chichkov, B.; Skottman, H. Human stem cell based corneal tissue mimicking structures using laser-assisted 3D bioprinting and functional bioinks. *Biomaterials* **2018**, *171*, 57–71. [CrossRef]

132. Knowlton, S.; Yenilmez, B.; Anand, S.; Tasoglu, S. Photocrosslinking-based bioprinting: Examining crosslinking schemes. *Bioprinting* **2017**, *5*, 10–18. [CrossRef]

133. Bártolo, P.J. *Stereolithography: Materials, Processes and Applications*; Springer: New York, NY, USA, 2011; ISBN 978-0-387-92903-3.

134. Drzewiecki, K.E.; Malavade, J.N.; Ahmed, I.; Lowe, C.J.; Shreiber, D.I. A thermoreversible, photocrosslinkable collagen bio-ink for free-form fabrication of scaffolds for regenerative medicine. *Technology* **2017**, *5*, 185–195. [CrossRef]

135. Goldstein, T.A.; Epstein, C.J.; Schwartz, J.; Krush, A.; Lagalante, D.J.; Mercadante, K.P.; Zeltsman, D.; Smith, L.P.; Grande, D.A. Feasibility of bioprinting with a modified desktop 3D printer. *Tissue Eng. Part C Methods* **2016**, *22*, 1071–1076. [CrossRef]

136. Murphy, S.V.; Skardal, A.; Atala, A. Evaluation of hydrogels for bio-printing applications. *J. Biomed. Mater. Res. Part A* **2013**, *101A*, 272–284. [CrossRef] [PubMed]

137. Hollister, S.J.; Murphy, W.L. Scaffold translation: Barriers between concept and clinic. *Tissue Eng. Part B. Rev.* **2011**, *17*, 459–474. [CrossRef]

138. Pashuck, E.T.; Stevens, M.M. Designing regenerative biomaterial therapies for the clinic. *Sci. Transl. Med.* **2012**, *4*, 160sr4. [CrossRef] [PubMed]

139. Fernandez-Moure, J.S. Lost in Translation: The gap in scientific advancements and clinical application. *Front. Bioeng. Biotechnol.* **2016**, *4*, 43. [CrossRef] [PubMed]

140. Ratcliffe, A. Difficulties in the translation of functionalized biomaterials into regenerative medicine clinical products. *Biomaterials* **2011**, *32*, 4215–4217. [CrossRef]

141. Ng, W.L.; Wang, S.; Yeong, W.Y.; Naing, M.W. Skin bioprinting: Impending reality or fantasy? *Trends Biotechnol.* **2016**, *34*, 689–699. [CrossRef] [PubMed]

142. Hourd, P.; Medcalf, N.; Segal, J.; Williams, D.J. A 3D bioprinting exemplar of the consequences of the regulatory requirements on customized processes. *Regen. Med.* **2015**, *10*, 863–883. [CrossRef]

143. Furth, M.E.; Atala, A.; van Dyke, M.E. Smart biomaterials design for tissue engineering and regenerative medicine. *Biomaterials* **2007**, *28*, 5068–5073. [CrossRef]

 *bioengineering*

Article

Formulation and Characterization of Alginate Dialdehyde, Gelatin, and Platelet-Rich Plasma-Based Bioink for Bioprinting Applications

Lakshmi T. Somasekharan [1,†]**, Naresh Kasoju** [2,†] **, Riya Raju** [1] **and Anugya Bhatt** [1,*]

1   Division of Thrombosis Research, Department of Applied Biology, Biomedical Technology Wing, Sree Chitra Tirunal Institute for Medical Sciences and Technology, Thiruvananthapuram, Kerala 695012, India; s.lakshmit@gmail.com (L.T.S.); riyarajud@gmail.com (R.R.)
2   Division of Tissue Culture, Department of Applied Biology, Biomedical Technology Wing, Sree Chitra Tirunal Institute for Medical Sciences and Technology, Thiruvananthapuram, Kerala 695012, India; naresh.kasoju@sctimst.ac.in
*   Correspondence: anugyabhatt@sctimst.ac.in; Tel.: +91-471-252-0219
†   These authors contributed equally to the work and thus are recognized as equal first authors.

Received: 18 August 2020; Accepted: 7 September 2020; Published: 9 September 2020

Abstract: Layer-by-layer additive manufacturing process has evolved into three-dimensional (3D) "bio-printing" as a means of constructing cell-laden functional tissue equivalents. The process typically involves the mixing of cells of interest with an appropriate hydrogel, termed as "bioink", followed by printing and tissue maturation. An ideal bioink should have adequate mechanical, rheological, and biological features of the target tissues. However, native extracellular matrix (ECM) is made of an intricate milieu of soluble and non-soluble extracellular factors, and mimicking such a composition is challenging. To this end, here we report the formulation of a multi-component bioink composed of gelatin and alginate -based scaffolding material, as well as a platelet-rich plasma (PRP) suspension, which mimics the insoluble and soluble factors of native ECM respectively. Briefly, sodium alginate was subjected to controlled oxidation to yield alginate dialdehyde (ADA), and was mixed with gelatin and PRP in various volume ratios in the presence of borax. The formulation was systematically characterized for its gelation time, swelling, and water uptake, as well as its morphological, chemical, and rheological properties; furthermore, blood- and cytocompatibility were assessed as per ISO 10993 (International Organization for Standardization). Printability, shape fidelity, and cell-laden printing was evaluated using the RegenHU 3D Discovery bioprinter. The results indicated the successful development of ADA–gelatin–PRP based bioink for 3D bioprinting and biofabrication applications.

Keywords: biofabrication; bioink; hydrogels; growth factor cocktail; bioactive scaffold; printability

1. Introduction

With the demand for innovations and technologies to generate biomimetic organs in the field of medical technology comes the need for generation of novel three-dimensional (3D) objects that can change the face of medical science. One such innovation that has found an eminent place in tissue engineering and regenerative medicine is the 3D bioprinting technology [1]. As the name suggests, 3D bioprinting is an additive manufacturing process that uses a 3D bioprinter and biocompatible biomaterials to generate 3D tissues through layer-by-layer extrusion [2]. The resultant tissues can be used to replace, repair, or reconstruct damaged tissue/organ in the human body and fabricate 3D tissues for in vitro toxicological testing applications. Such material that incorporates cells and existing hydrogel biomaterial components to fabricate scaffolds for 3D bioprinting application is called "bioink". Bioinks are generated from biocompatible polymers that can be tuned for their printability, biodegradability, and better mechanical property. They are physical scaffolds to which the cell attaches and proliferates to form a tissue construct [3]. Many commonly available bioinks that have been used for 3D bioprinting include collagen, alginate, gelatin, chitosan, and tissue-specific decellularized extracellular matrix. They are used either alone or in different combinations to improve overall performance in terms of cell proliferation, metabolic activity, and tissue-specific functions [4]. There have been approaches using extracellular matrices and biological components, including growth factors to develop bioinks that can support cellular growth and can be used for various tissue engineering approaches [5,6]; however, poor mechanical properties and printability limit their uses [7].

Alginate is a widely used biopolymer for the generation of scaffolds for tissue engineering applications, due to its availability, low cost, biocompatibility, and one-step gelation process [5]. Hydrogels based on an oxidized form of alginate (ADA: alginate dialdehyde) offer more reactive groups compared to native alginate, and thus were explored in combination with other polymers in a variety of cell and tissue engineering applications [6]. One of the widely used polymers in combination with ADA is gelatin (Gel), which is a thermoresponsive biopolymer derived from collagen. Gelatin is a biocompatible, bioresorbable biopolymer rich in arginine, glycine, and aspartic acid (RGD) motifs that help in cell attachment, and is therefore widely explored as a scaffolding biomaterial in tissue engineering [7]. Although the gelatin component in ADA–Gel offers cell adhesion motifs, it does not provide any other bioactive cues. One of the attractive sources of bioactive cues is platelet-rich plasma (PRP), which is enriched by a range of plasma proteins and growth factors, the most prominent being the platelet-derived growth factor, transforming growth factor, vascular endothelial growth factor, epidermal growth factor, insulin-like growth factor, and fibroblast growth factor [8]. Numerous growth factors, cytokines, and thrombin–fibrin in PRP are capable of enhancing angiogenesis, stem cell recruitment, and tissue regeneration of bone, tendon, skin, and cartilage, including cell proliferation, differentiation, and improved synthesis of the extracellular matrix. PRP has been successfully used as a therapeutic agent in the field of dermatology—for instance, in wound healing and cosmetic medicine [9]. PRP has been explored for various tissue engineering applications, as a culture supplement for enhancing cell proliferation or in tissue regeneration therapy, such as in orthopedic applications [10]. Another advantage of PRP is the autologous nature, which makes it an inexpensive and immunologically safe source in different tissue engineering applications. Its properties of enhancing angiogenesis, stem cell recruitment, and tissue regeneration are now being explored in regard to generating biocompatible bioink for cell proliferation and development [10–13].

In the current study, inspired by bioactive properties of PRP, we aim to prepare and characterize PRP supplemented ADA–Gel bioink formulation for potential 3D bioprinting applications. We followed previously reported protocols to synthesize ADA and an ADA–Gel conjugate, and subsequently performed systematic characterization studies. PRP from healthy volunteers was isolated and mixed with ADA–Gel to formulate ADA–Gel–PRP bioink. The resultant bioink was systemically characterized for its rheological, mechanical, chemical, and physical properties. Consequently, the feasibility of using ADA–Gel–PRP as a bioink for 3D bioprinting applications was verified by assessing its printability using a 3D bioprinter (RegenHu3D Discovery). Lastly, the cytocompatibility of the formulation was assessed by encapsulating the model cell line (L929, mouse fibroblast cell line), followed by a cell viability check by microscopy and CCK-8 (cell counting kit 8) assay.

2. Materials and Methods

2.1. Materials

Alginic acid sodium salt from brown algae was purchased from Sigma-Aldrich (Bangalore, India), gelatin was purchased from Gelita (Eberbach, Germany), sodium tetraborate (borax) was purchased from Fisher Scientific (United Kingdom), and sodium metaperiodate (EMSURE) was purchased from Merck (Mumbai, India). PRP was isolated from blood samples taken from healthy volunteers after

Institutional Ethics Committee 1 approval (IEC number: SCT/IEC/1366/APRIL-2019), and L929 cell line was obtained from American Type Culture Collection (Manassas, VA, United States). All cell culture related reagents and consumables were obtained from Thermo-Scientific (Bangalore, India).

2.2. Preparation of ADA–Gel–PRP Bioink Formulation

Alginate di-aldehyde was prepared by controlled oxidation of sodium alginate by metaperiodate in the ethanol–water mixture, as per the earlier method described by Balakrishnan et al. [14]. Gelatin was used as received without any processing or modification. To prepare PRP, stored/fresh blood samples collected from healthy human volunteers were subjected to centrifugation at 750× g for 5 min. The optimization of bioink formulation was done by varying ratios of ADA and gelatin. Typically, 12% (w/v) gelatin solution was prepared in DMEM-F12 (Dulbecco's Modified Eagle Medium/Nutrient Mixture F-12, supplemented with 10% (v/v) fetal bovine serum (FBS) and kept at 40 °C until complete dissolution. It was then mixed with a 12% (w/v) ADA solution, prepared by dissolving lyophilized ADA in 0.05 M borax. Bioink was formulated by mixing ADA/Gelatin/PRP in a ratio of 1.0:1.0:0.2 (v/v). The resulted formulation was characterized for gelation time, swelling properties, and rheological parameters.

2.3. Characterization of ADA–Gel–PRP Bioink Formulation

2.3.1. Physico-Chemical Properties

The gelation time of ADA–Gel–PRP was determined using the tube inversion method [15]. Briefly, ADA, gelatin, and PRP solutions were mixed in a vial, incubated at room temperature, and at regular intervals, the vials were inverted to check sol-gel transition. To determine swelling index and water uptake (%), pre-weighed, freeze-dried disc samples (15 mm × 5 mm) were immersed in 2 mL phosphate-buffered saline (PBS) at 37 °C for 24 h under static conditions, and the weight change was recorded to determine swelling index and water uptake (%), as per an earlier report [16]. Samples were freeze-dried (Edwards Modulyo 4K, Pharma Bioteck, United Kingdom) at −55 °C for 12 h. Subsequently, the microstructure of freeze-dried ADA–Gel–PRP gels was investigated using a scanning electron microscope (SEM, Hitachi, Model S-2400, Tokyo, Japan) and micro-CT (Micro-computed tomography40, Scanco, Bruttisellen, Switzerland). Finally, successful completion of Schiff's reaction and the formation of covalent bonds within ADA–Gel–PRP in presence of borax was investigated with attenuated total reflection Fourier-transform infrared spectroscopy (ATR-FTIR; 4200, JASCO FT/IR).

2.3.2. Rheological Properties

Rheological property is one of the critical parameters to be considered while formulating any novel bioink for 3D bioprinting. In the current study, the rheological properties of ADA–Gel–PRP (12% w/v ADA in 0.05 M Borax, 12% w/v gelatin in DMEM-F12, PRP in a volume ratio of 1.0:1.0:0.2) were carried out by a modular compact rheometer (MCR 102, Anton Paar). ADA–Gel was also analyzed for comparison purposes. A cone plate with a cone diameter of 24 mm and a cone angle of 2.009° was used, and the measurement gap was fixed at 0.105 mm. All experiments were performed at 25 °C. The viscosity of the hydrogel was measured at a constant shear rate of 100/s. Storage modulus (G′) and loss modulus (G″) were measured at an angular frequency from 100.0 to 0.1 rads/s at an amplitude gamma of 1%.

2.3.3. Biocompatibility Properties

For the hemolysis assay, the blood compatibility of the ADA–Gel and ADA–Gel–PRP hydrogels was analyzed by estimating hemolysis (%) test, as per ISO 10993-4, wherein the hydrogel discs of known size were placed in 2 mL of blood in a Petri plate. Samples were kept for agitation at 70 ± 5 rpm at 37 °C for 30 min. Subsequently, the whole blood from the sample was drawn and centrifuged for plasma separation at 1000× g for 15 min. From the supernatant, 100 μL of the plasma was taken and mixed with 1 mL of 0.1% (*w/v*) sodium bicarbonate. The absorbance of the liberated plasma hemoglobin was measured at 380 nm, 415 nm, and 450 nm in a UV-Vis spectrophotometer, and hemolysis (%) was calculated from the following Equation (1), where *free hemoglobin* is the level of hemoglobin liberated in the plasma, and the *total hemoglobin* was that from the initial whole blood count.

$$\% \; Hemolysis \; = \left(\left[\frac{Free \; haemoglobin}{Total \; haemoglobin} \right] \div 1000 \right) \times 100 \tag{1}$$

The cytotoxicity of the fabricated hydrogels (ADA–Gel and ADA–Gel–PRP) was evaluated using a test on extracts followed by an MTT (3-[4-C-dimethylthiazol-2-yl]-2,5-diphenyl tetrazolium bromide) assay, as per ISO 10993-5 [17]. Briefly, the hydrogel discs, of a known size having a surface area of about 1.25 cm^2, were incubating in 1 mL culture medium at 37 °C for 24 h. After incubation, the spent culture medium containing potential leachables from the hydrogel, or otherwise termed as extracts, were collected. This was considered as 100% (*v/v*) extract of the test material, and was subsequently diluted with fresh culture medium to prepare 50%, 25%, and 12.5% (*v/v*) extracts. Ultra-high molecular weight polyethylene (UHMWPE) samples extracted in similar conditions were considered as a negative control (the one which does not harm the cells), and freshly prepared dilute phenol (1.3% *w/v*) in culture medium was considered as a positive control (the one that harms the cells). Different dilutions of test and control extracts were placed on a monolayer of L929 cells in a 96-well plate at 100 μL/well, and incubated in a CO$_2$ incubator at 37 °C for 24 h. Subsequently, the spent medium was exchanged with freshly prepared 50 μL/well MTT reagent (1 mg/mL in medium without serum), and the cells were incubated at 37 °C for 2 h. The spent medium was discarded, and the formazan crystals were dissolved in 100 μL/well isopropanol. Cells without any treatment were considered as cell control. Absorbance was read at 570 nm in a spectrophotometer, and the metabolic activity % was calculated as per Equation (2):

$$Metabolic \; activity \; \% \; = \frac{Absorbance \; of \; treated \; wells}{Absorbance \; of \; cell \; control} \times 100 \tag{2}$$

2.4. Assessment of Printability

The printability of ADA–Gel–PRP formulation was assessed by using a state of the art bioprinting platform (Regen HU 3D Discovery). Typically, 1 mL of ADA solution (12% *w/v* in 0.05 M borax, kept at room temperature), 1 mL of gelatin solution (12% *w/v* in DMEM-F12, kept at 40 °C), and 0.2 mL of PRP (kept at room temperature) were mixed in a 35 mm culture dish to form a bioink. The bioink was then loaded into a 3 cc cartridge, as per the manufacturer guidelines. A 410 μm nozzle was attached to the cartridge tip, and it was then fixed onto print head 1 of the bioprinter. The needle height and stage were calibrated using the software, as per manufacturer instructions. A design template was prepared using the software provided with the bioprinter (size of the construct 1.5 × 1.5 cm^2). Approximately 15 min after mixing of ADA-Gel-PRP components, the printing was initiated as per the design drawn earlier, using a pneumatically controlled extrusion print head at a feed rate of 7.5 mm/sec. The versatility of printing and shape fidelity of the construct was examined by printing multiple shapes (three shapes) and multiple layers (up to 10 layers).

2.5. Cell-Laden Bioprinting

A model cell line, i.e., L929 mouse fibroblast cell line, was used to assess the cell-laden bioprinting using ADA-Gel-PRP formulation. Overall, the protocol for cell-laden bioprinting was the same as described in the previous section. However, for cell-laden bioprinting, about 1 million cells in the pellet form were mixed with 0.2 mL of PRP. This was then mixed with 1 mL of ADA (12% *w/v* in 0.05 M borax) and 1 mL of gelatin (12% *w/v* in DMEM-F12) solutions to prepare the cell-laden bioink formulation. The said bioink was loaded into a sterile cartridge, and the printing was started as described in the previous section. The 3D bioprinted, cell-laden constructs, collected in 12-well plates, were fed with DMEM-F12 supplemented with FBS (10% *v/v*) and incubated at 37 °C for 24 h. The cell viability in the cell-laden constructs was examined by CCK-8 (cell counting kit 8) assay, as per the kit manual. Briefly, after the incubation period, the spent medium was discarded, and 500 μL of CCK-8 reagent (5% *v/v* in serum-free DMEM) was added to each well. The constructs were further incubated for 4 h in the dark at 37 °C. About 100 μL of spent medium was collected into a fresh 96-well culture plate. The absorbance of the solution was measured at 450 nm (against a reference at 650 nm), and the cell viability (%) was calculated as per Equation (2). Cell viability was further confirmed with SEM for qualitative assessment, and was also evaluated by live dead staining fluorescein diacetate (FDA) and propidium iodide (PI) using confocal microscopy.

2.6. Statistical Analysis

The qualitative data shown was a representative of a group of replicates ($n = 3$). The quantitative values were averaged and expressed as mean ± standard deviation ($n = 3$). Statistical significance among the test and the control values were determined by one-way ANOVA, and the values were considered significant at $p < 0.05$.

3. Results and Discussion

The development of tissue construction using 3D bioprinting technology has become an attractive option in the field of tissue engineering, as it offers an exciting therapeutic alternative to numerous patients. Several industries are coming forward to invest in making tissue substitutes for potential applications in the biomedical field and beyond. The bioink, a hydrogel used for 3D printing, shall meet several criteria required for an efficient tissue fabrication, such as shear thinning, mechanical properties, biodegradability, and biocompatibility. Various formulations of bioinks are being synthesized using one or more biocompatible biomaterials by following several crosslinking strategies. Here we demonstrate the formulation of ADA–Gel–PRP hydrogel and the feasibility of using it as a bioink for 3D bioprinting applications. Alginate–gelatin-based hydrogels have been used as scaffolds for cell attachment and proliferation in several studies [18]. The ADA–Gel–PRP hydrogel proposed in the current study has the inclusion of more reactive groups that enhance crosslinking and provide a better environment for cell growth. ADA is an oxidized form of alginate that has a reactive aldehyde group, which facilitates covalent crosslinking with the amine groups of gelatin and PRP through Schiff's base reactions (Figure 1).

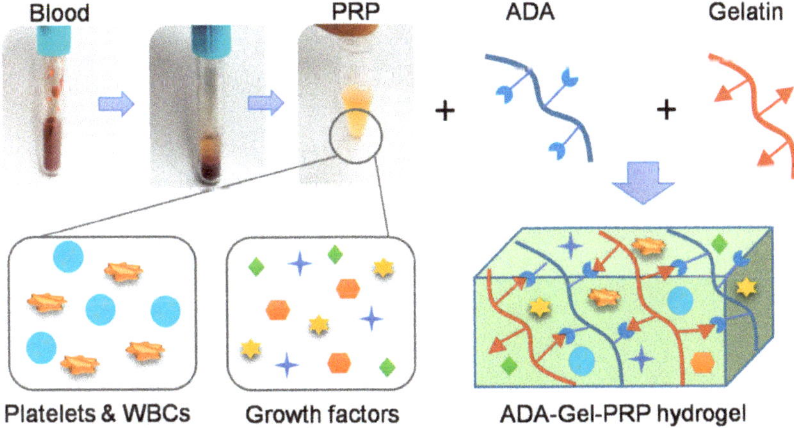

Figure 1. Schematic of ADA–Gel–PRP (dialdehyde–gelatin–platelet-rich plasma) hydrogel-based bioink formulation: the hydrogel network forms by covalent interaction of the aldehyde group of ADA with amine groups of gelatin through Schiff base reaction, incorporating components of PRP.

Gelation (gel transition) time is the time taken for a solution to become a gel. It is typically optimized to adjust for a bioink's rheological properties. The characteristic property of hydrogel in forming a gel helps in cell encapsulation and perfects the printability of hydrogel for the generation of a 3D construct. In practical terms, we found that the mixing of ADA, gelatin, and PRP along with cells, loading this bioink into a cartridge, assembling the print head, and instrument calibration took about 3 min. Based on this, amongst various combinations and permutations of ADA, gelatin, and PRP, 12% (*w/v*) ADA in 0.05 M borax, 12% (*w/v*) gelatin in PBS, and 200 µL of PRP in a 1.0:1.0:0.2 volume ratio was found to be optimal, with a gelation time of about 4 min (rationale for the selection of optimal composition is described in Supplementary Materials). Subsequently, we investigated the swelling behavior and water uptake capacity of the hydrogel made from this optimal concentration. The swelling behavior is a critical criterion to consider, as it alters the pore volume of a hydrogel and affects the properties and performance of the gel [19]. The water uptake capacity is also an important criterion to consider, since the encapsulated cells absorb nutrients from media to maintain cell growth, mobility, and spreading. In the current study, ADA–Gel–PRP-based bioink formulation was found to have a swelling index of 0.59 ± 0.02 (or in other words, the swelling ratio of final weight/initial weight was 1.59 ± 0.02) and a water uptake capacity of ~40%, thus indicating that the hydrogel formulation does not swell much, yet holds enough media to sustain cellular activity [20].

The highly interconnected porous structure is a prime requirement for any scaffold to promote proper cell seeding, attachment, and migration [21]. A considerable amount of hydrogel porosity is required for the better diffusion of nutrients and oxygen in the 3D construct, particularly in the absence of a functional vasculature system [22]. The cross-sectional morphology of the optimized hydrogel was investigated by SEM and micro-CT (Figure 2). SEM analysis showed the interconnecting porous nature of ADA–Gel–PRP hydrogels. Furthermore, micro-CT imaging was performed to analyze the pore size and porosity of hydrogel. The 3D morphology of the ADA–Gel–PRP revealed that the hydrogel was highly porous, and shows an even distribution of pores, with the pore size found to be 150 ± 50 µm and a porosity of 89% ± 5% (as analyzed through micro-CT software). However, the pore properties showed here represent the freeze-dried form of the hydrogel. They may or may not represent the wet form of the hydrogel in its absolute sense, perhaps due to the belief that the pore network would be altered upon sample swelling. Yet, since mean pore size was 150 µm in dried form, we believe that even after moderate shrinking these pores could allow efficient gas/nutrient exchange.

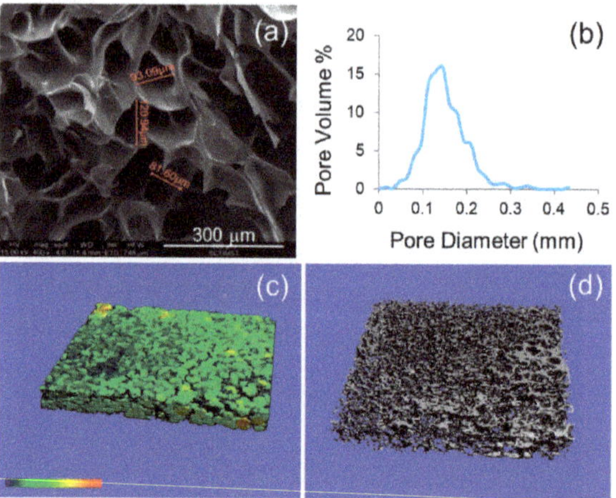

Figure 2. Morphological analysis of ADA–Gel–PRP hydrogel: (**a**) scanning electron microscope (SEM) and (**b–d**) micro-CT analysis reveal the highly interconnected porous nature of the hydrogel, with a uniform pore size and pore size distribution.

The chemical/structural features of ADA–Gel–PRP formulation and their ingredients as analyzed by FTIR are presented in Figure 3. The characteristic FTIR peaks of gelatin were seen at 1629 cm^{-1} and 1523 cm^{-1}, representing C=O stretching vibration of amide I and N–H/C–N stretching vibration of amide II, respectively. Similarly, PRP was rich in proteins and exhibited similar primary and secondary amine groups at 1633 cm^{-1} and 1531 cm^{-1}, respectively. ADA showed a characteristic peak at 1715 cm^{-1}, representing the C=O symmetric vibration. Hydrogel formation of ADA with gelatin occurs by Schiff's base reaction between the aldehyde groups of ADA and amine groups present in gelatin and PRP. In ADA–Gel–PRP hydrogel, the amide I band of PRP and gelatin was shifted to 1548 cm^{-1}, which denotes its involvement in gelation, and thus confirms the Schiff's base reaction within the hydrogel [23].

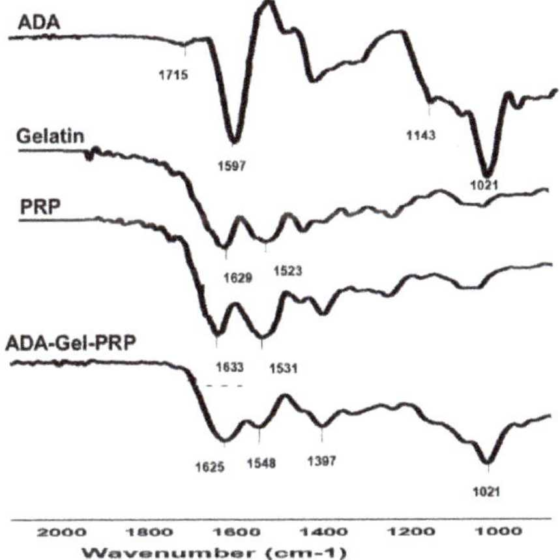

Figure 3. FTIR spectra of ADA–Gel–PRP formulation: a comparative observation of FTIR spectra of ADA, Gel, PRP, and ADA–Gel–PRP suggests successful Schiff's base reactions, wherein aldehyde and amine group interaction (CH=N) was shown by broadening of the peak at 1633 and 1548 cm^{-1}.

The rheological properties of any hydrogel are critical determinants for its ability to withstand shear force while being extruded through the needle [20,24]. A material is said to be thixotropic when its viscosity decreases with time at constant shear stress [25,26]. Thus, thixotropic nature allows highly viscous hydrogels to be extruded out from the printing nozzle effectively. Consequently, the viscosity of the hydrogel was estimated with time at a constant shear rate of 100/s. Furthermore, a frequency sweep test was performed within the linear viscoelastic region of hydrogel to determine the storage and loss modulus of the hydrogel in the angular frequency range of 0.1 to 100.0 rad/s. As shown in Figure 4, the viscosity of the hydrogel decreases with time at a constant shear rate, which showed the thixotropic or shear-thinning property of ADA–Gel–PRP hydrogel. The frequency sweep test performed on ADA–Gel–PRP hydrogel exhibited a storage modulus (G′) higher than the loss modulus (G″), indicating that the elastic character was always higher when a load is applied. Furthermore, both ADA–Gel and ADA–Gel–PRP samples showed a tan δ less than 1, and thus indicated the gel-like characteristic of both these samples. However, tan δ of ADA–Gel–PRP was relatively less than ADA–Gel, and thus indicated that the presence of PRP led to stiffer gel. Perhaps the addition of PRP improves the mechanical properties, printability, and stability of hydrogel, which is an important parameter for the bioink.

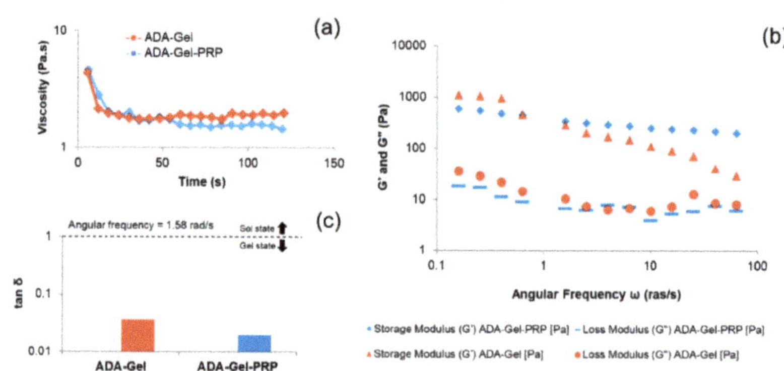

Figure 4. Rheological properties of ADA–Gel–PRP hydrogel: (**a**) viscosity plot, (**b**) storage and loss modulus plot, and (**c**) tan δ data show shear-thinning properties and stiffness of ADA–Gel–PRP bioink formulation.

Hemocompatibility of a bioink is an important aspect, as 3D scaffolds materials are used to treat wounds and injuries to the patients, thus having a short- or long-term exposure of the material to the blood. Hence, ADA–Gel–PRP bioink formulation was evaluated for hemolysis as per ISO 10993-4. The presence of free hemoglobin in the plasma is caused by the lysis of red blood cells (RBCs) in the blood when in contact with the test sample. We found that ADA–gel–PRP formulation exhibited a hemolysis of 0.04%, which was found to be less than the normal range of <0.1%, as per the ISO 10993-4 standard. Besides, since 3D-printed scaffolds come into contact with the body or wound, as in the case of a skin construct, it becomes necessary to evaluate their cytotoxic nature. The key components of our bioink formulation, i.e., ADA and gelatin, were reported be non-cytotoxic and cytocompatible [14]. However, the use of borax to increase the efficiency of crosslinking may cause toxicity, depending on its concentration. In the current study, we found that ADA in 0.05 M borax mixed with Gel–PRP was found to be the optimal non-cytotoxic concentration, showing nearly 100% cell viability (Figure 5), as per ISO 10993-5 (cell viability of 60% or lower was typically considered as cytotoxic). In a recent study, Tilman et al., showed that a plasma–alginate-based bioink promotes cell growth and proliferation. Plasma contains growth factors and proteins, which helps cellular adhesion; these proteins also play an important role in cell–matrix interactions [27].

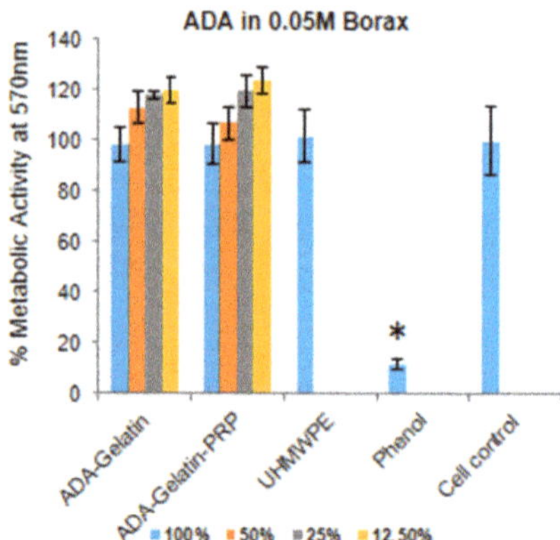

Figure 5. Cytotoxicity analysis of ADA–Gel–PRP bioink formulation, as per ISO 10993-5. Compared to cell control without any treatment, L929 cells treated with various dilutions of extracts of ADA–Gel and AD–Gel–PRP hydrogels exhibited ≥100% cell viability. Ultra-high molecular weight polyethylene

(UHMWPE; negative cytotoxic control) showed ≥100% cell viability, and diluted phenol (positive cytotoxic control) showed <10% cell viability, as anticipated (100% extracts: undiluted culture medium with extracts of the hydrogels, at 50%, 25%, and 12.5%; *v/v* extracts: undiluted culture medium with extracts of the hydrogels mixed with fresh culture medium in 1:2, 1:4, and 1:8 *v/v* ratios, respectively). * Cell viability (%) in phenol sample was statistically significant when compared to the cell viability % in test, UHMWPE, and cell control samples, $p < 0.05$).

Evaluation of bioink printability—in particular, the ability to form smooth and continuous filament and to form and sustain 3D structure—is of prime importance in bioprinting and biofabrication [28–30]. In the current study, we have systematically evaluated the bioprinter parameters, such as nozzle diameter, print head temperature, and feed rate, as well as the bioink formulation composition, in the process of developing a printable formulation. A formulation with 1 mL of ADA solution (12% *w/v* in 0.05 M borax, kept at ambient temperature), 1 mL of gelatin solution (12% *w/v* in DMEM-F12, kept at 40 °C), and 0.2 mL of PRP (kept at ambient temperature) were found to be yielding smooth and continuous filaments with a 410 μm nozzle and a feed rate of 7.5 mm/sec followed at ambient temperature. As shown in Figure 6, the morphological features of a single stack, as well as multiple stacks (10 layered constructs), were observed through stereo zoom microscope images. The lateral view of printed constructs having one, three, and five stacks suggests the shape fidelity of the printed constructs. Furthermore, we found that bioink formulation can be bioprinted into multiple shapes, therefore suggesting the versatility of the formulation and bioprinting process.

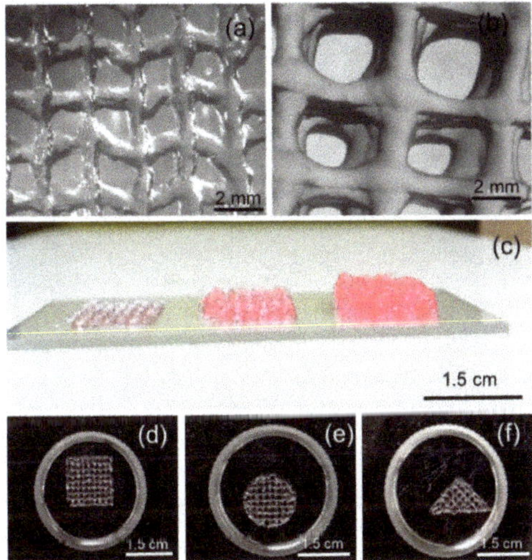

Figure 6. Printability aspects of ADA–Gel–PRP formulation. Stereo zoom microscope images suggested smooth and continuous filament formation in (**a**) single and (**b**) five-stack (or 10-layer) constructs. (**c**) Lateral view of single and multi-stacked constructs and (**d–f**) constructs in various shapes indicates the versatility of the formulation and bioprinting process.

It is known that the composition of any hydrogel formulation would influence cell viability, owing to the cytotoxic nature of the components involved. However, there is emerging evidence that suggests that printing parameters would also influence and seriously affect cell viability in bio-plotted constructs [31]. Therefore, in the current study, we have assessed the viability of L929 cells bioprinted with ADA–Gel–PRP formulation. After 24 h of incubation, the cell-laden constructs were subjected to a CCK-8 assay, and as evident from Figure 7a, the cell viability in 3D bioprinted constructs was nearly 80%, which was similar to the viability observed in the manually cast constructs. Perhaps, in two-dimensional (2D) culture, the cells have a treated surface with abundant cell anchoring moieties to attach, have no diffusion barrier of any kind for nutrient/gas exchange, and have abundant space to expand, whereas in 3D culture, cell viability and growth would be influenced by many factors, including but not limited to chemical composition of the ink, cell–matrix interactions, diffusion barrier to nutrient exchange, restricted freedom to expand due to gel stiffness, etc. The cell-laden construct was fixed and observed under an SEM for qualitative assessment of the cell adhesion, and as can be seen from Figure 7b, the cells were seen on/in the printed construct. The cell viability in the cell-laden construct was also evaluated by live dead staining (FDA/PI) and confocal microscopy analysis, and as presented in Figure 7c, an abundant number of viable cells (stained in green) were seen in the construct. The qualitative and quantitative analysis of cell-laden constructs confirmed that the ADA–Gel–PRP was biocompatible, and the printing parameters proposed in this study were optimal for the preparation of viable cell-laden constructs. However, the formulation composition and the printing parameters may have to be fine-tuned for specific cells and applications of interest [31].

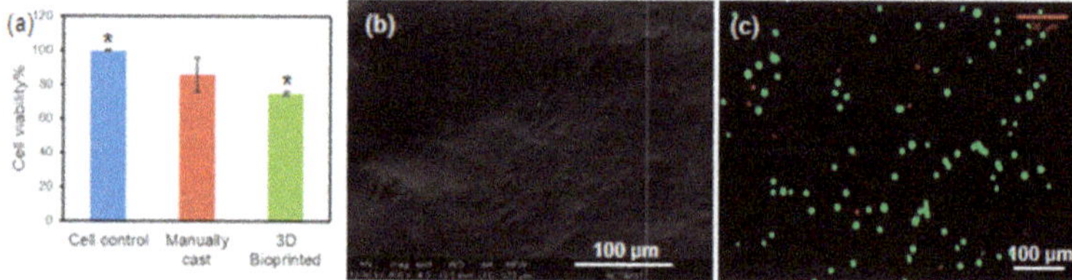

Figure 7. Cell viability in 3D bioprinted constructs: (**a**) CCK-8 assay, (**b**) SEM imaging, and (**c**) fluorescent live dead staining collectively suggest that the ADA–Gel–PRP formulation and the set bioprinting parameters were optimal in yielding a viable construct. * Denotes that the differences in cell viability % were statistically significant ($p < 0.05$).

4. Conclusions

Formulation of a printable bioink that has bioactive ingredients is a significant challenge, and to this end, here we report ADA–Gel–PRP-based hydrogel formulation as a novel bioink for potential bioprinting and biofabrication applications. Various combinations of ADA, Gel, and PRP in varying concentrations were studied, and a 1:1 (*v/v*) ratio of 12% (*w/v*) ADA and 12% (*w/v*) Gel, along with 200 µL of PRP, was found to yield a stable formulation. The gelation time was 4.0 ± 0.5 min, the swelling index was 0.59 ± 0.02, and water uptake was 37% ± 0.8%. SEM and micro-CT analysis indicated that the hydrogel has highly interconnected porous morphology with a mean pore size of 150 µm. Covalent crosslinking between ADA–Gel–PRP in the presence of borax was confirmed by FTIR spectroscopy. The rheological analysis revealed shear-thinning properties of the formulation; furthermore, it also suggested increased gel stiffness in ADA–Gel in the presence of PRP. In vitro cytotoxicity, as per ISO 10993-5, and hemolysis, as per ISO 10993-4, suggest the non-cytotoxic and non-hemolytic nature of the formulation. The printability aspect was assessed in a RegenHu 3D Discovery bioprinter, and the parameters were optimized to yield structurally stable constructs with smooth and continuous filaments. Finally, cell culture studies suggest that the cells encapsulated in the 3D bio-printed, cell-laden constructs were viable, with more than 80% cell viability. Further studies exploring this formulation in the development of multi-cellular tissue constructs are ongoing.

Supplementary Materials: Additional methods and results are available online at https://www.mdpi.com/2306-5354/7/3/108/s1.

Author Contributions: Conceptualization, N.K. and A.B.; methodology, L.T.S. and R.R.; data curation, L.T.S., N.K. and A.B.; writing—original draft preparation, L.T.S. and R.R.; writing—review and editing, N.K. and A.B.; funding acquisition, N.K. and A.B. All authors have read and agreed to the published version of the manuscript.

Funding: This research was funded by Department of Science and Technology, Government of India, TRC 8137.

Acknowledgments: Authors acknowledge the Department of Science and Technology, Government of India, and, Sree Chitra Tirunal Institute for Medical Sciences and Technology, Trivandrum for funding this work (TRC P8137). We thank our colleagues Rashmi R., Jimna M. Ameer, and Suvanish Kumar for technical assistance.

Conflicts of Interest: The authors declare no conflict of interest. The funders had no role in the design of the study; in the collection, analyses, or interpretation of data; in the writing of the manuscript, or in the decision to publish the results. An Indian patent application (no. 201941031304) has been submitted based on this study.

References

1. Chen, S.; Shi, Y.; Luo, Y.; Ma, J. Layer-by-layer coated porous 3D printed hydroxyapatite composite scaffolds for controlled drug delivery. *Colloids Surf. B Biointerfaces* **2019**, *179*, 121–127. [CrossRef]
2. Gu, Q.; Hao, J.; Lu, Y.; Wang, L.; Wallace, G.G.; Zhou, Q. Three-dimensional bio-printing. *Sci. China Life Sci.* **2015**, *58*, 411–419. [CrossRef] [PubMed]
3. Cui, H.; Zhu, W.; Nowicki, M.; Zhou, X.; Khademhosseini, A.; Zhang, L.G. Hierarchical fabrication of engineered vascularized bone biphasic constructs via dual 3d bioprinting: Integrating regional bioactive factors into architectural design. *Adv. Health Mater.* **2016**, *5*, 2174–2181. [CrossRef] [PubMed]
4. Roehm, K.D.; Madihally, S.V. Bioprinted chitosan-gelatin thermosensitive hydrogels using an inexpensive 3D printer. *Biofabrication* **2017**, *10*, 015002. [CrossRef]
5. Ter Horst, B.; Chouhan, G.; Moiemen, N.; Grover, L.M. Advances in keratinocyte delivery in burn wound care. *Adv. Drug Deliv. Rev.* **2018**, *123*, 18–32. [CrossRef]
6. Sarker, B.; Papageorgiou, D.G.; Silva, R.; Zehnder, T.; Gul-E-Noor, F.; Bertmer, M.; Kaschta, J.; Chrissafis, K.; Detsch, R.; Boccaccini, A.R. Fabrication of alginate-gelatin crosslinked hydrogel microcapsules and evaluation of the microstructure and physico-chemical properties. *J. Mater. Chem. B* **2014**, *2*, 1470–1482. [CrossRef]
7. Davidenko, N.; Schuster, C.F.; Bax, D.V.; Farndale, R.W.; Hamaia, S.; Best, S.M.; Cameron, R.E. Evaluation of cell binding to collagen and gelatin: A study of the effect of 2d and 3d architecture and surface chemistry. *J. Mater. Sci. Mater. Med.* **2016**, *27*, 148. [CrossRef] [PubMed]
8. Samberg, M.; Ii, R.S.; Natesan, S.; Kowalczewski, A.; Becerra, S.; Wrice, N.; Cap, A.; Christy, R.J. Platelet rich plasma hydrogels promote in vitro and in vivo angiogenic potential of adipose-derived stem cells. *Acta Biomater.* **2019**, *87*, 76–87. [CrossRef]
9. Gil Park, Y.; Lee, I.H.; Park, E.S.; Kim, J.Y. Hydrogel and platelet-rich plasma combined treatment to accelerate wound healing in a nude mouse model. *Arch. Plast. Surg.* **2017**, *44*, 194–201. [CrossRef]
10. Faramarzi, N.; Yazdi, I.K.; Nabavinia, M.; Gemma, A.; Fanelli, A.; Caizzone, A.; Ptaszek, L.M.; Sinha, I.; Khademhosseini, A.; Ruskin, J.N.; et al. Patient-specific bioinks for 3d bioprinting of tissue engineering scaffolds. *Adv. Health Mater.* **2018**, *7*, e1701347. [CrossRef]
11. Irmak, G.; Gümüşderelioğlu, M. Photoactivated platelet rich plasma (PRP) based patient-specific bio-ink for cartilage tissue engineering. *Biomed. Mater.* **2020**. [CrossRef]
12. Li, Z.; Zhang, X.; Yuan, T.; Zhang, Y.; Luo, C.; Zhang, J.; Liu, Y.; Fan, W. Addition of platelet-rich plasma to silk fibroin hydrogel bioprinting for cartilage regeneration. *Tissue Eng. Part A* **2020**, *26*, 15–16. [CrossRef]
13. Hedegaard, C.L.; Mata, A. Integrating self-assembly and biofabrication for the development of structures with enhanced complexity and hierarchical control. *Biofabrication* **2020**, *12*, 032002. [CrossRef]
14. Balakrishnan, B.; Jayakrishnan, A. Self-cross-linking biopolymers as injectable in situ forming biodegradable scaffolds. *Biomaterials* **2005**, *26*, 3941–3951. [CrossRef] [PubMed]
15. Wei, H.-L.; Yang, Z.; Zheng, L.-M.; Shen, Y.-M. Thermosensitive hydrogels synthesized by fast Diels–Alder reaction in water. *Polymer* **2009**, *50*, 2836–2840. [CrossRef]
16. Kasoju, N.; Hawkins, N.; Pop-Georgievski, O.; Kubies, D.; Vollrath, F. Silk fibroin gelation via non-solvent induced phase separation. *Biomater. Sci.* **2016**, *4*, 460–473. [CrossRef] [PubMed]
17. Ameer, J.M.; Venkatesan, R.B.; Damodaran, V.; Nishad, K.V.; Arumugam, S.; Asari, A.K.P.R.; Kasoju, N.; Mohamed, A.J.; Babu, V.R.; Vinod, D.; et al. Fabrication of co-cultured tissue constructs using a dual cell seeding compatible cell culture insert with a clip-on scaffold for potential regenerative medicine and toxicological screening applications. *J. Sci. Adv. Mater. Devices* **2020**, *5*, 207–217. [CrossRef]
18. Sarker, B.; Singh, R.; Silva, R.; Roether, J.A.; Kaschta, J.; Detsch, R.; Schubert, D.W.; Cicha, I.; Boccaccini, A.R. Evaluation of fibroblasts adhesion and proliferation on alginate-gelatin crosslinked hydrogel. *PLoS ONE* **2014**, *9*, e107952. [CrossRef]

19. Ganji, F.; Vasheghani-farahani, S. Theoretical description of hydrogel swelling: A review. *Iran. Polymer J.* **2010**, *19*, 375–398.

20. Gupta, N.V.; Shivakumar, H. Investigation of swelling behavior and mechanical properties of a pH-sensitive superporous hydrogel composite. *Iran. J. Pharm. Res. IJPR* **2012**, *11*, 481–493.

21. Kasoju, N.; Kubies, D.; Fábryová, E.; Kříž, J.; Kumorek, M.M.; Sticová, E.; Rypáček, F. In vivo vascularization of anisotropic channeled porous polylactide-based capsules for islet transplantation: The effects of scaffold architecture and implantation site. *Physiol. Res.* **2015**, *64*, S75–S84. [CrossRef] [PubMed]

22. Khademhosseini, A.; Langer, R. Microengineered hydrogels for tissue engineering. *Biomaterials* **2007**, *28*, 5087–5092. [CrossRef] [PubMed]

23. Marin, L.; Simionescu, B.; Barboiu, M. Imino-chitosan biodynamers. *Chem. Commun.* **2012**, *48*, 8778. [CrossRef] [PubMed]

24. Tang, G.; Du, B.; Stadler, F.J. A novel approach to analyze the rheological properties of hydrogels with network structure simulation. *J. Polym. Res.* **2017**, *25*, 4. [CrossRef]

25. A Barnes, H. Thixotropy—A review. *J. Non-Newtonian Fluid Mech.* **1997**, *70*, 1–33. [CrossRef]

26. Li, H.; Liu, S.; Lin, L. Rheological study on 3D printability of alginate hydrogel and effect of graphene oxide. *Int. J. Bioprinting* **2016**, *2*. [CrossRef]

27. Ahlfeld, T.; Cubo-Mateo, N.; Cometta, S.; Guduric, V.; Vater, C.; Bernhardt, A.; Akkineni, A.R.; Lode, A.; Gelinsky, M. A novel plasma-based bioink stimulates cell proliferation and differentiation in bioprinted, mineralized constructs. *ACS Appl. Mater. Interfaces* **2020**, *12*, 12557–12572. [CrossRef]

28. Paxton, N.; Smolan, W.; Böck, T.; Melchels, F.; Groll, J.; Jungst, T. Proposal to assess printability of bioinks for extrusion-based bioprinting and evaluation of rheological properties governing bioprintability. *Biofabrication* **2017**, *9*, 044107. [CrossRef]

29. Ouyang, L.; Yao, R.; Zhao, Y.; Sun, W. Effect of bioink properties on printability and cell viability for 3D bioplotting of embryonic stem cells. *Biofabrication* **2016**, *8*, 035020. [CrossRef]

30. Gao, T.; Gillispie, G.J.; Copus, J.S.; Kumar, P.R.A.; Seol, Y.-J.; Atala, A.; Yoo, J.J.; Lee, S.J. Optimization of gelatin–alginate composite bioink printability using rheological parameters: A systematic approach. *Biofabrication* **2018**, *10*, 034106. [CrossRef]

31. Zhao, Y.; Li, Y.; Mao, S.; Sun, W.; Yao, R. The influence of printing parameters on cell survival rate and printability in microextrusion-based 3D cell printing technology. *Biofabrication* **2015**, *7*, 045002. [CrossRef] [PubMed]

 *bioengineering*

Article

Advanced Bioink for 3D Bioprinting of Complex Free-Standing Structures with High Stiffness

Yawei Gu [1], Benjamin Schwarz [1], Aurelien Forget [1], Andrea Barbero [2], Ivan Martin [2] and V. Prasad Shastri [1,*]

[1] Institute for Macromolecular Chemistry, University of Freiburg, 79104 Freiburg, Germany; yawei.gu@makro.uni-freiburg.de (Y.G.); benjaminschwarz02@gmail.com (B.S.); aurelien.forget@makro.uni-freiburg.de (A.F.)
[2] Tissue Engineering Laboratory, Department of Biomedicine, University Hospital Basel, University of Basel, 4031 Basel, Switzerland; andrea.barbero@usb.ch (A.B.); ivan.martin@usb.ch (I.M.)
* Correspondence: prasad.shastri@gmail.com or prasad.shastri@makro.uni-freiburg.de; Tel.: +49(0)761-203-6268

Received: 30 September 2020; Accepted: 3 November 2020; Published: 7 November 2020

Abstract: One of the challenges in 3D-bioprinting is the realization of complex, volumetrically defined structures, that are also anatomically accurate and relevant. Towards this end, in this study we report the development and validation of a carboxylated agarose (CA)-based bioink that is amenable to 3D printing of free-standing structures with high stiffness at physiological temperature using microextrusion printing without the need for a fugitive phase or post-processing or support material (FRESH). By blending CA with negligible amounts of native agarose (NA) a bioink formulation (CANA) which is suitable for printing with nozzles of varying internal diameters under ideal pneumatic pressure was developed. The ability of the CANA ink to exhibit reproducible sol-gel transition at physiological temperature of 37 °C was established through rigorous characterization of the thermal behavior, and rheological properties. Using a customized bioprinter equipped with temperature-controlled nozzle and print bed, high-aspect ratio objects possessing anatomically-relevant curvature and architecture have been printed with high print reproducibility and dimension fidelity. Objects printed with CANA bioink were found to be structurally stable over a wide temperature range of 4 °C to 37 °C, and exhibited robust layer-to-layer bonding and integration, with evenly stratified structures, and a porous interior that is conducive to fluid transport. This exceptional layer-to-layer fusion (bonding) afforded by the CANA bioink during the print obviated the need for post-processing to stabilize printed structures. As a result, this novel CANA bioink is capable of yielding large (5–10 mm tall) free-standing objects ranging from simple tall cylinders, hemispheres, bifurcated 'Y'-shaped and 'S'-shaped hollow tubes, and cylinders with compartments without the need for support and/or a fugitive phase. Studies with human nasal chondrocytes showed that the CANA bioink is amenable to the incorporation of high density of cells (30 million/mL) without impact on printability. Furthermore, printed cells showed high viability and underwent mitosis which is necessary for promoting remodeling processes. The ability to print complex structures with high cell densities, combined with excellent cell and tissue biocompatibility of CA bodes well for the exploitation of CANA bioinks as a versatile 3D-bioprinting platform for the clinical translation of regenerative paradigms.

Keywords: carboxylated agarose; bioink; 3D printing; free-standing; human nasal chondrocytes; clinical translational

1. Introduction

Currently, there are three major challenges in 3D bioprinting: (1) reproducibility of printing process (i.e., printed structures), (2) cell viability in printed structures, and (3) the printing of large, free standing complex structures without the need for post-processing or support material (FRESH process). Few polymers possess the necessary physicochemical characteristics to simultaneously address the aforementioned challenges. Biopolymers like collagen, gelatin, fibrin, and hyaluronic acid are biocompatible materials that can mimic various aspects of the extracellular matrix (ECM), however, they possess poor printability. While synthetic thermoplastic polymers like poly(glycolic acid), poly(lactic acid), poly(ε-caprolactone) can address some of the challenges stated earlier [1], they cannot be processed in a traditional bioprinter and cannot accommodate cells during printing. Poly (ethylene glycol) (PEG), a water-soluble polymer, has been applied in tissue engineering for decades but it lacks desired physiochemical characteristics such as stability in aqueous medium to be considered as printing medium for bioprinting, without chemical modifications. Injectable hydrogels for example, Pluronic, and di, tri and multi-block copolymers of PEG with hydroxy acids, and self-assemble amphiphilic peptide hydrogels, have been explored extensively in cell encapsulation and tissue engineering, but are not suited for bioprinting as they either lack sufficient solubility in water to yield aqueous solutions that can be printed and/or do not yield mechanically stable structures that can be handled post-printing [2–4]. So far, common solutions to enhance the processability have been blending with viscous components such as nanocellulose and Laponite clay [5,6] and/or post-printing chemical modification using crosslinkers to improve the stability of the bioink and printed structures. Therefore, materials applied in 3D bioprinting are now dominated by methacrylated polymers like gelatin methacryloyl (GelMA) [7,8], collagen methacrylate (ColMA) [9], or bioink blended with alginate, a polymer capable of undergoing ionic crosslinking in presence of divalent cations [10]. While the strategy of introducing viscous components in a bioink is simple and effective and can promote the integrity of the extruding bioink which is necessary for microextrusion printing, this comes at the expense of increase in the general viscosity and changes to shear-viscosity both which can have undesired outcomes on cell viability. The chemical transformation of gelatin to GelMA provides a means to overcome a key limitation of gelatin namely, the dependence of its physical state on temperature; by covalently crosslinking the acrylate groups to realize a permanent shape. Therefore, most reported 3D bioprinting of complex structures require either fugitive phases, like poly(ε-caprolactone), carbohydrate glass, alginate, or Pluronic F127 [11–16], or post-processing like photo-crosslinking [17,18] or chemical crosslinking [19–21], which makes the fabrication complicated and highly demanding. Moreover, the addition of fugitive materials, crosslinking reagents and UV-light also introduces more variables that produce complexity, which might hinder the translation of these technologies to the clinical space, and additionally, residual fugitive phase and crosslinking chemistries can also present cytotoxicity.

Agarose, a naturally occurring polysaccharide derived from red seaweed has been used as a 3D matrix to culture chondrocytes for over three decades [22,23], but has limited use in tissue engineering per se. The gelling behavior of solutions of native agarose (NA) is characterized by a hysteresis, meaning the transition from solution-to-gel occurs at lower temperature than the gel-to-solution transition [24]. This characteristic of agarose on one hand makes it an exceptional gel forming agent that needs no crosslinking reagent, and is stable under physiological temperature after gelation and as result has aroused interest in 3D bioprinting over the past decade [25,26]. However, few studies on printing pure agarose bioink have been reported so far. On the other hand, the thermal gelation and hysteresis behavior confines the processing of agarose to a narrow concentration range and temperature range. As a result, NA tends to be used as an assistive material such as non-adhesive moulds [27] or Supplementary Materials to increase viscosity for other bioink components [28,29].

Recently, we reported the development of carboxylated agarose, a derivative of NA bearing carboxylic acid groups in the backbone [30] and the exploitation of CA as a bioink [31]. The significant difference between NA and CA lies in introduction of β-sheet and β-strand structures [32] in the polymer backbone and the concomitant reduction of a-helices which promotes physical gelation through β-sheet/β-strand-β-sheet/β-strand interactions thus decreasing the complex viscosity compared to NA and lower sol-gel transitions [30]. These new physiochemical characteristics make CA more easily processed at or under physiological temperature. Furthermore, unlike NA, where mechanical strength of gel phase is highly dependent on concentration, the mechanical properties of CA gels can be altered independent of concentration through varying the degree of carboxylation. Furthermore, mechanically defined hydrogels of CA have been shown to recruit and promote the stabilization of new blood vessels (vasculature) through a mechanobiology paradigm [33]. All these exceptional properties make CA a promising bioink for 3D bioprinting, and for the first time, using CA, we were able to print structures with a range of mechanically discrete microdomains in a single print while simultaneously encapsulating human mesenchymal stem cells at high viability within these structures [31]. Building on these findings and taking advantage of the mechanically stability of CA gels, we recently also demonstrated printing hollow structures of complex shape mimicking vascular tree within solid matrices of CA [34]. However, to date, the printing of free-standing complex structures without the need for post-processing such as crosslinking, or a support matrix has not been realized using biocompatible bioinks.

In this paper, we report a novel bioink formulation for printing free-standing complex structures. By combining CA with a negligible amount of NA (less than 0.30 *w/w* % with respect to CA concentration), a single phase bioink system CANA that possesses nuanced sol-gel transition behavior, and is suitable for printing with nozzles of various diameters was realized. Using CANA bioink on a commercial 3D bioprinter equipped with in-house modifications for precise temperature control at the print head assembly and print platform, and precisely rendered printing models; the printing of both simple and complex (curved and bifurcating), high-aspect ratio free-standing structures is demonstrated. The printed objects show structural integrity, elasticity, and fusion between layers, verifying the exceptional processability of the novel bioink formulation. CANA bioink is also amenable to the incorporation of high density of cells without appreciable loss in cell viability. The CANA bioink platform represents a step forward in the transformation of 3D-cell bioprinting in a clinically viable platform.

2. Materials and Methods

2.1. Synthesis of Carboxylated Agarose and Bioink Preparation

In this study, CA with 40% carboxylation (CA40) was used. CA40 was synthesized as previously described [30]. Briefly, 10 g of native agarose type 1 (GeneOn, Germany) was transferred into a three-necked round bottom flask, equipped with a mechanical stirrer and pH meter. The reaction vessel was heated up to 90 °C to dissolve the agarose and then cooled down to 0 °C in an ice bath under mechanical stirring. The reactor was then charged with 300 mg TEMPO (Abcr, Germany), 1.5 g NaBr (0.9 mmol), and 37.5 mL NaOCl (15% *v/v* solution) under vigorous stirring. The pH of the solution was adjusted to pH 10.8 throughout the duration of the reaction, and the degree of carboxylation was controlled by the addition of predetermined volumes of NaOH solution (0.5 M). At the end of the reaction 1.5 g $NaBH_4$ was added, and the solution was acidified to pH 8 and stirred for 1 h. The CA was precipitated by sequential addition of 150 g NaCl and 500 mL ethanol, and the solid was collected by vacuum filtration and extracted using ethanol. Residual ethanol was removed by extensive dialysis against water and the CA was obtained as a white solid upon lyophilization overnight. The degree of carboxylation was verified by the appearance of peaks associated with aliphatic carboxylic acid groups via FTIR (KBr) (vc = o:1750 cm^{-1}) (Bruker Optics, Germany) and NMR 300 MHz (13C: 180 ppm) (Bruker BioSpin, Germany). The CANA bioink formulation was prepared as follows. Lyophilized CA (78 mg) and NA (2 mg) were added into 1 mL Dulbecco's phosphate-buffered saline (DPBS, Gibco, Germany) and the mixtures was heated up to 95°C until a clear solution was obtained to yield a bioink composed of 8% *w/v* (7.8% CA + 0.2% NA) solids (CA + NA). For the NA samples with different concentrations, a pre-define amount of NA was added to DPBS and prepared similarly to the CANA bioink. All the concentration used in NA samples are % *w/v*.

2.2. Rheological Tests and Compress Tests

Rheological characterizations were performed on a Kinexus Pro+ rotary rheometer (Malvern Instruments, Malvern, United Kingdom) using a cone and plate assembly comprising an upper 4 cone plate 40 mm in diameter. Sample for rheological testing was prepared as follows: The sample was first heated to 95 °C until a clear solution was obtained before transferring to the stage set to a desired temperature. For the thermal-dynamic rheological characterization, samples were loaded on the lower plate at 60 °C, and then heated or cooled to the first temperature, and then maintained at that temperature for 5 min to equilibrate. Then the sample was heated up or cooled down to the target temperature at a rate of 5 °C/min at a constant frequency of 0.1 Hz and a constant shear strain of 1%. For the frequency sweeping, samples were loaded on the lower plate at 37 °C, equilibrated for 5 min before a frequency sweep from 10 Hz to 0.1 Hz.

For the compression tests, a parallel plate configuration comprising an upper parallel plate with 40 mm in diameter and a lower plate were used. All samples were loaded on to the lower plate at 37 °C, and an initial gap was set at 0.3 mm. The samples were then trimmed to fit the plate diameter. Sample was cooled down to 4 °C, before heating up to 37 °C again. After equilibration at 37 °C for 5 min, the upper plate was lowered down to the prescribed gap to initiate compression of the hydrogel. The compression was terminated when upper plate reached the prescribed gap or when the detected normal force reached 50 N.

The radial elastic deformation tests on the printed cylinders were also performed with a Kinexus Pro+ rotary rheometer. The diameter of all the samples were measured before tests. The samples were then placed in the center of the lower plate horizontally. The upper plate was lowered at a speed of 0.1 mm/s to compress the cylinders (Figure S1) and the normal force and distance between upper and lower plate were recorded.

2.3. 3D Printing

Structures were printed on an Inkredible-2 (Cellink, Sweden) 3D printer with several custom in-house modifications which included an aluminum temperature-control module to heat the print nozzle and a water-cooled print bed both of which were machined and built by the machine shop at the Institute for Macromolecular Chemistry at the University of Freiburg. The g-code was created on Slic3r (GNU Affero General Public License) using the parameters provided by the bioprinter manufacturer. The g-code was then edited on Cellink HeartWare (Cellink, Sweden). In preparation for printing, CANA bioink was first heated to 90 °C into a clear solution, and then transferred into a 42 °C oven for at least 10 min. Then the bioink was loaded into the cartridge at 37 °C for another 10 min to reach equilibrium before mounting on the printer for printing.

2.4. Scanning Electron Microscopy (SEM) and Optical Microscopy

SEM's were obtained on an environmental scanning electron microscopy Quanta 250 FEG (Oxford Instrument, United Kingdom). The samples were prepared as follows. The printed structure was first cut in the desired direction in the wet state and then frozen in liquid nitrogen, followed by lyophilization overnight. Optical microscopy images were obtained on a Zeiss Observer A1 (Carl Zeiss, Germany). The samples for optical microscopy were prepared as follows. Freshly printed structures were immersed in 30% sucrose solution for 48 h before embedding in Optimal cutting temperature compound (VWR, Germany). The embedded samples were then cryo-sectioned and imaged without further staining.

2.5. Cell Culture, Bioprinting, and Live/Dead Assay

Human nasal chondrocytes (NCs) were isolated from patients undergoing reconstructive surgery after informed consent and in accordance with the local ethical committee (University Hospital Basel, Ref Number 78/07). <u>Isolation of NCs</u>: The biopsy was dissected to remove the perichondrium [35] and the pure cartilage was cut into pieces and incubated overnight in complete medium for sterility control. Complete medium (CM) consisted of Dulbecco's modified Eagle's medium (DMEM, Gibco, UK), 10% fetal bovine serum (FBS, Gibco, UK), 4.5 mg/mL d-glucose, 0.1 mM nonessential amino acids, 1 mM sodium pyruvate, 100 mM HEPES buffer, 100 U/mL penicillin, 100 mg/mL streptomycin, and 29.2 mg/mL L-glutamine (Gibco, UK). Nasal chondrocytes (NCc) were isolated from the cartilage samples by enzymatic digestion as previously described [36] with 0.15% collagenase II (Worthington) for 22 h at 37 °C. <u>Expansion of NCs</u>: isolated NCs were expanded in CM supplemented with 5 ng/mL FGF (R&D Systems), and 1 ng/mL TGF β-1 (R&D Systems) for two passages (corresponding to 81–0 population doublings). Passage 2 NCs were trypsinized and centrifuged before mixing with CANA bioink equilibrated at 42 °C at a concentration of 30 million per mL bioink. The NCs-laden bioink was then transferred to a 37 °C cartridge for printing. The printed structures were then cultured in media immediately after printing. NC viability was ascertained using Live/Dead Assay (Life Technologies, Germany) immediately after printing, on day 1, day 4, and day 7.

2.6. Statistical Analysis

Statsitical analysis was carried using statistical package embedded in Origin graphing software and statistical signifcance was determined using Tukey's multiple comparion test. A p-value of ≤ 0.05 was considered statistically significant.

3. Results

3.1. Rheological Properties

As the temperature of the printing bed would be set at 4 °C to induce rapid and homogenous physical crosslinking of the printed structure, the storage modulus (G') of the 8% *w/v* of CANA bioink gel was measured at 4 °C and was determined to be 212 ± 7.2 kPa (Table 1). In order to place a metric on this value, a relationship between storage modulus of NA over a concentration range of 0.5% to 5% was established and this was used to estimate the concentration of NA that could yield gels of similar modulus (Figure 1A). From this comparative analysis the stiffness of CANA was estimated to lie between that of 3% NA and 3.2% NA gel. Therefore, the rheological behavior of CANA was compared to these two concentrations of NA. To gain further insights into the dynamic changes to rheology of the bioink as a function of temperature, the viscosity of the bioink during a cooling process from 90 °C–4 °C was examined. From this temperature sweep curve, it is clear that the gelling point of CANA bioink is shifted to the left compared to NA to a temperature that is lower than the printing temperature of 37 °C (Figure 1B). While the viscosity of CANA solution at 37 °C was less than 1 Pa·S, in comparison the viscosities of both 3% and 3.2% NA solutions were two-orders of magnitude (100×). When comparing the storage and loss moduli at 37 °C it is clear that the moduli of NA sols at both concentrations in general are magnitudes higher and show a large fluctuation compared to CANA (Table 1). Importantly, only CANA can reach the target shear strain (1%). This indicates that gel domains have already formed in NA at 37 °C while the CANA is still in a fluid-like state and this exceptional property of CANA makes it extrudable under 37 °C. It is also noted that this kind of special fluid-like state is not permanent and shows dependency on shear rate (Figure 1C). Since, the physical crosslinks in CA are dominated by β-sheet and β-strands which have low H-bonding propensity [30] one could expect changes to gel storage modulus as a function of temperature. Therefore, we investigated the rheological behavior of CANA when the gel recovers from 4 °C to physiological temperature of 37 °C. It was found that the storage modulus of CANA showed a sharp decrease from 180 kPa to 15 kPa, which is consistent with the breaking of β-sheet and β-strands interactions resulting in a softer gel at body temperature (Figure 1D). Furthermore, this is indicative of a fluidic behavior in the CANA gels even at physiological conditions. This is beneficial as the migration of cells within softer gels is facilitated than in stiffer gels [37]. In contrast, NA gels show very negligible changes suggesting that they are in a stable gel state.

Table 1. Comparison of the moduli and shear strain of different formulations in the sol state at 37 °C and in the gel state at 4 °C (n = 3).

	G′ at 37 °C (Pa)	G″ at 37 °C (Pa)	Shear Strain at 37 °C (%)	G′ at 4 °C (kPa)
CANA (8 *w/v* %)	0.85 ± 0.41	0.48 ± 0.04	0.10 ± 0.00	212.18 ± 7.24
3% NA	168.70 ± 121.32	41.22 ± 43.01	0.30 ± 0.13	140.63 ± 3.38
3.2% NA	1006.82 ± 380.78	253.78 ± 34.66	0.41 ± 0.23	237.30 ± 3.26

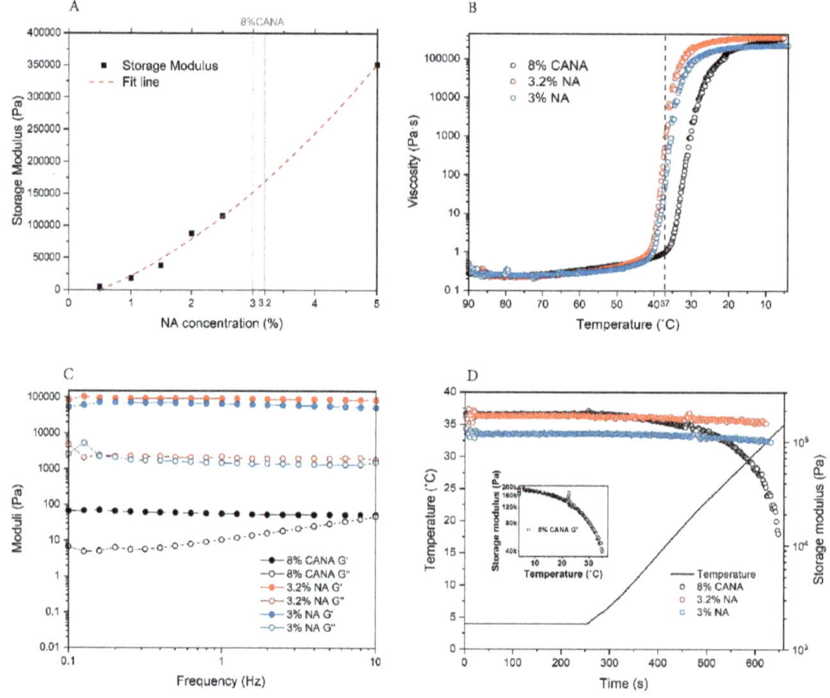

Figure 1. Rheological properties of the CANA bioink compared to NA. (**A**) Estimation of the concentration of NA necessary to achieve the stiffness of the CANA bioink formulation at 4 °C determined using a NA standard curve. The storage modulus of the CANA bioink is depicted by the grey rectangle area. (**B**) Comparison of the viscosity of CANA bioink, 3% NA, and 3.2% NA under a temperature sweep. Dash line represents 37 °C. (**C**) Comparison of the storage modulus and loss modulus of CANA bioink, 3% NA, and 3.2% NA under a frequency sweep from 0.1 Hz to 10 Hz under 37 °C. (**D**) Thermal recovering properties of the CANA bioink, 3% NA, and 3.2% NA. The inset shows the storage modulus of CANA bioink as function of temperature. (n = 3).

3.2. 3D Printing of CANA Bioink into Complex High-Aspect Ratio Structures and Mechanical Properties

The key parameters in microextrusion printing of temperature-sensitive materials are temperature, pressure, and synergy between printing speed and gelation kinetics. To effectively address these variables we modified the printer, as illustrated in Figure 2A. In addition to the built-in temperature-controlling cartridge provided by the manufacturer, we incorporated two more temperature controlling modules, namely, a metal heating module for the printing nozzle and a print bed with water cooled platform. During printing the temperatures of both cartridge (T1) and nozzle module (T2) were set to 37 °C, while the temperature of print bed (T3) was set to 4 °C.

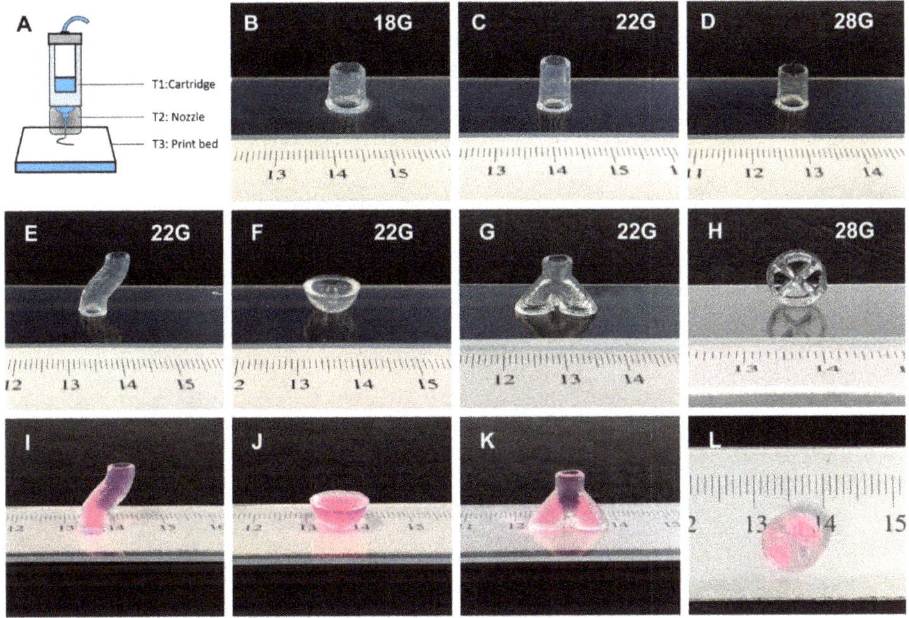

Figure 2. (**A**) Schematic of in-house-modified Cellink Inkredible-2 3D printer. T1, T2, and T3 represent the temperatures of bioink cartridge, nozzle, and print bed respectively. (**B–D**) show the constructs printed using 18G, 22G, and 28G nozzles, respectively. (**E–H**) show structures (s-shape cylinder, hemisphere, a bifurcated tube, and a cylinder with compartments) with various degrees of curvature and complexity printed using the CANA bioink system, and (**I–L**) show the same structures filled with cell culture media demonstrating the structural integrity of the printed structures.

In order to examine the printability of CANA bioink, metal flat-head nozzles of different sizes (18G, 22G, and 28G) were explored. Based on an exhaustive optimization study, pneumatic pressures ideal for extrusion of CANA ink through the various nozzles was identified as follows: 7 kPa for 18G, 22.3 ± 7.1 kPa for 22G, and 44 ± 2.9 kPa for 28G. Using these parameters hollow cylinders 8–10 mm in height were successfully printed using all the nozzles (Figure 2B–D).

As CANA bioink can work well with nozzles of different sizes, we inquired whether it can yield high aspect ratio and complex structures. Towards this objective only 22G and 28G nozzles were used, so that the details of the structures could be more accurately built. As shown in Figure 2 structures ranging from S-shaped hollow cylinders, hemispherical surfaces, bifurcated tubes, and cylinders with compartments were successfully printed (Figure 2E–H). After printing, the cylinders were examined for leaks by adding culture media inside the cylinders—"the leak test". None of the cylinders were found to leak media (Figure 2I–L). Furthermore, from the leak test, it could be seen that both the S-shaped cylinder and the bifurcated tube have even lumen interior, demonstrating the excellent capacity of CANA bioink to render structures. Finally, all the above-mentioned structures could be picked up and handled 30 s after the printing (Supplementary Materials, Video S1).

The elastic modulus of CANA hydrogel was measured under compression at 4 °C to simulate conditions immediately after printing and at 37 °C. And as a reference, compressive moduli of 3% NA, 3.2% NA, and 1% NA were also determined 37 °C (Figure 3A). The modulus of the CANA hydrogel at 37 °C (110 ± 15.89 kPa) was comparable to NA hydrogels. However, at 4 °C the gels were significantly stiffer (133.44 ± 7.48 kPa). In comparison the compressive modulus of a printed sheet (dimension: 1 cm × 1 cm × 0.2 cm) of the CANA gel at 37 °C was found to be 92.64 ± 8.19 kPa (n = 3) and not statistically different from cast samples ($p > 0.1$). This value is only marginally lower than the bulk modulus of the CANA hydrogel suggesting that the bulk properties of printed structures are not significantly altered during printing. The statistically significantly lower modulus of the CANA hydrogel at 37 °C maybe attributed to thermal recovery of the physically crosslinked network through chain relaxation events associated with reduced H-bonding between β-beta sheet structures in CA backbone and also increased mobility of the chains. The softening of structures at physiological conditions maybe be beneficial from cell encapsulation and survival standpoint.

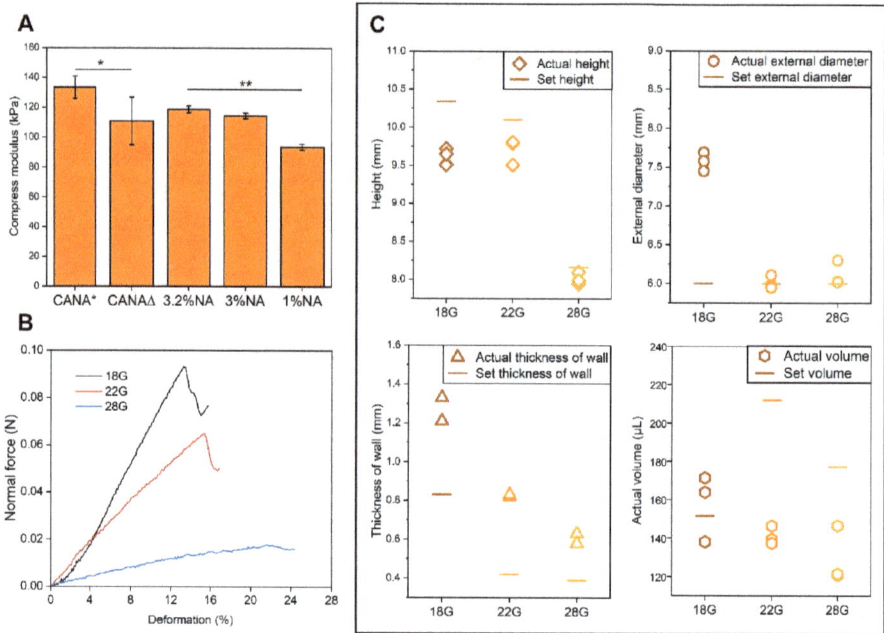

Figure 3. (**A**) Compressive modulus of different hydrogels at 4 °C and 37 °C (n = 3). (CANA* represents the compressive modulus of bulk CANA hydrogel at 4 °C, and CANAΔ represents the compressive modulus of bulk CANA hydrogel at 37 °C. All samples were tested at 37 °C after equilibration for 5 min at 4 °C, except that CANA* was tested at 4 °C after equilibration for 5 min at 4 °C. The compression modulus of 1 *w/v* % NA gels are presented for comparison purposes as it represents the concentration at which NA can be printed through microextrusion printing based on literature (* = $p \leq 0.05$, ** = $p \leq 0.01$, n = 3). (**B**) Force-displacement curves in response to imposition of normal force (n = 3). (**C**) Comparison between prescribed dimension in the g-code and measured dimension (height, external diameter, thickness of wall, and volume contained within the printed cylinder, n = 3) of hollow cylinders printed using 18G, 22G, and 28G nozzles.

In order to gain further insights into the mechanical properties of the printed objects, the elastic deformation of the cylinders under radial-compression was analyzed (Figure 3B). All the printed hollow cylinders exhibited an elastic deformation region and could withstand 12% to 20% deformation depending on the nozzle used to print the structure (Table 2). This is a very important outcome, as it clearly demonstrates that the unique physicochemical characteristics of CA, which is its shear thinning behavior and physical crosslinking, can be exploited in conjunction with print nozzle parameters to produce materials of different bulk properties from the same bioink. Among the cylinders, those printed with 28G nozzles had best elastic deformability, and this might be because finer nozzles yield better fidelity.

Table 2. Initial diameter and degree of deformation of cylinders printed by different nozzles (n = 3).

	Initial Diameter	**Height at Failure Point**	**Degree of Deformation**
18G	7.70 ± 37 mm	6.78 ± 0.41 mm	12.02%
22G	6.51 ± 0.06 mm	5.51 ± 0.21 mm	15.37%
28G	6.02 ± 0.02 mm	4.70 ± 0.11 mm	21.00%

Crosslinking of polymer chains is accompanied by reduction in volume. Therefore, one aspect that needs to be considered in the 3D printing of hydrogels is the fidelity of printed structures vis-à-vis the prescribed g-code. Therefore, we measured the actual dimensions of the printed cylinders, and compared them to design parameters. As shown in Figure 3C, 28 G nozzles yielded the best performance in terms of reproduction of the set height, external diameter, and wall thickness, and 22G nozzles yielded better outcomes than 18 nozzles. As the size of nozzle increases, the extruded volume per unit time is larger, and thus, the cold flow phenomenon is more prominent, resulting in a shortened height after printing. Both 18G and 22G nozzles yielded structures with volumes reasonably close to that expected from calculation.

3.3. Characterization of The Microscopic Structures of Printed Structures

In order to again further insights into the microstructure of the printed objects freeze-dried cylinders were analyzed using scanning electron microscopy (SEM). SEM of the surface of the cylinder revealed good integration of the printed layers with discernible boundaries between layers (Figure 4A). Comparison of layer height to prescribed height in g-code from SEM images revealed that the height of the printed layers was with 80–90% of the prescribed values and also showed not noticeable dependence on nozzle diameter, although the layers in structures printed with 28G nozzle were very close to prescribed values (Table 3). The interior of the wall of the printed cylinder possessed porous microstructure (Figure 4B) with a pore size between 10–30 μM (Figure 4B, inset). The layer-by-layer fusion was further confirmed in the vertical cross sections (Figure 4C), although some delamination between layers was evident. This could be partially attributed to the freeze-drying process, and an incomplete fusion between the entire length of the layers during the printing. Interestingly, the size of the nozzle had an impact on delamination. In structures printed with a finer nozzle the delamination was more obvious (Figure S2). This could be due to the longer printing times with the finer nozzle for a given volume of bioink in comparison to a larger nozzle. Since fusion (or bonding) between layers requires that the layers have some liquid-like characteristics in order to promote polymer chain entanglement at the layer boundaries, increasing the print time (due to reduced dispensed volume at smaller nozzles) can lead to a premature gel-state in the preceding layer, thus diminishing fusion and hence delamination. However, this partial delamination was not sufficient to promote entire separation of layers (Figure S2). This led to exploring an alternative explanation that focuses on the conditions necessary to promote annealing of layers. Since temperature can influence mobility of polymer chains and fluidity of hydrogels, an experiment was undertaken wherein printed samples were incubated in the media either at the ambient temperature, or at 4 °C overnight. Samples were then sectioned in a wet state to visualize the relationship between layers. As shown in Figure S3 separation between layers occurred more often in the samples incubated in cold media than in media equilibrated to ambient-temperature. This phenomenon is also consistent with the thermal recovering observed in CANA (Figure 1D), where the softening of the gel with increase in temperature alludes to higher mobility in chains due to loss of H-bond promoted physical crosslinks between beta sheet structures in the CA.

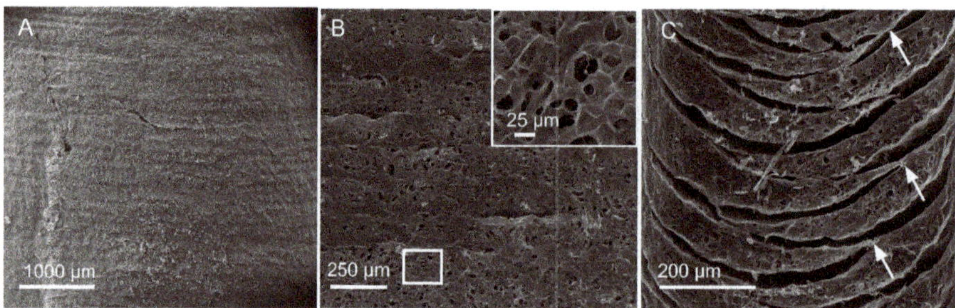

Figure 4. SEM images of a printed cylinder. (**A**) View of the external surface of the cylinder, showing integrity with clear boundaries. (**B**) Interior surface of cylinder wall showing the presence of porous microstructure clearly visible in the inset, which is a magnification of the region bounded by the white rectangle. (**C**) View of the vertical cross section of the cylinder. Arrows show the regions of fusion between layers.

Table 3. The set layer height in printing and the layer height obtained from SEM (n = 3).

	18G	22G	28G
Set height	0.45 mm	0.33 mm	0.16 mm
Height in SEM	0.36 ± 0.18 mm	0.29 ± 0.16 mm	0.14 ± 0.18 mm

3.4. Cellular Biocompatibility of CANA Bioink—Printing of Structures Containing Human Ncs

In past studies, we had shown that CA hydrogels have excellent cellular [30,38] and tissue biocompatibility [33]. Notwithstanding, mechanical properties (stiff versus soft) can impact cell viability. A study to ascertain the suitability of CANA bioinks for printing cell-laden structures was therefore undertaken. Based on past efforts in the encapsulation of chondrocytes in agarose gels showing that densities in excess of 10 million chondrocytes is necessary to have functional constructs [39], in this study, we explored the printing of 30 million NCs per mL of bioink and used cell viability (dead/live assay) as first approximation to ascertain the cellular biocompatibility of the CANA bioink (Figure 5). The viability of the printed NCs immediately after printing was around 83 ± 5%, which is an acceptable and expected outcome, due to shear-induced effects on the cells during the printing process. After 24 h of culture, the viability was 75 ± 5%, which dropped slightly to 71 ± 4% on day 4. However, on day 7, the viability of NCs increased slightly (74 ± 3%) and more importantly, diploid cells, i.e., cells undergoing mitosis (cell division), could be frequently observed in the samples (Figure 5, pointed out by white arrows). Based on these observations, it can be concluded that CANA bioinks can support high-density cell printing; and structures printed using CANA bioink are capable of supporting primary human cell culture and proliferation.

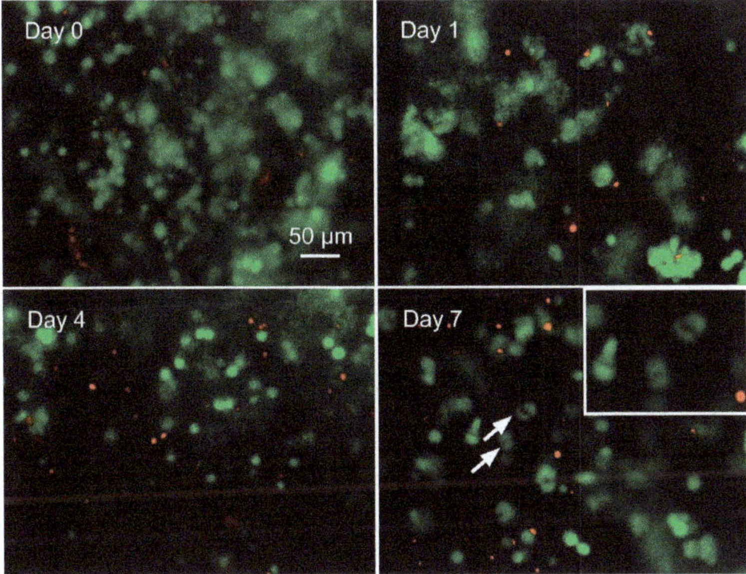

Figure 5. Viability of printed human nasal chondrocytes within cylinders (7 mm diameter × 2mm in height, wall thickness 1.5 mm) printed with CANA bioink instantly after print (day 0), on day 1, day 4, and day 7, as determined by live/dead assay. All experiments were carried out in triplicate. (n = 3). The inset shows a higher magnification of the area pointed out by white arrows and identifies chondrocytes with a typical morphological profile of cells in the process of undergoing cell division (mitosis), indicating that the CANA bioinks are permissive to proliferation of human primary cells.

4. Discussion

Native agarose was used as a component of a bioink for inkjet bioprinting first in the pioneering work by Xu et al., [40] and more recently in 3D bioprinting by Campos et al., who reported the bioprinting cell-laden agarose gel in a hydrophobic high-density fluid [25]. However, beyond these reports there are no notable studies on bioprinting of NA. Due to the poor processability of NA its application in 3D bioprinting is confined to printing with solutions of low concentration (1–1.5 *w/v* %) [25,41]. Due to the rapid gelation of agarose above physiological temperature, it is not capable of yielding high-aspect ratio structures with good resolution, as the printed layers are incapable of fusing with preceding and subsequent layers. Recently, López-Marcial et al., reported bioprinting of a higher concentration of NA albeit blended with alginate, in which the concentration of NA approaches 3% [28]. It is known that the gelation of NA is a result of physical association of double- and single-stranded α helixes via H-bonding [32]. The addition of alginate essentially disrupts helical-helical interaction necessary for gelation of agarose, thus making the bioink more fluidic during printing. However, in that study López-Marcial et al., only showed the printing of a simple geometry and the general printability of blended bioink remains to be further explored.

In comparison to NA, polymer chains of CA—carboxylated derivative of NA, interact with one another, additionally, via β-sheet and β-strand structures [27,29], and therefore gels formed from CA possess many interesting attributes including tunable mechanical properties and sol-gel transition and shear thinning behavior. In previous studies, we had presented the advantages of CA in 3D bioprinting [31,34]. However, whether CA can yield free-standing complex structures of ideal heights has never been explored. Furthermore, whether one can print constructs with similar or higher stiffness compared to printable NA was also an open question. Based on our preliminary experiments the printable range of CA40 (40% carboxylation) was determined to between 7.5 *w/v* % and 10 *w/v* % and based on this information a blend predominantly composed of CA40 with negligible amounts of NA was explored as a potential bioink. The viscosity curve of CANA shared several positive attributes in comparison to NA, including (1) gelation below physiological temperature, (2) a gel phase with fluid-like characteristics that was conducive to microextrusion below 37 °C, and (3) storage modulus comparable to gels of high NA concentrations. These characteristics make CANA an optimal bioink that can yield structures of higher stiffness than the printable NA (~1.5 *w/v* %), and of high-aspect ratio, as the liquid-like characteristics of the gel phase promotes adhesion of layers post-printing, which is very critical to ensure that tall structures do not kink or collapse under their own weight. Using the CANA bioink several complex structures including hemisphere, bifurcated hollow cylinders and S-shape cylinders were printed with high fidelity and layer resolution and bonding. Moreover, we have demonstrated that large number of cells (~30 million/mL) can be dispersed in CANA bioink without impacting its printability, and cells within printed structures remain largely viable and also undergo cell division. Therefore, cell-laden structures possessing accurate anatomical attributes can be realized without the need for scaffolding or supporting phase. Furthermore, Since, CA can be readily modified with cell adhesion sequence such as arginine-glycine-aspartic acid (RGD) [33,38], one can envision the customization of cellular microenvironments within CANA bioinks for 3D-bioprinting-assisted engineering of tissues such as bone, cartilage or cardiovascular components.

5. Conclusions

Excellent printability is crucial and necessary in 3D bioprinting especially in generating advanced complex structures. Development of a novel bioink should follow a "simple" approach to ensure wide-scale adoption. In this study, through blending of modified agarose (CA) with NA a novel bioink—CANA—that possesses an ideal rheological behavior for printing at physiological temperatures was developed. The unique rheological properties of the CANA bioink make it suitable for printing on any microextrusion printer. The superior gelling behavior of CANA bioink leads to exceptional fusion between layers during and after printing, obviating the need of post-processing to stabilize printed structures. Thus, CANA bioink is suitable for printing using nozzles of different diameters' and lengths sizes and require nominal pneumatic pressure. Using CANA bioink large free-standing high aspect ratio objects such as, straight hollow cylinders, S-shaped hollow cylinders, Y-shaped bronchi-like structures, hemispheres, and cylinders with compartments, possessing good elasticity were successfully printed. The stiffness of the printed structures was comparable to those of gels obtained by casting procedure, and from high concentration of NA. In spite of the high stiffness, CANA bioink is also amenable to incorporation of cells at high densities without compromising printability and cell viability. This when combined with the excellent cytocompatibility and in vivo biocompatibility of CA paves the way for its further development as a translational bioink platform in laboratory and clinical research.

Supplementary Materials: The following are available online at http://www.mdpi.com/2306-5354/7/4/141/s1, Figure S1. View of the compressing tests on cylinders. A, B, and C show the initial diameters of cylinders printed by 18G, 22G, and 28G nozzles before compressing, respectively. D, E, and F show the process of compressing tests; Figure S2. SEM images of the view of the interior surface and the vertical cross section of a printed cylinder. The upper row shows the internal surface of the cylinders printed by 18G (A), 22G (B), and 28G (C). The lower row shows the vertical section of the cylinders; Figure S3. Bright field optical microscopic images of a vertical section of a cylinder after immersion in cold media (A,C), and a cylinder after immersion in media at ambient temperature (B,D). Panels B and D are magnified images of the parts within the dashed rectangles in A and C. The cylinders were printed using a 28G nozzle; Video S1: Bifurcated Structure.

Author Contributions: Conceptualization, Y.G. and V.P.S.; methodology, Y.G., B.S., A.F. and A.B.; formal analysis, Y.G. and V.P.S.; investigation, Y.G., B.S. and A.F.; resources, I.M. and V.P.S.; data curation, Y.G.; writing—original draft preparation, Y.G.; writing—review and editing, A.B., I.M. and V.P.S.; supervision, V.P.S.; project administration, V.P.S.; funding acquisition, V.P.S. All authors have read and agreed to the published version of the manuscript.

Funding: This was work was supported in part by the excellence initiative of the German federal and state governments (EXC 294).

Acknowledgments: The authors wish to thank MAPTECH HOLDINGS UG for giving generous access to the Rheometer. The authors also wish to thank Lucas Ahrens and Markus Heiny for technical assistance with the bioprinter, Vincent Ahmadi for technical help with scanning electron microscopy, and Florian Miessmer for the synthesis of the CA. YG would like to thank the China Scholarship Council for support through a Doctoral Fellowship.

Conflicts of Interest: The authors declare no conflict of interest.

References

1. Valino, A.D.; Dizon, J.R.C.; Espera, A.H.; Chen, Q.; Messman, J.; Advincula, R.C. Advances in 3D Printing of Thermoplastic Polymer Composites and Nanocomposites. *Prog. Polym. Sci.* **2019**, *98*, 101162. [CrossRef]
2. Yu, L.; Ding, J. Injectable Hydrogels as Unique Biomedical Materials. *Chem. Soc. Rev.* **2008**, *37*, 1473–1481. [CrossRef] [PubMed]
3. Yang, J.-A.; Yeom, J.; Hwang, B.W.; Hoffman, A.S.; Hahn, S.K. In Situ-Forming Injectable Hydrogels for Regenerative Medicine. *Prog. Polym. Sci.* **2014**, *39*, 1973–1986. [CrossRef]
4. Hendricks, M.P.; Sato, K.; Palmer, L.C.; Stupp, S.I. Supramolecular Assembly of Peptide Amphiphiles. *Acc. Chem. Res.* **2017**, *50*, 2440–2448. [CrossRef] [PubMed]
5. Peak, C.W.; Stein, J.; Gold, K.A.; Gaharwar, A.K. Nanoengineered Colloidal Inks for 3D Bioprinting. *Langmuir* **2018**, *34*, 917–925. [CrossRef] [PubMed]
6. Chimene, D.; Lennox, K.K.; Kaunas, R.R.; Gaharwar, A.K. Advanced Bioinks for 3D Printing: A Materials Science Perspective. *Ann. Biomed. Eng.* **2016**, *44*, 2090–2102. [CrossRef]
7. Gungor-Ozkerim, P.S.; Inci, I.; Zhang, Y.S.; Khademhosseini, A.; Dokmeci, M.R. Bioinks for 3D Bioprinting: An Overview. *Biomater. Sci.* **2018**, *6*, 915–946. [CrossRef]
8. Gu, Y.; Zhang, L.; Du, X.; Fan, Z.; Wang, L.; Sun, W.; Cheng, Y.; Zhu, Y.; Chen, C. Reversible Physical Crosslinking Strategy with Optimal Temperature for 3D Bioprinting of Human Chondrocyte-Laden Gelatin Methacryloyl Bioink. *J. Biomater. Appl.* **2018**, *33*, 609–618. [CrossRef]
9. Drzewiecki, K.E.; Malavade, J.N.; Ahmed, I.; Lowe, C.J.; Shreiber, D.I. A Thermoreversible, Photocrosslinkable Collagen Bio-Ink for Free-Form Fabrication of Scaffolds for Regenerative Medicine. *Technology* **2017**, *5*, 185–195. [CrossRef]
10. Axpe, E.; Oyen, M.L. Applications of Alginate-Based Bioinks in 3D Bioprinting. *Int. J. Mol. Sci.* **2016**, *17*, 1976. [CrossRef]
11. Kang, H.-W.; Lee, S.J.; Ko, I.K.; Kengla, C.; Yoo, J.J.; Atala, A. A 3D Bioprinting System to Produce Human-Scale Tissue Constructs with Structural Integrity. *Nat. Biotechnol.* **2016**, *34*, 312–319. [CrossRef]
12. Miller, J.S.; Stevens, K.R.; Yang, M.T.; Baker, B.M.; Nguyen, D.-H.T.; Cohen, D.M.; Toro, E.; Chen, A.A.; Galie, P.A.; Yu, X.; et al. Rapid Casting of Patterned Vascular Networks for Perfusable Engineered Three-Dimensional Tissues. *Nat. Mater.* **2012**, *11*, 768–774. [CrossRef]
13. Compaan, A.M.; Christensen, K.; Huang, Y. Inkjet Bioprinting of 3D Silk Fibroin Cellular Constructs Using Sacrificial Alginate. *ACS Biomater. Sci. Eng.* **2017**, *3*, 1519–1526. [CrossRef]
14. Suntornnond, R.; Tan, E.Y.S.; An, J.; Chua, C.K. A Highly Printable and Biocompatible Hydrogel Composite for Direct Printing of Soft and Perfusable Vasculature like Structures. *Sci. Rep.* **2017**, *7*, 16902. [CrossRef]
15. Christensen, K.; Xu, C.; Chai, W.; Zhang, Z.; Fu, J.; Huang, Y. Freeform Inkjet Printing of Cellular Structures with Bifurcations. *Biotechnol. Bioeng.* **2015**, *112*, 1047–1055. [CrossRef]
16. Gao, G.; Park, J.Y.; Kim, B.S.; Jang, J.; Cho, D.-W. Coaxial Cell Printing of Freestanding, Perfusable, and Functional In Vitro Vascular Models for Recapitulation of Native Vascular Endothelium Pathophysiology. *Adv. Healthc. Mater.* **2018**, *7*, e1801102. [CrossRef]
17. Skardal, A.; Zhang, J.; McCoard, L.; Xu, X.; Oottamasathien, S.; Prestwich, G.D. Photocrosslinkable Hyaluronan-Gelatin Hydrogels for Two-Step Bioprinting. *Tissue Eng. Part A* **2010**, *16*, 2675–2685. [CrossRef]
18. Yue, K.; Trujillo-de Santiago, G.; Alvarez, M.M.; Tamayol, A.; Annabi, N.; Khademhosseini, A. Synthesis, Properties, and Biomedical Applications of Gelatin Methacryloyl (GelMA) Hydrogels. *Biomaterials* **2015**, *73*, 254–271. [CrossRef] [PubMed]
19. Nishiyama, Y.; Nakamura, M.; Henmi, C.; Yamaguchi, K.; Mochizuki, S.; Nakagawa, H.; Takiura, K. Development of a Three-Dimensional Bioprinter: Construction of Cell Supporting Structures Using Hydrogel and State-of-the-Art Inkjet Technology. *J. Biomech. Eng.* **2009**, *131*, 035001. [CrossRef]

20. Tabriz, A.G.; Hermida, M.A.; Leslie, N.R.; Shu, W. Three-Dimensional Bioprinting of Complex Cell Laden Alginate Hydrogel Structures. *Biofabrication* **2015**, *7*, 045012. [CrossRef]

21. Afghah, F.; Altunbek, M.; Dikyol, C.; Koc, B. Preparation and Characterization of Nanoclay-Hydrogel Composite Support-Bath for Bioprinting of Complex Structures. *Sci. Rep.* **2020**, *10*, 5257. [CrossRef] [PubMed]

22. Seyedin, S.M.; Thompson, A.Y.; Rosen, D.M.; Piez, K.A. In Vitro Induction of Cartilage-Specific Macromolecules by a Bone Extract. *J. Cell Biol.* **1983**, *97*, 1950–1953. [CrossRef] [PubMed]

23. Thompson, A.Y.; Piez, K.A.; Seyedin, S.M. Chondrogenesis in Agarose Gel Culture. A Model for Chondrogenic Induction, Proliferation and Differentiation. *Exp. Cell Res.* **1985**, *157*, 483–494. [CrossRef]

24. Arnott, S.; Fulmer, A.; Scott, W.E.; Dea, I.C.; Moorhouse, R.; Rees, D.A. The Agarose Double Helix and Its Function in Agarose Gel Structure. *J. Mol. Biol.* **1974**, *90*, 269–284. [CrossRef]

25. Duarte Campos, D.F.; Blaeser, A.; Weber, M.; Jäkel, J.; Neuss, S.; Jahnen-Dechent, W.; Fischer, H. Three-Dimensional Printing of Stem Cell-Laden Hydrogels Submerged in a Hydrophobic High-Density Fluid. *Biofabrication* **2013**, *5*, 015003. [CrossRef]

26. Gu, Q.; Tomaskovic-Crook, E.; Lozano, R.; Chen, Y.; Kapsa, R.M.; Zhou, Q.; Wallace, G.G.; Crook, J.M. Functional 3D Neural Mini-Tissues from Printed Gel-Based Bioink and Human Neural Stem Cells. *Adv. Healthc. Mater.* **2016**, *5*, 1429–1438. [CrossRef]

27. Norotte, C.; Marga, F.S.; Niklason, L.E.; Forgacs, G. Scaffold-Free Vascular Tissue Engineering Using Bioprinting. *Biomaterials* **2009**, *30*, 5910–5917. [CrossRef]

28. López-Marcial, G.R.; Zeng, A.Y.; Osuna, C.; Dennis, J.; García, J.M.; O'Connell, G.D. Agarose-Based Hydrogels as Suitable Bioprinting Materials for Tissue Engineering. *ACS Biomater. Sci. Eng.* **2018**, *4*, 3610–3616. [CrossRef]

29. Mirdamadi, E.; Muselimyan, N.; Koti, P.; Asfour, H.; Sarvazyan, N. Agarose Slurry as a Support Medium for Bioprinting and Culturing Freestanding Cell-Laden Hydrogel Constructs. *3D Print. Addit. Manuf.* **2019**, *6*, 158–164. [CrossRef]

30. Forget, A.; Christensen, J.; Lüdeke, S.; Kohler, E.; Tobias, S.; Matloubi, M.; Thomann, R.; Shastri, V.P. Polysaccharide Hydrogels with Tunable Stiffness and Provasculogenic Properties via α-Helix to β-Sheet Switch in Secondary Structure. *Proc. Natl. Acad. Sci. USA* **2013**, *110*, 12887. [CrossRef]

31. Forget, A.; Blaeser, A.; Miessmer, F.; Köpf, M.; Campos, D.F.D.; Voelcker, N.H.; Blencowe, A.; Fischer, H.; Shastri, V.P. Mechanically Tunable Bioink for 3D Bioprinting of Human Cells. *Adv. Healthc. Mater.* **2017**, *6*. [CrossRef]

32. Rüther, A.; Forget, A.; Roy, A.; Carballo, C.; Mießmer, F.; Dukor, R.K.; Nafie, L.A.; Johannessen, C.; Shastri, V.P.; Lüdeke, S. Unravelling a Direct Role for Polysaccharide β-Strands in the Higher Order Structure of Physical Hydrogels. *Angew. Chem. Int. Ed. Engl.* **2017**, *56*, 4603–4607. [CrossRef]

33. Forget, A.; Gianni-Barrera, R.; Uccelli, A.; Sarem, M.; Kohler, E.; Fogli, B.; Muraro, M.G.; Bichet, S.; Aumann, K.; Banfi, A.; et al. Mechanically Defined Microenvironment Promotes Stabilization of Microvasculature, which Correlates with the Enrichment of a Novel Piezo-1(+) Population of Circulating CD11b(+)/CD115(+) Monocytes. *Adv. Mater.* **2019**, *31*, e1808050. [CrossRef]

34. Forget, A.; Derme, T.; Mitterberger, D.; Heiny, M.; Sweeney, C.; Mudili, L.; Dargaville, T.R.; Shastri, V.P. Architecture-Inspired Paradigm for 3D Bioprinting of Vessel-like Structures Using Extrudable Carboxylated Agarose Hydrogels. *Emergent Mater.* **2019**, *2*, 233–243. [CrossRef]

35. Asnaghi, M.A.; Power, L.; Barbero, A.; Haug, M.; Köppl, R.; Wendt, D.; Martin, I. Biomarker Signatures of Quality for Engineering Nasal Chondrocyte-Derived Cartilage. *Front. Bioeng. Biotechnol.* **2020**, *8*, 283. [CrossRef]

36. Jakob, M.; Démarteau, O.; Schäfer, D.; Stumm, M.; Heberer, M.; Martin, I. Enzymatic Digestion of Adult Human Articular Cartilage Yields a Small Fraction of the Total Available Cells. *Connect. Tissue Res.* **2003**, *44*, 173–180. [CrossRef] [PubMed]

37. Engler, A.J.; Sen, S.; Sweeney, H.L.; Discher, D.E. Matrix Elasticity Directs Stem Cell Lineage Specification. *Cell* **2006**, *126*, 677–689. [CrossRef]

38. Arya, N.; Forget, A.; Sarem, M.; Shastri, V.P. RGDSP Functionalized Carboxylated Agarose as Extrudable Carriers for Chondrocyte Delivery. *Mater. Sci. Eng. C Mater. Biol. Appl.* **2019**, *99*, 103–111. [CrossRef]

39. Mauck, R.L.; Seyhan, S.L.; Ateshian, G.A.; Hung, C.T. Influence of Seeding Density and Dynamic Deformational Loading on the Developing Structure/Function Relationships of Chondrocyte-Seeded Agarose Hydrogels. *Ann. Biomed. Eng.* **2002**, *30*, 1046–1056. [CrossRef]
40. Xu, T.; Jin, J.; Gregory, C.; Hickman, J.J.J.J.; Boland, T. Inkjet Printing of Viable Mammalian Cells. *Biomaterials* **2005**, *26*, 93–99. [CrossRef] [PubMed]
41. Duarte Campos, D.F.; Blaeser, A.; Korsten, A.; Neuss, S.; Jäkel, J.; Vogt, M.; Fischer, H. The Stiffness and Structure of Three-Dimensional Printed Hydrogels Direct the Differentiation of Mesenchymal Stromal Cells toward Adipogenic and Osteogenic Lineages. *Tissue Eng. Part A* **2015**, *21*, 740–756. [CrossRef]

Publisher's Note: MDPI stays neutral with regard to jurisdictional claims in published maps and institutional affiliations.

Perspective

Biofabrication of Bacterial Constructs: New Three-Dimensional Biomaterials

Amin Shavandi [1,*]  **and Esmat Jalalvandi** [2]

[1] BioMatter-Biomass transformation Lab (BTL), École interfacultaire de Bioingénieurs (EIB), Université Libre de Bruxelles, Avenue F.D. Roosevelt, 50 - CP 165/61, 1050 Brussels, Belgium

[2] School of Engineering and Physical Sciences, Heriot-Watt University, Edinburgh EH14 4AS, UK; e.jalalvandi@hw.ac.uk

* Correspondence: amin.shavandi@ulb.ac.be

Received: 16 April 2019; Accepted: 10 May 2019; Published: 14 May 2019

Abstract: An enormous number of bacteria live in almost every environment; from deep oceans to below the surface of the earth or in our gastrointestinal tract. Although biofabrication is growing and maturing very quickly, the involvement of bacteria in this process has not been developed at a similar pace. From the development of a new generation of biomaterials to green bioremediation for the removal of hazardous environmental pollutants or to develop innovative food products in a recent trend, researchers have used cutting-edge biofabrication techniques to reveal the great potential of 3D structured bacterial constructs. These 3D bacterial workhouses may fundamentally change our approach toward biomaterials.

Keywords: bacteria biofabrication; 3D printing; tissue engineering; probiotic food

---

We are a giant symbiont organism composed of Homo sapiens and microbial cells, while the microbes in our body have a larger genome size than us [1,2]. We are learning more and more about the significance of bacteria in our body and their communication with organs such as liver and brain. An incredible recent study found a key role of bacteria in the development of schizophrenia [3]. Biofabrication in recent years progressed toward the concept of an automated development of structured materials with biological function. Living cells, bioactive molecules and hybrid cell-material structures have developed through biofabrication, among others [4]. However, bacteria traditionally have not been largely considered in biofabrication, and their enormous potential for the development of functional 3D biomaterials has mainly remained unknown. In an early study, Weible and co-workers [5] printed patterns of bacteria on the agar surface in a petri dish using soft lithography. For this purpose, agarose stamps were fabricated by casting agarose solutions on the Polydimethylsiloxane (PDMS) moulds. By applying cell suspension, the cells were deposited on the surface of the stamp which was then used for transferring the pattern of bacteria to agar plates containing culture media. The authors could use one single stamp to print a bacterial pattern at least 250 times on agar culture. The reported agarose stamp technique is capable of patterning various strains of bacteria in a simultaneous, reproducible and rapid process. The proposed stamping technique can be useful for different scientists interested in developing a pattern of bacteria on cell culture media in order to study the organisms' interaction with each other, molecules or the material surface (Figure 1A).

To better understand the cell–cell interactions in a complex microbial environment besides gaining insight into the role of geometry on the bacterial pathogenicity, Connell et al. [6] proposed a new 3D printed cellular model using a laser-based lithography method. Applying this technique, selected bacteria were trapped and sealed within the cavities formed by the crosslinked chains of gelatin. The authors showed the interaction between human pathogens of *Staphylococcus aureus* and *Pseudomonas aeruginosa* in a 3D structure indicating the survival of *S. aureus* from antibiotic treatment with β-lactam when enclosed in 3D shell communities composed of *P. aeruginosa*. Given that a bacterial community thrives in a 3D structure in the human body, the proposed technique can be useful to study the role of geometry in pathogenicity.

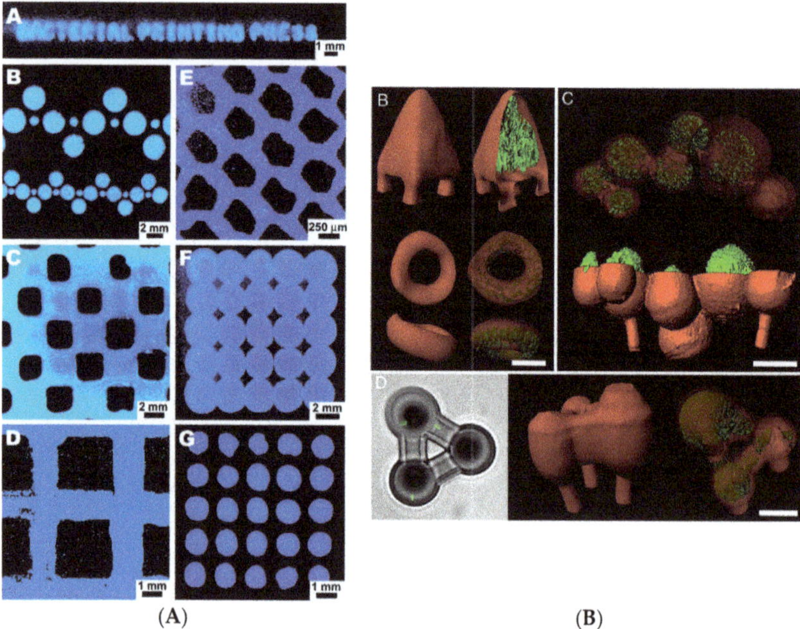

(A)   (B)

Figure 1. (**A**) Various patterns of *Vibrio fischeri* colonies printed on the agar surface in a petri dish using soft lithography. Adapted with permission from [5]. Copyright (2005) American Chemical Society. (**B**) Confocal fluorescence images show gelatin-based micro-3D printing in the presence of Pseudomonas aeruginosa microcolonies [6].

The incorporation of desired proteins such as bovine serum albumin (BSA) to the gel can enhance the mechanical and chemical properties of the 3D structure, and it is possible to print various cell types using different fabrication gels (Figure 1B). The ability of bacteria to generate new materials was combined with the tunable properties of the 3D printing method by Lehner and co-workers [7], who fabricated the 3D structure of bacterial cultures leading to the development of new sustainable materials. Using a modified commercial 3D printer, a mixture of bacteria and alginate was extruded, cross-linked and formed a gel upon contact with a calcium ion containing surface resulting in the preparation of 3D microbial structures in a reproducible manner. A high printing resolution was achieved using this technique, and the rate of extrusion and print head speed were reported as two major parameters affecting the printing resolution. The developed system allowed printing multilayer bacterial structures. Two different strains of *Escherichia coli* able to express proteins at distinct colours were printed in a bilayer structure (Figure 2A). The analysis indicated a good separation in addition to bacterial survival and metabolic activity in the gel layers up to 48 hours of incubation. The proposed printing system can be used for the preparation of different bacteria containing materials in a patterned format within millimetre resolution. Nevertheless, limitations of the system include the production of 3D printed structures with internal bridges or hollow spaces. In addition, it is not yet possible to directly process bacteria in a complex 3D structure. The development of biofilm is also not controlled, and the chemistry of the matrix polymer can be the limiting factor regarding the stability of the developed bacterial structure.

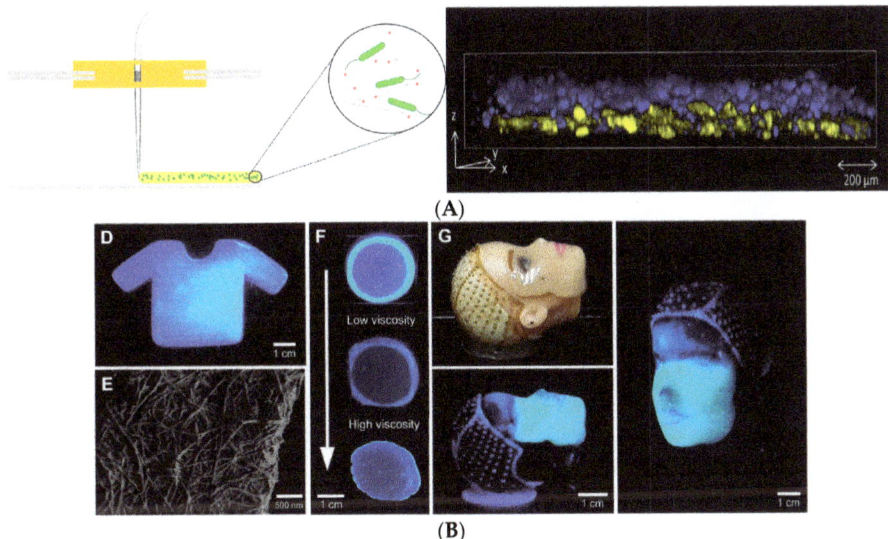

Figure 2. (**A**) Modified strains of *Escherichia coli* contain blue fluorescent and yellow cells printed in a layered structure using a modified commercial 3D printer. (ACS Author Choice—This is an open access article published under a Creative Commons Non-Commercial No Derivative Works (CC-BY-NC-ND) Attribution License). (**B**) 3D-printed complex structures containing bacteria for various bioremediation and biomedical applications [8]; Open-access article distributed under the terms of the Creative Commons Attribution-Non Commercial license.

By combining genetic engineering and 3D printing, the same group in a recent study [8] developed a standardised and reproducible method for the production of 3D biofilm structures. A low-cost 3D printer "The Biolinker" was utilised to print bacterial suspension in an alginate solution which turns into a gel on a substrate containing calcium. The authors printed engineered *E. coli* that in the presence of an inducer produces biofilm and the biofilm formation could be controlled by the genetic control of a gene (csgA). These 3D printed biofilms could have diverse functions and applications, such as sequestration of metal ions or water filtration. Schaffner and co-workers [9] proposed a 3D printing system to develop cell-laden hydrogels called "Flink" with the ability to control the cells' concentration and their spatial distribution in the 3D structure. The developed biocompatible hydrogel synthesised from nontoxic substances of k-carrageenan, hyaluronic acid and fumed silica had the viscoelastic properties suitable for the immobilisation of the cells and the production of 3D printed structures through multilateral direct ink writing (DIW). The authors showed the capability of the designed system with two examples of 3D structures for bioremediation and biomedical applications. In the first example, in order to benefit from phenol degrading capability of *Pseudomonas putida*, this bacterium was immobilised and 3D-printed. The 3D bacterial lattice structure with the high surface area could degrade the phenol as a major and toxic substance without the need for a supporting material. The degradation of phenol was found to be caused by bacteria that have been released from the 3D structure in the phenol-containing medium as well as the cells immobilised on the surface of the 3D structure. In the second example, Flink was loaded with *acetobacter xylinum* for the in situ production of cellulose in the form of a 3D structure with good mechanical properties, making it suitable for biomedical applications. In this case, once the bacteria have produced cellulose, the ink residue was washed away, leaving a cellulose network with a specific geometry and topography (Figure 2B).

In line with the potential of 3D printed bacteria for bioremediation applications, Qian et al. [10] developed living inks using freeze-dried baker's yeast and printed catalytically active structures at a high resolution (100 μm) through direct ink writing techniques. The cell-containing ink had shear-thinning rheological behaviour suitable for extrusion 3D printing. Printed cell structures can ferment glucose and produce ethanol and CO_2. The proposed ink system in this study can be used for printing different microbes with catalytic activities for a broad range of biotechnological applications. The food industry can also benefit from the 3D fabrication of bacteria. In a different direction, in a recent study, Zhang and colleagues [11] investigated the possibility of manufacturing 3D printed cereal-based food, loaded with probiotic bacteria *Lactobacillus plantarum* WCFS1. Like hydrogel composition—which affects viscosity and printability—the dough formulation, the flour type and water content determine the feasibility of printing the dough in 3D structures. The probiotic loaded bacterial structures were printed using a fused deposition modelling method in two honeycomb and concentric structures (Figure 3) and baked at different temperature of 145, 175 and 205 °C. One major obstacle to produce bakery products containing probiotic bacteria is the survival of the organism at the baking temperature. In this regard, increasing the rate of drying of the products can help shorten the baking time. Therefore, food structures with high surface-to-volume ratios produced by the 3D printing approach can be beneficial for this purpose. The authors have incorporated sodium caseinate to increase the viscoelastic properties of the dough to enhance its printability. The probiotics could survive the baking process of 6 min at 145 °C in honeycomb structures, and by having more than 106 CFU/g, the baked product could be defined as a probiotic food. The results reported in this study may offer a new avenue to the development of innovative bakery products containing probiotics.

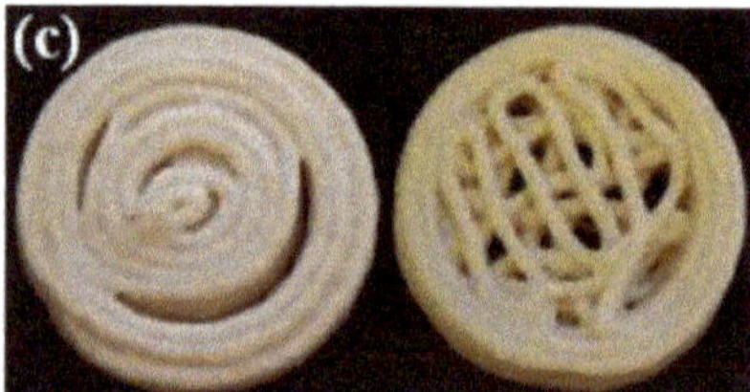
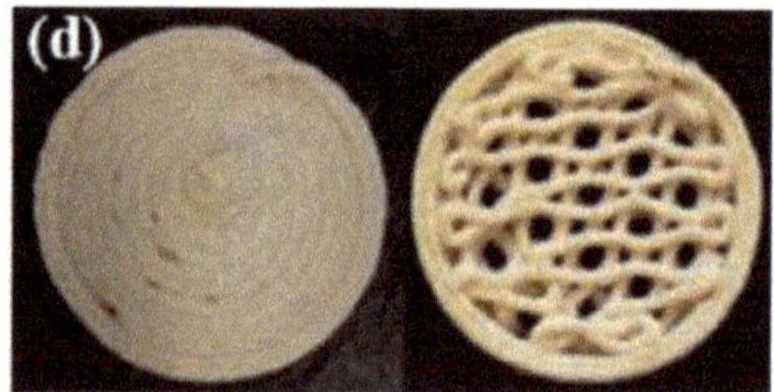

Figure 3. 3D printing dough formulations with different nozzle sizes of 1.6 mm (**left**) and 1.2 mm (**right**) containing probiotic bacteria [9].

Microbial biofilm models are the major form of microbial life, which are composed of a microorganism and extra polysaccharides. These 3D structures can be used to evaluate microbial metabolism and their interaction with the surrounding media. Recent research at Montana State University focused on engineering 3D printed biofilms of methanotroph bacteria with specific characteristics in order to convert methane into various organic materials, such as bioplastics. This can result in the reduction of the methane emitted into the environment and at the same time result in the development of bio-based materials as an intersection of technology and nature [10]. In another study, a microporous 3D bacterial cellulose foam was developed through foaming, and bacterial cellulose was formed directly at the air–water interfaces of the air bubble (Figure 4A) [11]. The authors used Cremodan as a surfactant for foaming and Xanthan as a thickener to obtain the stable foam. The biocompatible 3D structured bacterial cellulose may have potential applications in skin tissue engineering and wound healing which addresses the lack of porosity of the bacterial cellulose conventionally produced.

In a recent creative study [12], Manoor's lab 3D printed colonies of cyanobacteria and graphene nanoribbons on biotic and abiotic (polysiloxane) mushroom pileus and created bionic mushrooms (Figure 4B). The highly packed 3D printed bacteria could nourish from the mushroom for generating photosynthetic electricity which was collected using graphene nanoribbons. Interestingly, the biotic mushroom provided bacteria a suitable environment (e.g., temperature, pH and moisture) helping toward their viability, while these conditions were absent in abiotic mushroom resulting in less viability of the bacteria.

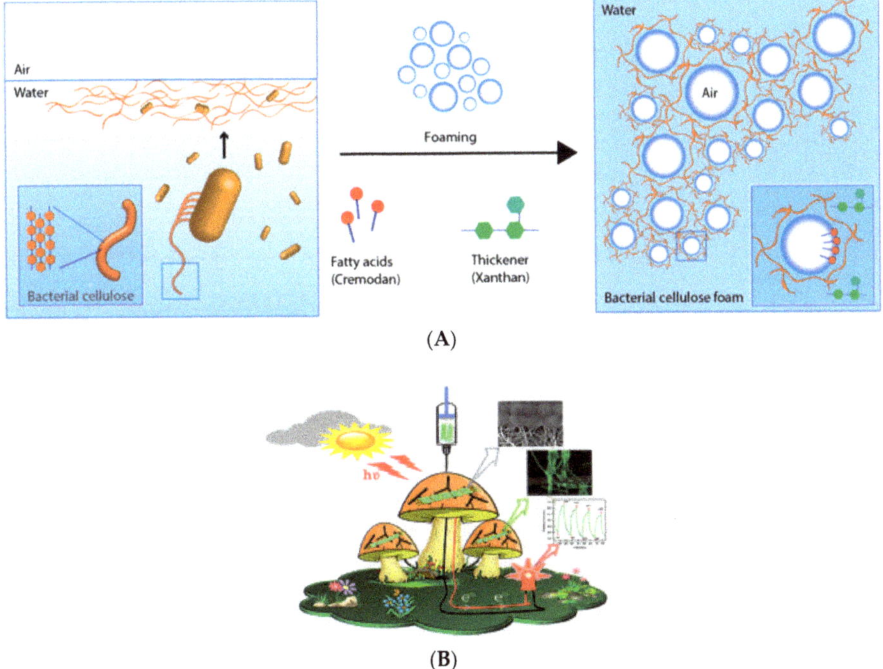

(**A**)

(**B**)

Figure 4. (**A**) Schematic of the formation of bacterial cellulose foam through foaming of a mannitol-based media with a bacterial suspension of *Gluconoacetobacter xylinus*. Cremodan as a surfactant stabilises the foam. Xanthan provides stability to the foam [11] (Open access). (**B**) Schematic drawing of mushrooms with 3D printed layer of cyanobacteria and graphene nanoribbons for the production of bioelectricity. Reprinted with permission from [12]. Copyright (2018) American Chemical Society.

The work highlighted the possibility to harness the benefit of 3D printed bacteria to realise an environmentally friendly source of photosynthetic electricity [12].

In conclusion, in this perspective, we have summarised the state-of-the-art biofabrication of bacterial constructs, highlighting the progress and unmet challenges. The potential application of 3D printed bacterial constructs is diverse, ranging from studying the development of infection in vivo to producing 3D structured probiotic foods, converting methane into bioplastics, producing photosynthetic electricity and biomedical applications. Despite these intriguing studies and reports, the current 3D bioprinters are slow and operate at small scales. Future studies are required to develop new 3D bioprinters which are affordable, scalable and able to print different types of bacterial inks with diverse viscosities at a short time and in a controlled fashion. Considering the high and growing demand for green products and the potential applications of 3D printed bacterial constructs, it is highly predictable that those barriers will soon be resolved.

Author Contributions: A.S. and E.S. contributed equally.

Funding: This research received no external funding.

Conflicts of Interest: The authors declare no conflict of interest.

References

1. Ley, R.E.; Peterson, D.A.; Gordon, J.I. Ecological and evolutionary forces shaping microbial diversity in the human intestine. *Cell* **2006**, *124*, 837–848. [CrossRef] [PubMed]

Bioengineering **2019**, *6*, 44

2. Shanahan, F. The host-microbe interface within the gut. *Best Pract. Res. Clin. Gastroenterol.* **2002**, *16*, 915–931. [CrossRef] [PubMed]

3. Zheng, P.; Zeng, B.; Liu, M.; Chen, J.; Pan, J.; Han, Y.; Liu, Y.; Cheng, K.; Zhou, C.; Wang, H.; et al. The gut microbiome from patients with schizophrenia modulates the glutamate-glutamine-GABA cycle and schizophrenia-relevant behaviors in mice. *Sci. Adv.* **2019**, *5*, eaau8317. [CrossRef] [PubMed]

4. Moroni, L.; Boland, T.; Burdick, J.A.; De Maria, C.; Derby, B.; Forgacs, G.; Groll, J.; Li, Q.; Malda, J.; Mironov, V.A.; et al. Biofabrication: A Guide to Technology and Terminology. *Trends Biotechnol.* **2018**, *36*, 384–402. [CrossRef] [PubMed]

5. Weibel, D.B.; Lee, A.; Mayer, M.; Brady, S.F.; Bruzewicz, D.; Yang, J.; DiLuzio, W.R.; Clardy, J.; Whitesides, G.M. Bacterial Printing Press that Regenerates Its Ink: Contact-Printing Bacteria Using Hydrogel Stamps. *Langmuir* **2005**, *21*, 6436–6442. [CrossRef] [PubMed]

6. Connell, J.L.; Ritschdorff, E.T.; Whiteley, M.; Shear, J.B. 3D printing of microscopic bacterial communities. *Proc. Natl. Acad. Sci. USA* **2013**, *110*, 18380. [CrossRef] [PubMed]

7. Lehner, B.A.E.; Schmieden, D.T.; Meyer, A.S. A Straightforward Approach for 3D Bacterial Printing. *ACS Synth. Biol.* **2017**, *6*, 1124–1130. [CrossRef] [PubMed]

8. Schaffner, M.; Rühs, P.A.; Coulter, F.; Kilcher, S.; Studart, A.R. 3D printing of bacteria into functional complex materials. *Sci. Adv.* **2017**, *3*, eaao6804. [CrossRef] [PubMed]

9. Zhang, L.; Lou, Y.; Schutyser, M.A.I. 3D printing of cereal-based food structures containing probiotics. *Food Struct.* **2018**, *18*, 14–22. [CrossRef]

10. Marshall, S. MSU Researchers Receive $1.8 Million to Study Methane-Converting Microbes. Available online: https://www.montana.edu/news/17214/msu-researchers-receive-1-8-million-to-study-methane-converting-microbes (accessed on 11 May 2019).

11. Rühs, P.A.; Storz, F.; López Gómez, Y.A.; Haug, M.; Fischer, P. 3D bacterial cellulose biofilms formed by foam templating. *NPJ Biofilms Microbiomes* **2018**, *4*, 21. [CrossRef] [PubMed]

12. Joshi, S.; Cook, E.; Mannoor, M.S. Bacterial Nanobionics via 3D Printing. *Nano Lett.* **2018**, *18*, 7448–7456. [CrossRef] [PubMed]

Article

Additive Manufacturing of Prostheses Using Forest-Based Composites

Erik Stenvall [1], Göran Flodberg [2], Henrik Pettersson [2], Kennet Hellberg [3],
Liselotte Hermansson [4,5], Martin Wallin [4] and Li Yang [2,*]

[1]   Stora Enso AB, Sommargatan 101A, 65009 Karlstad, Sweden; erik.stenvall@storaenso.com
[2]   RISE—Research Institutes of Sweden, Drottning Kristinas väg 61, 11486 Stockholm, Sweden;
      goran.flodberg@ri.se (G.F.); henrik.pettersson@ri.se (H.P.)
[3]   Embreis AB, Tumstocksvägen 11 A, 18766 Täby, Sweden; kennet.hellberg@embreis.com
[4]   Department of Prosthetics and Orthotics, Faculty of Medicine and Health, Örebro University,
      70185 Örebro, Sweden; liselotte.hermansson@regionorebrolan.se or liselotte.hermansson@oru.se (L.H.);
      martin.wallin@regionorebrolan.se (M.W.)
[5]   University Health Care Research Center, Faculty of Medicine and Health, Örebro University,
      70185 Örebro, Sweden
*   Correspondence: li.yang@ri.se; Tel.: +46-76-8767-134

Received: 14 August 2020; Accepted: 28 August 2020; Published: 1 September 2020

Abstract: A custom-made prosthetic product is unique for each patient. Fossil-based thermoplastics are the dominant raw materials in both prosthetic and industrial applications; there is a general demand for reducing their use and replacing them with renewable, biobased materials. A transtibial prosthesis sets strict demands on mechanical strength, durability, reliability, etc., which depend on the biocomposite used and also the additive manufacturing (AM) process. The aim of this project was to develop systematic solutions for prosthetic products and services by combining biocomposites using forestry-based derivatives with AM techniques. Composite materials made of polypropylene (PP) reinforced with microfibrillated cellulose (MFC) were developed. The MFC contents (20, 30 and 40 wt%) were uniformly dispersed in the polymer PP matrix, and the MFC addition significantly enhanced the mechanical performance of the materials. With 30 wt% MFC, the tensile strength and Young´s modulus was about twice that of the PP when injection molding was performed. The composite material was successfully applied with an AM process, i.e., fused deposition modeling (FDM), and a transtibial prosthesis was created based on the end-user's data. A clinical trial of the prosthesis was conducted with successful outcomes in terms of wearing experience, appearance (color), and acceptance towards the materials and the technique. Given the layer-by-layer nature of AM processes, structural and process optimizations are needed to maximize the reinforcement effects of MFC to eliminate variations in the binding area between adjacent layers and to improve the adhesion between layers.

Keywords: biocomposite; forest-based MFC; fibrils; additive manufacturing; artificial limb; fused deposition modeling (FDM)

---

1. Introduction

Fossil-based thermoplastics are the dominant raw materials in both prosthetic and industrial applications; there is a general demand for reducing their use and replacing them with renewable biobased materials [1,2]. Forests hold the biggest share of renewable biomaterials on earth, which are inevitably central to the biobased economy. Various biocomposite materials have been studied and also commercialized to some extent in recent years. Biocomposites, also called natural fiber composites, can be made from many different types of plant fibers originating from bast, leaf, straw, grass, or wood [3]. Natural fibers from wood are normally pretreated by grinding and after this they are treated again by mechanical or chemical means to release fibrils from the wood fiber. Instead of having only fibrils of a narrow size distribution, microfibrillated cellulose (MFC) can contain fibers, fiber fragments, fibrillar fines, and nanofibrils [4,5]. Then, the MFC is melt blended with thermoplastic polymers in an extrusion process to form biocomposites. MFC can significantly improve the strength and stiffness when used as reinforcement in thermoplastics [6]. In additive manufacturing (AM), industry fiber reinforcement is often needed to meet the strength requirement or other functions [7–9] to replace traditional materials such as aluminum in many applications [10]; in this context, MFC becomes an interesting renewable-source reinforcement.

Prosthetic and orthotic products are important for healthcare and well-being. Today, approximately 3000 prosthetic limbs are manufactured by different companies in Sweden, with an estimated value of 100 million Swedish kronor (MSEK) per year. The materials used are fossil-based thermoplastics and reinforcement materials, for example, glass or carbon fibers, metals, etc. According to a market study, there are more than 100 million individuals worldwide with limb loss and more than one million amputations annually. In addition to four million people living with limb loss in 2014, there are 400,000 amputations/year in the USA and EU countries. The global prosthetics and orthotics market size has been estimated to be USD 9.2 billion in 2019 with an annual growth rate of 4.6% [11].

A prosthetic or orthotic product is unique and custom made for each patient. Specialists make the customization to ensure a good result for the patient. It requires multiple fitting-and-adjusting cycles, which is both time consuming and expensive. The specialist uses many different types of materials for the customization, for instance, thermoplastics, such as PP, PE (polyethylene), PETG (polyethylene terephthalate glycol), and PA (polyamide), as they are easy to customize; carbon fiber composites and special high-strength titanium alloys and aluminum alloys are also commonly used to ensure that the prostheses are both lightweight and durable, and soft foam and gel materials from silicone, EVA (ethylene-vinyl acetate), PUR (polyurethane), PE and SEBS (styrene ethylene butylene styrene) are used to provide cushioning effects.

Additive manufacturing (AM), also known as three-dimensional (3D) printing, is rapidly progressing in industrial application prototyping. Unlike traditional manufacturing processes, in which tools are needed, and geometrical complexity and production volume are often limiting factors, AM needs no tools, hence, it is not limited by either geometrical complexity or production volume, which makes it a perfect technique for personalization from design and prototyping to production. Due to a strong interest worldwide, many AM techniques have been developed in recent decades. These techniques operate on different principles, from binding, fusing, and melting to polymerization, and with different materials including sand, polymer, metal, and biomaterials [12]. Nevertheless, the number of polymeric materials available for AM applications based on FDM (fused deposition modeling) and polymer bed fusion (scanning laser sintering (SLS) and digital light processing (DLP)) are rather limited, which prevents the adoption of the techniques. The qualities of FDM-produced products are affected by various process parameters, for example, layer thickness, build orientation, raster width, and print speed. The process parameter settings and their ranges depend on the type of FDM machine. The optimum parameters can improve the qualities of three-dimensional (3D) printed parts [13].

In the last few decades, AM techniques have been introduced into the prosthesis sector [14,15] and AM is now considered to be an emerging prosthetic production technology. The number of businesses that offer prostheses produced by 3D CAD (additive manufacturing) is increasing rapidly. The first prosthetic parts that were additively manufactured were cosmetic outer shells for lower limb prostheses, in which the AM technique made a perceptual impact on the prosthetic and orthotic products. From the first production of cosmetic shelves, this technique is now suggested, for example, for production of prosthetic sockets [16] and prosthetic hands [17]. However, there is a scarcity in the literature regarding 3D-printed prostheses for the lower limbs. A recent scoping review [18] only found 11 articles that covered the implications of 3D printing in the field of lower limb amputation, but they concluded that it was a promising advancement in modern prosthetic fabrication. Among the existing studies, the focus has mainly been on user experiences and improved functionality of the AM-fabricated prosthetic parts using commercial materials [19–21].

The transition from conventional manufacturing to 3D printing of entire prostheses is not easy. A unique prosthesis must be created in a digital format, and then produced through AM [22]. Moreover, the AM materials must be high strength, light weight, and produced at a reasonable cost. In their 2020 review, Barrios et al. concluded that one of the future paths in this field would be based on the design of new materials [14]. However, the additively manufactured products, for example, lower limb prostheses, must be able to tolerate high and dynamic loads and provide sufficient durability. The mechanical properties of prosthetic sockets produced with this technology have yet to be proven. The strength and durability of the sockets were identified as important aspects as early as 2005 [16]; a review in 2018 [23] further emphasized the need for quality control and testing of the sockets. However, to our knowledge, no lower limb prosthesis, socket or other product component, has been produced using renewable source MFC-reinforced materials and AM technologies.

Our hypothesis is that MFC from wood fiber for reinforcement of composites combined with AM techniques enables innovative and personalized prosthetic solutions. The focus of this work is proof of the concept, i.e., obtaining MFC-based composites that are appropriate for FDM and prosthetic applications. The results presented were generated in two research projects. The aim of the projects was to develop systematic solutions to combine biocomposites with AM techniques, for example, FDM and SLS, through close collaboration between partners along the value chain from material development to end-user applications. Prosthetic products were chosen because they offer excellent opportunities for developing both AM and biocomposites, as there are well-established testing methods and quality requirements and standards which are often higher than for ordinary consumable products. Hence, prosthetic products are well suited to our purpose of technical developments in biocomposites for 3D printing applications. Prior to this work, there have been two project publications. Pore characteristics of SLS-built parts of nylon 12 with or without the addition of carbon fibers were found to be responsible for the variation of the parts' mechanical performance [24]. It has also been demonstrated that topology optimization based on FEM (finite element modeling) could be used in prosthesis design and structural optimization [25].

2. Materials and Methods

This section describes the composite materials, the processing techniques for the test specimen and demonstrators, the mechanical and structural test and measurement techniques, and a survey of the end user's attitude towards biobased composites and 3D-printed prostheses.

2.1. Composite Materials

Extensive studies on material composition and characterizations have been carried out. First, multiple trials with different MFC and PP blends were undertaken at a lab scale. Then, two pilot-scale trials were conducted to further verify the results. Finally, the composite materials studied in this work were produced by Stora Enso in a dry compounding process. The composites, called DuraSense PP AM quality, were specially fabricated for this work and consisted of a polypropylene (PP) matrix with 20, 30, and 40 percent MFC made from chemical pulp fiber. It is worthwhile noting that the quality of the MFC used in this work was adopted to composite fabrication. According to Chinga-Carrasco [4], an MFC could contain several material components, for example, fibers, fiber fragments, fibrillar fines, and nanofibrils. Due to the fact that the composites having different MFC contents were produced on different dates and in different batches, there were possible variations in both materials and production conditions. To improve both the compatibility and the dispersion of MFC in the PP matrix, the following two main strategies were followed: (I) An adapted compounding process at Stora Enso, which facilitates the dispersion of MFC in the specific polymer matrix and (II) the use of a functional coupling agent, which enhances the compatibility between the hydrophobic polymer phase and the hydrophilic MFC. The composite materials were in granulate form and suitable for FDM-based AM applications.

2.2. Production of the Test Samples and the Demonstrators

The ISO 3167 multipurpose test samples (dumbbells) were produced on a BOY 25 EVH injection molding machine from Dr. BOY GmbH & Co. KG, Neustadt-Fernthal, Germany, using the parameters listed in Table 1. Similar test samples were created with the FDM technique. The FDM machine was an ErectorBot 644LX (ErectorBot, Inc. CA), equipped with a custom extruder printhead with a nozzle of diameter 0.8 mm. The test specimens were built with either longitudinal lines or perpendicular lines inside a perimeter frame, as shown in Figure 1. The other parameters applied were line width of 0.9 mm, layer height of 0.3 mm, print temperature of 240 °C, and print speed of 33 mm/s.

Table 1. Injection molding parameters.

Temperature Profile from Nozzle	200/200/195/190/180 °C
Injection pressure	Ramping down from 100 to 40 MPa
Injection speed	Ramping down from 20 to 3.8 cm^3/s
Hold pressure	60 MPa
Hold time	8 s
Mold temperature	40 °C
Cooling time	20 s

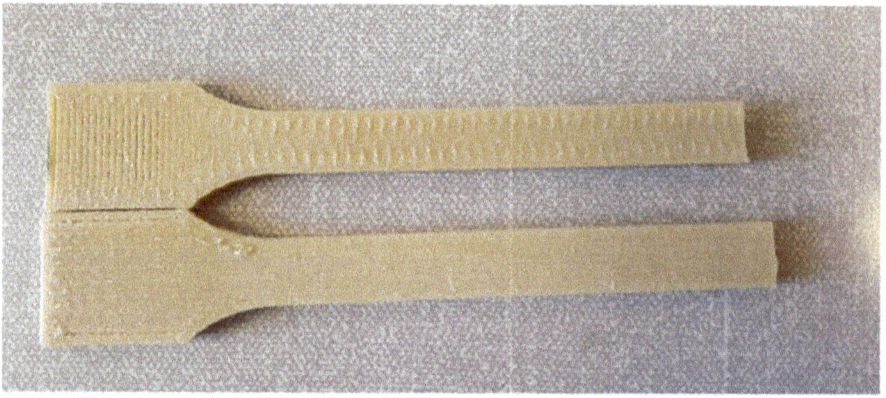

Figure 1. Standard test specimens built with fused deposition modeling (FDM) and with longitudinal lines (bottom) and perpendicular lines (top).

Figure 2a shows a transtibial (lower limb) prosthesis, which was the demonstrator for this work created using the FDM technique with biocomposite materials. To demonstrate AM's geometric precision, strength, and durability capabilities, a test socket (Figure 2b) was also 3D printed with similar, however, simpler geometry than the lower limb prosthesis. The socket was regarded as a building block of the lower limb prosthesis, and the simplified geometry enabled testing according to the ISO standards.

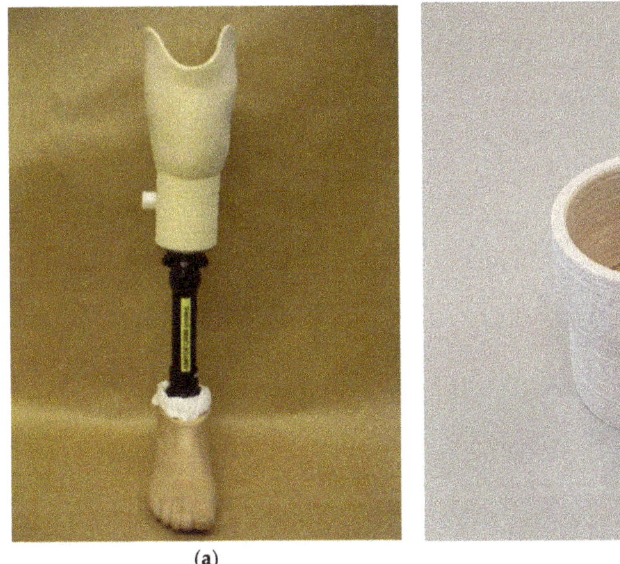

(a)   (b)

Figure 2. The 3D-printed demonstrators. (**a**) The lower limb prosthesis; (**b**) The test socket, a simplified variant of the prosthetic socket.

2.3. Mechanical Testing, SEM and X-ray Microtomography

The static ultimate test and the dynamic test were performed at Fillauer Europe AB and Embreis AB according to ISO 10328:2016 [26], a structural standard test for prosthetics regarding the structural testing of lower limb prostheses. Loading condition II was applied with P6 test geometry, which meant that the tested sockets experienced not only off-axis compression and bending but also twisting around a few axes, as illustrated in Figure 3a. The test equipment shown in Figure 3b include a Static Test Machine No. K1 Tinius Olsen 25ST from Tinius Olsen TMC, USA and a Sauter TVS 10KN100 from SAUTER GmbH, Balingen, Germany. The load was measured using a 10 kN load cell with a measurement accuracy of +/−5 N.

Electron scanning microscopy (SEM) was carried out using a Hitachi SU3500, Tokyo, Japan. The instrument is equipped with a backscattered electron detector (BSE), secondary electron detector (SE), and X-ray energy dispersive spectrometer (EDS). The pictures shown in this work were taken with the BSE detector at 13 kV and at 250X magnification.

In recent years, X-ray microtomography has become a valuable tool for material characterization [27]. The X-ray microtomography was performed with an Xradia MicroXCT-200 (Carl Zeiss X-ray Microscopy, Inc., Pleasanton, CA, USA). The scanning conditions were as follows: X-ray source (voltage 40 kV, power 4 W), the number of projections was 1.289, and the exposure time was 20 s/projection. The distances from the detector and from the X-ray source to the sample holder were 16 and 30 mm, respectively. The magnification was 20×, the pixel size of the image was 0.8149 µm, the pixel resolution was 1.07 µm, and the maximum analyzed volume was $1 \times 1 \times 1$ mm. The samples were examined in their X, Y, and Z directions.

The injection molded specimens for both tensile and impact testing were kept conditioned at 23 °C, 50% RH, for a minimum of 48 hours prior to testing. The tensile tests were performed using a hydraulic actuator with an MTS FlexTest™ 60 digital controller (MTS Systems Corporation, Eden Prairie, MN, USA). The specimens were fastened using clamps, and a contact extensometer was attached to the specimens with rubber bands. The load was measured with a 40 kN load cell, and the tests were performed at a rate of 50 mm/min. Data were recorded at a sampling rate of 500 Hz using MTS Series

793 control software. The impact tests were performed on a Zwick HIT5P pendulum impact tester (ZwickRoell GmbH & Co. KG, Ulm, Germany) equipped with a 5 Joule Charpy pendulum.

The dynamic viscosity of the composite material was measured using a Malvern Rosand RH 10 capillary rheometer with 15 mm twin bore at 190 °C.

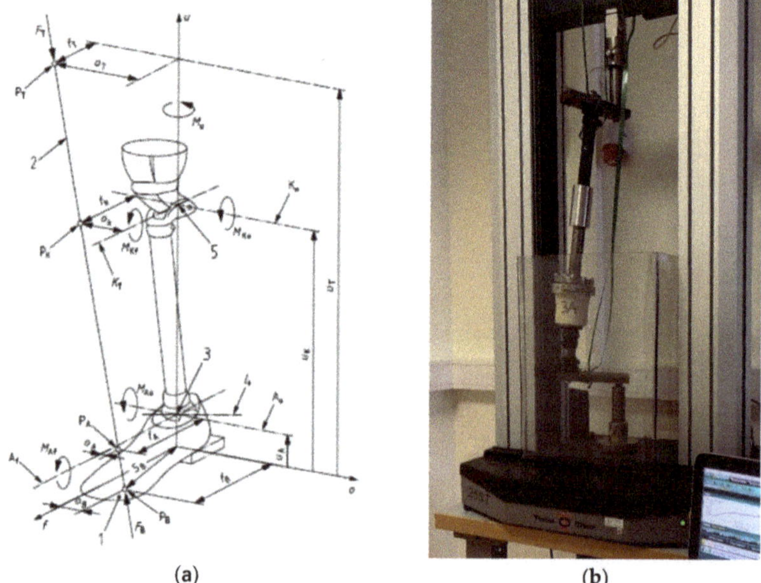

(a)　　　　(b)

Figure 3. Structural testing method for lower limb prostheses (ISO 10328:2016). (**a**) Illustration of loading condition II; (**b**) Test equipment with the test piece loaded.

2.4. User Attitude

Initially, a study-specific survey was constructed to inquire about the users' attitude for transferring production of orthotic and prosthetic devices to a more sustainable material. The users were asked how likely it was that they would choose a product produced with a biocomposite rather than fossil-based thermoplastics if it meant a compromise in design, color, or quality. The survey also asked about potential benefits or risks of using the new material, and what products the users would like to see made from this material. Outpatients visiting the department of Prosthetics and Orthotic, Örebro University Hospital, Örebro, Sweden, answered the waiting-room survey. The demonstrator was tested by a patient with a transtibial (lower limb, below-the-knee) prosthesis. The patient was 70 years old, a female with 52 years of experience using a prosthetic limb every day. The outcome was reported by the Satisfaction with Prosthesis subscales of the Trinity Amputation and Prosthesis Experience Scales, Revised (TAPES-R) [28,29]. These subscales have 3 (aesthetic) and 5 (functional) subscale items, scored on a 3-level rating scale in which a high score is positive, indicating satisfaction with the prosthesis. The patient responded to the questionnaire before and after wearing the demonstrator for a comparison with the conventional prosthesis.

3. Results

3.1. Mechanical Properties of the Biocomposites

To evaluate the characteristics of the MFC composite materials to be used for FDM, standard test specimens were injection molded, followed by tensile testing. The distribution of cellulose material in the PP matrix was examined by microtomography. Table 2 and Figure 4 show that the tensile strength (σ_m) and the tensile moduli (E_t) were significantly improved by the addition of MFC to the composites. Six samples were tested for each data point. Figure 4 shows that the ultimate tensile strength increases significantly with increasing MFC content. The tensile strength is about twice as much with 30 wt% MFC as compared with neat PP. The improvement in tensile strength becomes less significant with additional increases in MFC content. This could be because a higher fiber content resulted in more fiber damage and shorter fiber lengths. Fu and colleagues [30] showed that the mean glass and carbon fiber lengths in their PP composites after injection molding decreased with increasing fiber volume fractions. As a result, the reduction in mean fiber length offset the reinforcement effect by increasing fiber volume fraction. In addition, the E_t modulus values of the composites monotonically increased with MFC content. This was also observed by Fu et al. [30], for example, the modulus for both types of composites increased dramatically with increased fiber volume fraction. This indicated that the composite modulus depended on MFC volume fraction rather than length. Nevertheless, one should bear in mind that these composites, in Table 2, were produced in different batches with different raw materials and processing settings. Therefore, the differences in their mechanical properties are a collective reflection of several parameters rather than purely MFC content, and the comparison thus becomes indicative.

Table 2. Mechanical properties of the composite injection-molded test specimens.

MFC Content (wt%)	Tensile Strength σ_m (MPa)	Tensile Modulus E_t (MPa)	Elongation at Break ε_b (%)	U_N [kJ/m^2]
0	24.00 ± 0.81	1715 ± 97	>20	7.18 ± 0.40
20	32.10 ± 0.15	3112 ± 99	5.9 ± 0,7	3.79 ± 0.12
30	39.57 ± 0.24	3896 ± 57	4.5 ± 0,2	4.14 ± 0.14
40	40.81 ± 0,29	4525 ± 94	3.8 ± 0,2	3.41 ± 0.47

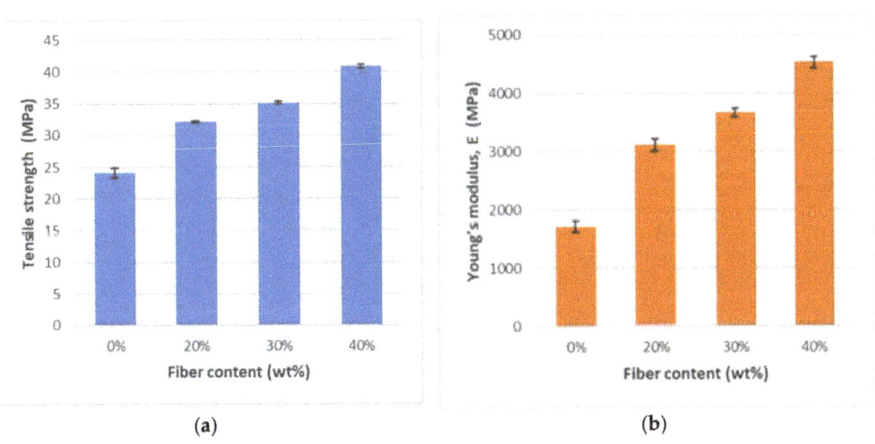

Figure 4. Relationships of the composites' mechanical properties by microfibrillated cellulose (MFC) content. (**a**) Ultimate tensile strength; (**b**) Tensile (Young's) modulus.

The elongation at break (tensile strain, ε_b) of the composites decreases with increasing MFC content, indicating more brittle-like fracture of the biocomposites as compared with neat PP. The impact strength was measured using the Charpy notch test method in which a notch was made in each test sample. Table 2 and Figure 5 show that the Charpy impact strength decreases with the addition of MFC, which is often the case due to reduced elasticity and flexibility with increased reinforcement as compared with neat PP. Similar observations have been reported by several authors [31]. The explanations were that the fibers induced a fracture change from ductile to brittle and increase fiber agglomeration, which also resulted in high-stress regions and non-uniform stress transfer.

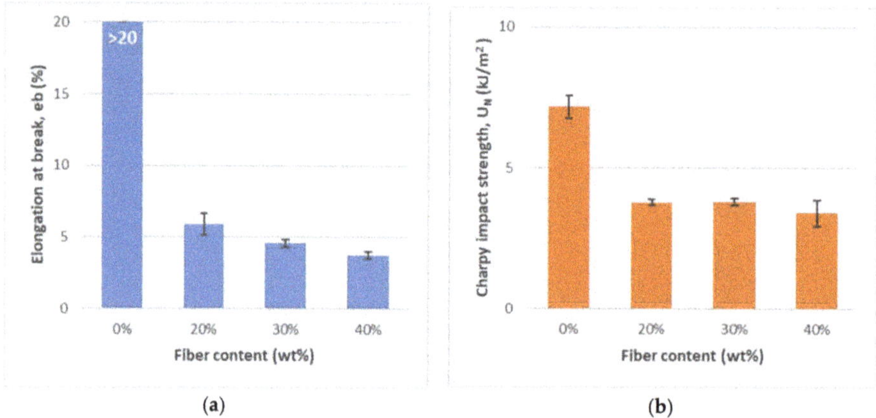

Figure 5. Relationships of the composites' mechanical properties by MFC content. (**a**)Elongation at break (%); (**b**) Charpy impact strength (kJ/m2).

The aspect ratio (length/diameter) of wood pulp fibers usually ranges between 44 and 75, depending on the processing method [32], whereas the aspect ratio of MFC can be up to several hundred due to its small diameter (<100 nm) and long fibril length (μm), depending on the fabrication method [33,34]. The higher the aspect ratio, the higher the reinforcement capacity when incorporated in composite materials [35]. The MFC, however, can consist of several inhomogeneous material components, for example, fibers, fiber fragments, fines, and fibrils [4]. A detailed analysis of the fiber dispersion is given in the next subsection.

Standard composite test specimens consisting of 20 wt% MFC were also created with the FDM technique. The mechanical properties of these test specimens are shown in Table 3. Three samples were tested for each data point. The tensile strength and modulus of the FDM test specimens built with longitudinal lines are approximately 68% of the injection molded values shown in Table 2. It was reported by Lay et al. [36] that the mechanical strengths of the FDM-built parts of three polymers, i.e., PLA (Polylactic acid), ABS (Acrylonitrile butadiene styrene), and nylon 6, were 48, 34 and 37% lower than their counterparts fabricated with injection moulding. In our case, the mechanical strength of the FDM built parts is about 32% lower than the injection moulding.

Table 3. Mechanical properties of the FDM-printed test specimens with the 20 wt% MFC content.

Sample	Tensile Strength σ_m (MPa)	Tensile Modulus E_t (MPa)	Elongation at Break ε_b (%)
Longitudinal print lines	21.96 ± 0.7	2100 ± 73	4.1 ± 1.3
Perpendicular print lines	16.11 ± 2.1	1885 ± 222	1.7 ± 0.4

It is well known that adding fibers in PP significantly increases the shear viscosity, and thus influences the processing behavior (flowability) of material deposition during the FDM process. Figure 6 shows that the biocomposites made of PP/MFC have a strong shear thinning behavior, resulting in a much lower viscosity and resistance to flow at high shear rates. It was also observed in this study that the processing temperature had a significant influence on the dynamic viscosity, which was similar to most thermoplastics.

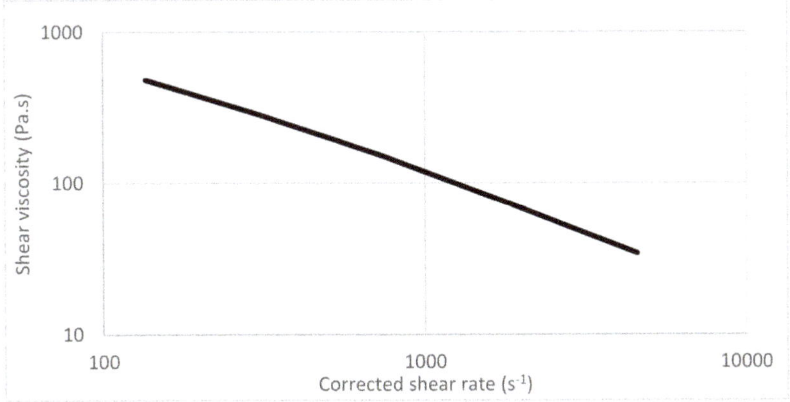

Figure 6. The dynamic shear viscosity versus corrected shear rate for the biocomposite polypropylene (PP)/MFC 70:30 measured with a capillary rheometer at 190 °C.

3.2. Fiber Dispersion and Pull-out for the Biocomposites

The uniform dispersion of MFC additives is crucial to the mechanical strength of composite materials. As stated in Section 2.1, the following two main strategies were followed to improve both the compatibility and the dispersion of MFC in the PP matrix: (I) An adapted compounding process which facilitates the dispersion of MFC material components in the polymer matrix and II) the use of a functional coupling agent, which enhances the compatibility between the hydrophobic polymer phase and the hydrophilic MFC components.

Figures 7 and 8 show that the cellulose material is uniformly dispersed in the PP matrix. However, single MFC fibrils are not visible in these images due to limited resolution. Much of the reinforcement effect observed in the tensile strength and Young's modulus can be attributed to MFC with a large L/D ratio. Another important factor is that a larger fiber specific surface area, such as in MFC, promotes binding of the fibers with the polymer matrices, resulting in improved mechanical strength.

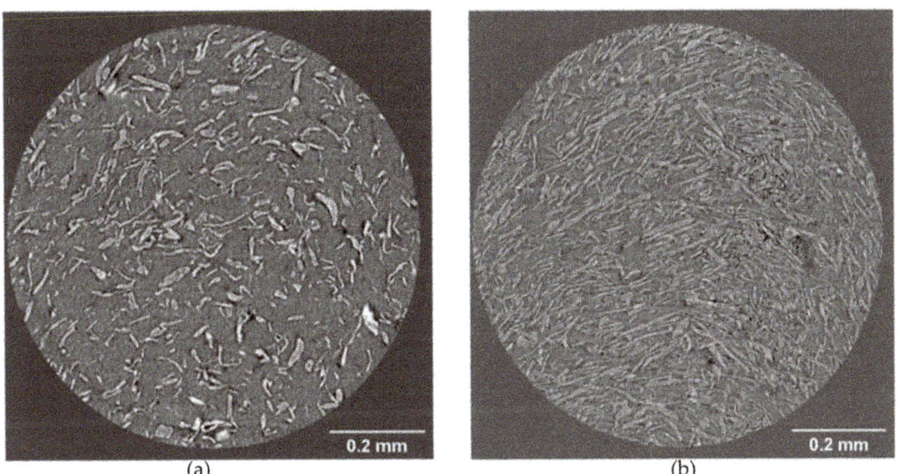

Figure 7. X-ray microtomographic images of the composite cross-section (x-y plane). (**a**) PP/MFC 80:20; (**b**) PP/MFC 60:40. The magnification is 20×.

Figure 7 shows that no cellulose material agglomerates are visible in the x-y cross-sections, even though they often occur in biocomposites due to strong fiber–fiber respective fibril–fibril interactions and the hydrophilic character of the MFC components, in contrast to the hydrophobic nature of polypropylene. Agglomeration results in regions where less energy is required to begin a crack [31].

Figure 7 also shows black spots or holes in the cross-sections in the images, which most likely originate from moisture. MFC holds water strongly, and it is difficult to obtain a completely dried cellulose material in the PP matrix, although it might be achievable in up-scaled production facilities. These holes occur randomly for all MFC composites and are visible in microtomography in both the x-y and y-z planes. They can affect the mechanical properties because they can form cracks.

The fiber-matrix interface property also affects the composite's mechanical behavior, of which the adhesion between fiber and polymer matrix is the key [37]. One common way to observe the interface effect is through a fiber pull-out investigation, which consists of a debonding process followed by a pull-out process. Since MFC contains fibers and fiber fragments, therefore, pull-out testing is possible. In this work, samples from a Charpy notch test were investigated, using an SEM technique, shown in Figure 8. As shown in the images, only a few sites could be connected to either fiber pull-out or gas formation in processing. The latter originated from the low water content in the hydrophilic fibers, as was also observed in the x-ray microtomographic images in Figure 7. Because these sites are few and well-separated from one another, they are not expected to be the major cause for the reduction in impact strength. In fact, impact strength between the biocomposites and the neat PP is difficult to compare since the materials have such different ductilities. The neat PP is very ductile and it is difficult for a crack to propagate, while the biocomposites are much stronger and much more brittle, allowing cracks to easily propagate inside the material. The crack propagation characteristics can, to some extent, be circumvented by using impact modifiers according to Thomason et al. [31], but this is beyond the scope of this work.

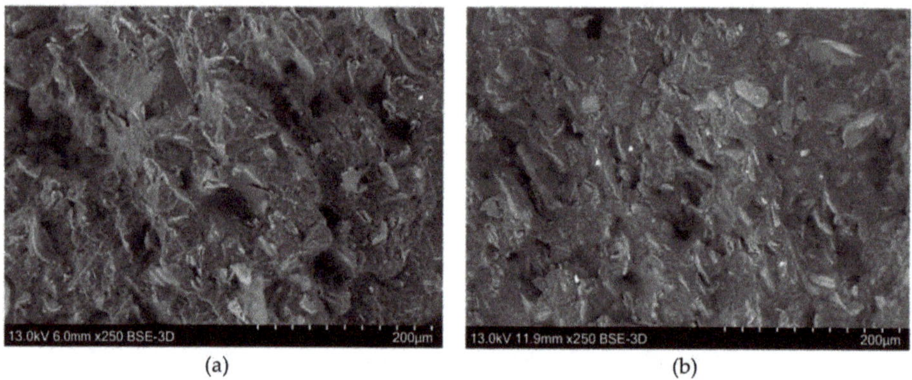

Figure 8. SEM images to identify fiber pull-out in cross-sections of injection-molded test specimens after the Charpy notch test of the composites. (**a**) PP/MFC 80:20; (**b**) PP/MFC 60:40, magnification 250×.

3.3. Mechanical Performance of the Test Socket

Polypropylene (PP) is a commonly used material for prosthetists and orthotists. Table 2 shows the test results of the new material with polypropylene (PP) as a matrix and 20% MFC made from chemical pulp fiber produced a tensile strength of 32.10 ± 0.15 MPa, more than 30% higher than that of PP. This material was used for additive manufacturing based on the FDM process.

Figure 9 shows the test result (loading history) of the FDM-manufactured socket with the biocomposite containing 20% MFC. The socket was broken at 2896 N, equivalent to 12.9 MPa. The test socket's mechanical performance was evaluated according to ISO 10328:2016. Loading condition II was applied using P6 test geometry, as shown in Figure 3a. This means that the socket was simultaneously subjected to compression and off-axis bending and also torsion along several axes, which simulates the worst-possible load situation during normal use of a prosthetic lower limb. Hence, the strength obtained from this test is not directly comparable with the tensile strength listed in Tables 2 and 3.

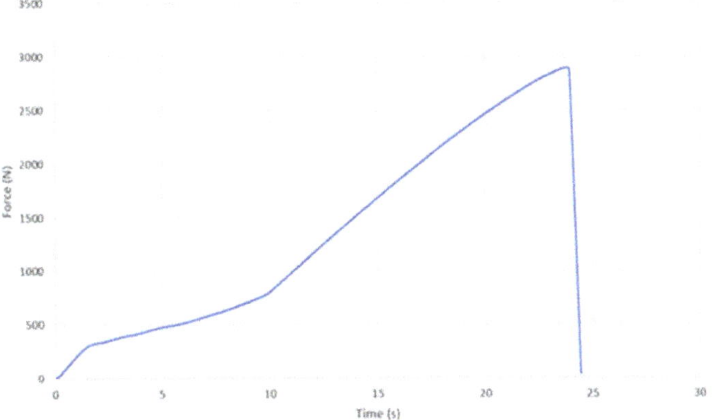

Figure 9. FDM-manufactured socket loading test curve in the sagittal plane. The testing method is illustrated in Figure 3.

Three sockets were tested at the end of this work. These sockets were manufactured using the FDM technique with the same composite material (PP/MFC 80:20) but in different build orientations, i.e., the transverse plane, the coronal (frontal) plane, and the sagittal plane. The test results are listed in Table 4, wherein "transverse plane" indicates the deposition layer-by-layer is in the transverse plane and so forth. The force at break (the strength of the test socket) is highly dependent on the build orientation. The strength of the socket built in the sagittal plane is twice that of the transverse plane. The underlying explanation for why the strength of the sockets built in the coronal plane and the sagittal plane was much higher than that of the sockets built in the transverse plane is that when printed in these directions, the MFC reinforcement was better realized. This reveals tremendous FDM process impacts on the final mechanical performance. The strength would be even higher if we could improve adhesion between the layers.

Table 4. Test results of the FDM-manufactured sockets with the same material (PP and 20 wt% MFC), but in different printing directions.

Build Orientation	Force at Break (N)	Strength (MPa)
Transverse plane	1343 ± 82	6.03 ± 0.27
Coronal (frontal) plane	2385	10.7
Sagittal plane	2896	12.9

Every test socket broke at the distal end where there was an internally sharp corner, and the load was the highest according to FEM simulation [25]. The sharp corner is a starting point for the fracture failure and the specific load application produces axial compression, shear forces, bending moments, and torque as a result of the load vector in load condition II. Therefore, the strength (MPa), listed in Table 4, is a combined strength caused by the force and the load geometry. The load geometry and the force that a prosthetic limb shall sustain according to ISO 10328:2016 comes from real-life measurements on prosthetic lower limb users. This makes it impossible to directly compare the strength of the material in the socket form (Table 4) with the tensile strengths obtained from the injection molded test samples (Table 2).

3.4. Analysis of Consumers' Attitude towards Biocomposites

The consumers (n = 27) reported that they most likely would choose a product made from biocomposite over a conventional plastic product unless it meant having to compromise quality (Figure 10a). The acceptance towards biocomposites was not affected by compromising in design or color, only 8% of the respondents were likely to avoid these products (Figure 10b). They reported no or low fear of risks related to the new material, and instead were interested in the environmental advantages such as biodegradability and recycling. When asked what products the consumers would like to see made from this new material, they responded "all parts that today are made from plastics". Example suggestions included all sorts of orthoses, i.e., insoles, corsets, and breast prostheses. The reasons stated included the environmental benefits, and also that these products came in close contact with the skin and were used quite extensively.

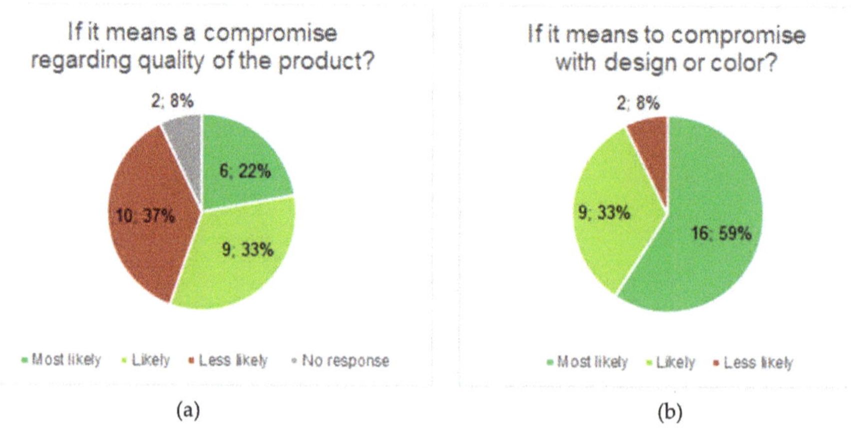

(a)   (b)

Figure 10. Consumers' attitudes towards more sustainable materials in orthotic and prosthetic applications. (**a**) Quality aspect; (**b**) Design and color.

3.5. Patient and Clinician Experiences with the 3D-Printed Prosthesis

Preparation of the product prior to client fitting showed that the biocomposite material was easy to cut with a jigsaw and easy to smooth the surface by machine grinding. Spontaneous comments from technicians were that working with the material in preparation for the socket created a pleasant wood smell. The socket demonstrated sufficient form stability to provide the correct pressure distribution in the socket and enough overall stability to make the prosthesis useable for gait in a lightweight female.

The patient reported that wearing the demonstrator felt no different from wearing the conventional prosthesis (Figure 11a). However, according to the patient, the demonstrator had a better color; and she would have chosen this demonstrator over a conventional prothesis had she had the opportunity to choose. This was reflected by the patient's score on the TAPES-R, where the overall score on the aesthetic satisfaction subscale was improved from six to seven; the satisfaction with appearance item score increased from two = satisfied to three = very satisfied; whereas on the functional satisfaction subscale the overall score was the same (10 = satisfied) before and after wearing the demonstrator. Moreover, despite the actual increase in weight (the conventional laminated socket without adaptors weighs 148 grams versus 393 grams for the experimental socket (the demonstrator)), the patient reported a sense of decreased weight of the demonstrator as compared with the conventional prosthesis during walking (Figure 11b).

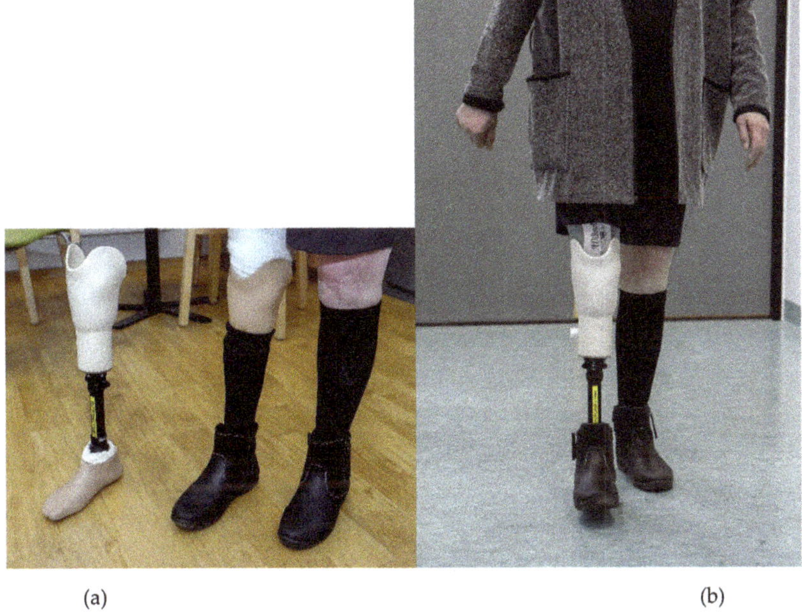

(a)   (b)

Figure 11. The demonstrator tested by a 70-year-old female with 52 years of experience using a transtibial prosthesis. (**a**) The patient wearing her conventional prosthesis, with the demonstrator to the left for comparison; (**b**) The patient walking with the demonstrator.

4. Discussion

From a material point of view, biocomposites offer a renewable and sustainable material alternative for AM-processed applications to traditionally used thermoplastic or thermosetting materials. However, biocomposites are new materials for this type of processing and there is still much to understand and develop in regard to MFC dispersion in the polymer matrix, MFC aspect ratio, MFC size and orientation, and interfacial strength between the MFC and the polymer, etc. Although the MFC materials clearly reinforce the polymer matrix, such as PP, it is very hard to transfer this reinforcement potential between layers in 3D printing. Thus, it is very important to choose the build orientation to maximize the effect of MFC reinforcement, as shown in Table 4. For the same reason, 3D printing processing conditions are very important for the dynamic behavior of the material in the FDM process, layer-layer adhesion, and the final properties of the built parts. In this study, PP was selected as the polymer matrix because it has not commonly been used in 3D printing but has been used extensively in the injection molding industry due to its beneficial mechanical properties in relation to its cost. It was observed that the MFC improved the melt-flow properties of PP composite and therefore enabled its use in 3D-printing. In the future, PP from renewable sources, which were not available when the project was started, could be used to further strengthen the sustainability aspect.

The overall mechanical strength of the additively manufactured object (AMO) depends on the composite material and also on the structural characteristics of the printed objects. The former determines the upper limit of the mechanical strength achievable with this material, whereas the latter determines the actual strength achieved with the AM technique. The overall mechanical strength of the AMO is defined by the weakest link in its structure. Hence, it is particularly important to identify the limiting factors related to the AM techniques.

There are several factors that can affect the overall mechanical strength of AMOs [13]. Voids and porosity in the bulk structure are known to be responsible for the variation in the mechanical strength of SLS processes [24]. They also exist in structures created by FDM, as the top and bottom filaments typically do not attach perfectly and bond to each other, forming air pockets and porous structures with large gaps between the strands [36]. Figure 12 clearly shows the voids at the start and end positions of each layer and even between the layers. Voids and pores are of particular concern to the mechanical strength when their locations are at or close to the AMO's surface. Variation in the effective binding area between adjacent material layers can also cause deteriorated mechanical performance. The area variation could have resulted from the dimension accuracy related to the precision of the print-head's movement in FDM, morphology of the underlying layer, the temperature gradient between the layers, etc. Moreover, there could have been a systematic area reduction due to the spatial gradient of the 3D model, as illustrated in Figure 13 (marked by red rings). This type of reduction is particularly severe with a thick building layer. Therefore, the overall mechanical strength can be improved using hardware and software improvements to reduce or even eliminate random and systematic binding-area variations. In addition, the mechanical strength of AMO can be further improved by testing different slicing, temperature, or printing speed settings, or in different ambient temperatures, etc. [13].

Figure 12. Illustration of defects (voids and pores) in an FDM structure.

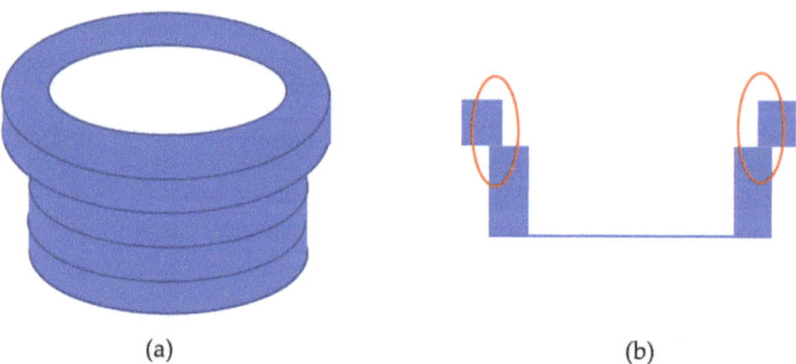

(a)   (b)

Figure 13. Illustration of reduced binding area between adjacent layers due to geometrical gradient. (**a**) The full structure; (**b**) The cross-section.

Additive manufacturing can potentially enable paradigm changes in the prosthetic value chain from design to terminal devices. The changes benefit patients as follows: They can receive their prosthesis faster; the digital prosthesis is easy to store and retrieve, enabling production of multiple devices from the same scan; and the digitalized prosthesis is easy to visualize, therefore, a patient can see what the prosthesis should look like prior to manufacturing. These lead to a more personal solution in which the patient's concerns and priorities can be addressed. There are many other benefits. With AM, material waste is reduced to a minimum, which together with biobased composites, reduces the imprint of the prosthesis on the environment. The consumers' view of the environmental aspects of production in this study showed that this is something that the market should consider.

Another important aspect of prosthetics is cost. The cost limits access to prosthetic devices in many parts of the world [38]. Costs relate to both production of the customized sockets and the finalizing of the product using prefabricated devices such as joints, feet, and hands. Very little attention has been given to research on these aspects of prosthetics. Biddiss et al. [38] discussed the implications of modular designs and rapid prototyping along with computer-aided design and manufacturing on the cost of prosthetic components. A review of 3D-printed upper-limb prostheses supported this suggestion [15], showing that the maximum material cost was $500. Another aspect of prosthetics, related to cost but rarely studied, is the implication for the patient's quality of life. There is a need for more high-quality research that reflects the effectiveness of different prosthesis interventions in terms of users' quality of life [39].

Although the perspective of combining MFC-reinforced composite with AM is promising for prosthetic applications, there are also challenges. From the prosthetic application point of view, the material looks very promising regarding the way it failed, however, it is unclear whether it is the polypropylene (PP) as matrix, the 20% MFC content, or both together that produces a more ductile break. Due to the layer-by-layer nature, MFC in the composites are less likely to over-bridge different layers. It is very important that the adhesion between each layer is optimal. To maximize the reinforcement effects and to obtain the best material tensile strength, process direction is of crucial importance. This is particularly true for FDM processes. Potentially, the stiffness and durability of a prosthetic socket made from this material could be improved by varying the thickness of the socket and making, for example, a honeycomb structure, especially in the distal parts of the socket. This is one of the advantages of the 3D printing technique. Future development should improve the adaptors to optimize transitioning into and out of the socket. Moreover, cost and education are required to introduce this procedure into industrial applications, both regarding the software and the hardware. Manufacturing with AM requires 3D scanning of the patient, data modulation of the 3D scan, designing the prosthesis in 3D CAD, additive manufacturing, mounting, adjusting, and finalizing the prosthesis.

There are also potential limitations in this study. The initial test of the demonstrator produced in this project showed promising results. However, the product was tested with only one type of suspension, the distal lock-pin attachment, and other systems could function differently with this material and production technique. The sockets produced with this new material and by FDM may not be impermeable to air. Therefore, future testing is required with other types of suspensions, for example, the sleeve-and-valve vacuum system or sealed-in suspension systems. In addition to the dynamic testing, cyclic mechanical testing is also important for prosthetic applications, which need to be complemented in the future. Moisture susceptibility, water absorption, as well as friction and wear behavior [40] could potentially be important aspects which have not yet been examined. Due to the hydrophilic nature of MFC, the composite can take up moisture in a humid environment and make the composite susceptible to microbial growth [41,42]. Espert et al. reported that when immersed in water, the composite's water absorption followed Fick's law and the mechanical properties were severely affected [43]. Modifying the cellulosic material to make them hydrophobic or using a cap layer are possible solutions [43].

5. Conclusions

Composite materials made of polypropylene reinforced by MFC were investigated. MFC materials (20, 30 and 40 wt%) were uniformly dispersed in the polymer PP matrix, and the mechanical performance of the materials was significantly enhanced by the addition of MFC. The ultimate tensile strength of the composite was about twice that of the PP when the MFC content was 30 wt%, while its Young's modulus more than doubled. The composite material with 20 wt% of MFC was successfully applied in an AM process using an FDM-based technique; a transtibial prosthesis was created based on the end-user's data. A clinic trial of the prosthesis was conducted with successful outcomes for wearing (walking) experiences, appearance (color), and acceptance of the materials and the technique. The 30 and 40% MFC composites have higher viscosity, therefore, their application requires a more powerful extruder, beyond the limit of the FDM machine used in this work.

The AMOs created with the FDM technique and the MFC-reinforced material exhibited strong dependence on build orientation. To utilize the full range of mechanical strength offered by the materials, structural and process optimizations are needed to transfer the strength of the material to the AMO's strength. Due to the layer-by-layer nature, MFC material components in the composites are less likely to over-bridge different layers, which leads to deteriorated tensile strength in the cross direction. Structural and process arrangements must be adapted to the terminal application to maximize the reinforcement effects of MFC, to eliminate variations in binding area between adjacent layers, and to improve adhesion between layers to create robust, durable prostheses using AM techniques.

This study finds that combining biocomposites with 3D printing offers a promising future for prostheses and orthosis solutions. However, further developments in both material and AM technology are needed to enable the industrial sector to achieve sustainability improvements as a result of using renewable-sourced 3D-printed materials.

Author Contributions: Conceptualization, G.F., K.H., L.H., H.P., E.S., and L.Y.; methodology, G.F., K.H., L.H., H.P., E.S., M.W., and L.Y.; validation, K.H., H.P., and M.W.; investigation, K.H, H.P., and M.W.; resources, E.S.; writing—original draft preparation, L.Y.; writing—review and editing, G.F., K.H., L.H., E.S., M.W., and L.Y.; visualization, K.H. and LY; supervision, L.Y.; project administration, L.Y.; funding acquisition, G.F., L.H., and L.Y. All authors have read and agreed to the published version of the manuscript.

Funding: This work was financially supported by national research agencies, Vinnova, Formas, and Energimyndigheten, through BioInnovation, a Strategic Innovation Program, through AMPOFORM and BioComp-PPS projects.

Acknowledgments: Fredrik Adås is acknowledged for his help with the microtomography of the test specimens. Anders Reiman is acknowledged for his help with scanning electron microscopy of the test specimens. We thank Fabian Rosén, CPO, for his contribution in the initial phase of the project.

Conflicts of Interest: The authors declare no conflict of interest.

References

1. Commission, E. BioBasedEconomy. 2012. Available online: https://www.biobasedeconomy.eu/policy/ (accessed on 1 August 2020).
2. Commission, E. Circular Economy. 2018. Available online: https://ec.europa.eu/environment/circular-economy/index_en.htm (accessed on 1 August 2020).
3. Abdul Khalil, H.P.S.; Bhat, A.H.; Ireana Yusra, A.F. Green composites from sustainable cellulose nanofibrils: A review. *Carbohydr. Polym.* **2012**, *87*, 963–979. [CrossRef]
4. Chinga-Carrasco, G. Cellulose fibres, nanofibrils and microfibrils: The morphological sequence of MFC components from a plant physiology and fibre technology point of view. *Nanoscale Res. Lett.* **2011**, *6*, 417–422. [CrossRef]
5. Miller, J. Nanocellulose State of the Industry. 2015. Available online: https://www.tappinano.org/media/1114/cellulose-nanomaterials-production-state-of-the-industry-dec-2015.pdf (accessed on 1 August 2020).
6. Suzuki, K.; Okumura, H.; Nakagaito, A.N.; Yano, H.H. Development of continuous process enabling nanofibrillation of pulp and melt compounding. *Cellulose* **2013**, *20*, 201–210. [CrossRef]

7. Xiao, X.L.; Chevali, V.S.; Song, P.G.; He, D.N.; Wang, H. Polylactide/hemp hurd biocomposites as sustainable 3D printing feedstock. *Compos. Sci. Technol.* **2019**, *184*, 107887. [CrossRef]

8. Song, P.A.; Yang, H.T.; Fu, S.Y.; Wu, Q.; Ye, J.W.; Lu, F.Z.; Jin, Y.M. Effect of carbon nanotubes on the mechanical properties of polypropylene/wood flour composites: Reinforcement mechanism. *J. Macromol. Sci. Part B* **2011**, *50*, 907–921. [CrossRef]

9. Fu, S.; Yang, H.; Jin, Y.; Lu, F.; Ye, J.; Wu, Q. Effects of carbon nanotubes and its functionalization on the thermal and flammability properties of polypropylene/wood flour composites. *J. Mater. Sci.* **2010**, *45*, 3520–3528. [CrossRef]

10. Parandoush, P.; Lin, D. A review on additive manufacturing of polymer-fiber composites. *Compos. Struct.* **2017**, *182*, 36–53. [CrossRef]

11. Prosthetics And Orthotics Market Size, Share & Trends Analysis Report By Type Orthotics (Upper Limb, Lower Limb, Spinal), Prosthetics (Upper Extremity, Lower Extremity), And Segment Forecasts, 2020–2027. 2020. Available online: https://www.grandviewresearch.com/industry-analysis/prosthetics-orthotics-market (accessed on 1 June 2020).

12. 3D Printing Technologies: An Overview. Available online: https://www.techpats.com/3d-printing-technologies-overview/ (accessed on 1 June 2020).

13. Dey, A.; Yodo, N. A Systematic survey of FDM process parameter optimization and their influence on part characteristics. *J. Manuf. Mater. Process* **2019**, *3*, 64–93. [CrossRef]

14. Barrios-Muriel, J.; Romero-Sánchez, F.; Alonso-Sánchez, F.J.; Rodríguez Salgado, D. Advances in Orthotic and prosthetic manufacturing: A technology review. *Materials* **2020**, *13*, 295. [CrossRef]

15. Ten Kate, J.; Smit, G.; Breedveld, P. 3D-printed upper limb prostheses: A review. *Disabil. Rehabil. Assist. Technol.* **2017**, *12*, 300–314. [CrossRef]

16. Herbert, N.; Simpson, D.; Spence, W.D.; Ion, W. A preliminary investigation into the development of 3-D printing of prosthetic sockets. *J. Rehabil. Res. Dev.* **2005**, *42*, 141–146. [CrossRef] [PubMed]

17. Zuniga, J.; Katsavelis, D.; Peck, J.; Stollberg, J.; Petrykowski, M.; Carson, A.; Fernandez, C. Cyborg beast: A low-cost 3d-printed prosthetic hand for children with upper-limb differences. *BMC Res. Notes* **2015**, *8*, 10. [CrossRef] [PubMed]

18. Ribeiro, D.; Cimino, S.R.; Mayo, A.L.; Ratto, M.; Hitzig, S.L. 3D printing and amputation: A scoping review. *Disabil. Rehabil. Assist. Technol.* **2019**. [CrossRef]

19. Goldstein, T.; Oreste, A.; Hutnick, G.; Chory, A.; Chehata, V.; Seldin, J.; Gallo, M.D.; Bloom, O. A pilot study testing a novel 3D printed amphibious lower limbprosthesis in a recreational pool setting. *PMRJ* **2020**, *12*, 783–793.

20. Abdelaal, O.; Darwish, S.; Elmougoud, K.A.; Aldahash, S. A new methodology for design and manufacturing of a customized silicone partial foot prosthesis using indirect additive manufacturing. *Int. J. Artif. Organs.* **2019**, *42*, 645–657. [CrossRef] [PubMed]

21. Fey, N.P.; Klute, G.K.; Neptune, R.R. The influence of energy storage and return foot stiffness on walking mechanics and muscle activity in below-knee amputees. *Clin. Biomech.* **2011**, *26*, 1025–1032. [CrossRef]

22. Hsu, L.H.; Huang, G.F.; Lu, C.T.; Hong, D.Y.; Liu, S.H. The development of a rapid prototyping prosthetic socket coated with a resin layer for transtibial amputees. *Prosthet. Orthot. Int.* **2010**, *34*, 37–45. [CrossRef]

23. Nguyen, K.T.; Benabou, L.; Alfayad, S. Systematic Review of Prosthetic Socket Fabrication using 3D printing. In Proceedings of the ACM 4th International Conference on Mechatronics and Robotics Engineering, Valenciennes, France, 7–11 February 2018.

24. Flodberg, G.; Pettersson, H.; Yang, L. Pore analysis and mechanical performance of selective laser sintered objects. *Addit. Manuf.* **2018**, *24*, 307–315. [CrossRef]

25. Lindberg, A.; Alfthan, J.; Pettersson, H.; Flodberg, G.; Yang, L. Mechanical performance of polymer powder bed fused objects – FEM simulation and verification. *Addit. Manuf.* **2018**, *24*, 577–586. [CrossRef]

26. ISO 10328:2016 Prosthetics—Structural Testing of Lower—Limb Prostheses—Requirements and Test Methods. 2016. Available online: https://www.iso.org/standard/70205.html (accessed on 1 June 2020).

27. Ali, M.A.; Umer, R.; Khan, K.A.; Cantwell, W.J. Application of X-ray computed tomography for the virtual permeability prediction of fiber reinforcements for liquid composite molding processes: A review. *Compos. Sci. Technol.* **2019**, *184*, 107828. [CrossRef]

28. Gallagher, P.; MacLachlan, M. Development and psychometric evaluation of the Trinity Amputation and Prosthesis Experience Scales (TAPES). *Rehabil. Psychol.* **2000**, *45*, 130–154. [CrossRef]

29. Gallagher, P.; Franchignoni, F.; Giordano, A.; MacLachlan, M. Trinity amputation and prosthesis experience scales: A psychometricssessment using classical Test Theory And Rasch Analysis (TAPES). *Am. J. Phys. Med. Reh.* **2010**, *89*, 487–496. [CrossRef] [PubMed]

30. Fua, S.F.; Lauke, B.; Mäder, E.; Yue, C.Y.; Hu, X. Tensile properties of short-glass-fibre- and short-carbon-fibre-reinforced polypropylene composites. *Compos. Part A Appl. Sci. Manuf.* **2000**, *31*, 1117–1125. [CrossRef]

31. Thomason, J.; Rudeiros-Fernández, J.L. A review of the impact performance of natural fiber thermoplastic composites. *Front. Mater.* **2018**, *5*, 1–18. [CrossRef]

32. Sjöholm, E.; Berthold, F.; Gamstedt, E.K.; Neagu, C.; Lindstrom, M. The Use of Conventional Pulped Wood Fibres as Reinforcement in Composites. In *Sustainable Natural and Polymeric Composites: Science and Technology*; Riso National Laboratory: Roskilde, Denmark, 2002.

33. Ankerfors, M. Microfibrillated cellulose: Energy-efficient preparation techniques and key properties in Department of Fibre and Polymer Technology. In *School of Chemical Science and Engineering*; KTH Royal Institute of Technology: Stockholm, Sweden, 2012.

34. Chen, Y.; Fan, D.B.; Han, Y.M.; Li, G.Y.; Wang, S.Q. Length-controlled cellulose nanofibrils produced using enzyme pretreatment and grinding. *Cellulose* **2017**, *24*, 5431–5442. [CrossRef]

35. Börjesson, M.; Westman, G. *Crystalline nanocellulose—Preparation, Modification, and Properties in Cellulose—Fundamental Aspects and Current Trends*; Poletto, M., Ornaghi Junior, H.L., Eds.; IntechOpen: London, UK, 2015; Available online: https://www.intechopen.com/books/cellulose-fundamental-aspects-and-current-trends/crystalline-nanocellulose-preparation-modification-and-properties (accessed on 1 April 2020).

36. Lay, M.; Thajudin, N.-L.N.; Hamid, Z.A.A.; Rusli, A.; Abdullah, M.K.; Shuib, R.K. Comparison of physical and mechanical properties of PLA, ABS and nylon 6 fabricated using fused deposition modeling and injection molding. *Compos. Part B: Eng.* **2019**, *176*, 107341. [CrossRef]

37. Wells, J.K.; Beaumont, P.W.R. Debonding and pull-out processes in fibrous composites. *J. Mater. Sci.* **1985**, *20*, 1275–1284. [CrossRef]

38. Biddiss, E.; McKeever, P.; Lindsay, S.; Chau, T. Implications of prosthesis funding structures on the use of prostheses: Experiences of individuals with upper limb absence. *Prosthet. Orthot. Int.* **2011**, *35*, 215–224. [CrossRef]

39. Samuelsson, K.; Töytäri, T.; Salminen, A.L.; Brandt, Å. Effects of lower limb prosthesis on activity, participation, and quality of life: A systematic review. *Prosthet. Orthot. Int.* **2018**, *36*, 145–158. [CrossRef]

40. Liu, L.; Wang, Z.Y.; Yu, Y.M.; Fu, S.Y.; Nie, Y.J.; Wang, H.; Song, P.G. Engineering interfaces toward high-performance polypropylene/coir fiber biocomposites with enhanced friction and wear behavior. *ACS Sustain. Chem. Eng.* **2019**, *7*, 18453–18462. [CrossRef]

41. Jin, S.; Matuana, L.M. Wood/plastic composites co-extruded with multi-walled carbon nanotube-filled rigid poly(vinyl chloride) cap layer. *Polym. Int.* **2009**, *59*, 648–657. [CrossRef]

42. Stark, N.M.; Gardner, D.J. 7—Outdoor durability of wood–polymer composites. In *Wood-Polymer Composites*; Woodhead Publishing: Shaston, UK, 2008; pp. 142–165.

43. Espert, A.; Vilaplana, F.; Karlsson, S. Comparison of water absorption in natural cellulosic fibres from wood and one-year crops in polypropylene composites and its influence on their mechanical properties. *Compos. Part A: Appl. Sci. Manuf.* **2004**, *35*, 1267–1276. [CrossRef]

 *bioengineering*

Article

3D Printing of Gelled and Cross-Linked Cellulose Solutions; an Exploration of Printing Parameters and Gel Behaviour

Tim Huber [1,2,*] **, Hossein Najaf Zadeh [2,3], Sean Feast [2,4], Thea Roughan [1] and Conan Fee [1,2]**

[1]   School of Product Design, University of Canterbury, Private Bag 4800, Christchurch 8020, New Zealand; tlr50@uclive.ac.nz (T.R.); conan.fee@canterbury.ac.nz (C.F.)

[2]   Biomolecular Interaction Centre, University of Canterbury, Private Bag 4800, Christchurch 8020, New Zealand; hossein.najafzadeh@pg.canterbury.ac.nz (H.N.Z.); sean.feast@pg.canterbury.ac.nz (S.F.)

[3]   Department of Mechanical Engineering, University of Canterbury, Private Bag 4800, Christchurch 8020, New Zealand

[4]   Department of Chemical and Process Engineering, University of Canterbury, Private Bag 4800, Christchurch 8020, New Zealand

*   Correspondence: tim.huber@canterbury.ac.nz

Received: 19 February 2020; Accepted: 25 March 2020; Published: 27 March 2020

Abstract: In recent years, 3D printing has enabled the fabrication of complex designs, with low-cost customization and an ever-increasing range of materials. Yet, these abilities have also created an enormous challenge in optimizing a large number of process parameters, especially in the 3D printing of swellable, non-toxic, biocompatible and biodegradable materials, so-called bio-ink materials. In this work, a cellulose gel, made out of aqueous solutions of cellulose, sodium hydroxide and urea, was used to demonstrate the formation of a shear thinning bio-ink material necessary for an extrusion-based 3D printing. After analysing the shear thinning behaviour of the cellulose gel by rheometry a Design of Experiments (DoE) was applied to optimize the 3D bioprinter settings for printing the cellulose gel. The optimum print settings were then used to print a human ear shape, without a need for support material. The results clearly indicate that the found settings allow the printing of more complex parts with high-fidelity. This confirms the capability of the applied method to 3D print a newly developed bio-ink material.

Keywords: bioprinting; cellulose; hydrogel; physical cross-linking

1. Introduction

3D printing, often also referred to as additive manufacturing, of biopolymers has attracted growing attention within the last few years. A large and increasing amount of 3D printing techniques is being developed, of which a large fraction fall within the categories of inkjet 3D printing, stereolithography printing and extrusion based 3D printing [1,2]. Currently, the most used polymers in 3D printing are polylactic acid (PLA), acrylonitrile butadiene styrene (ABS) and nylon, but a growing number of new materials have become available to be used on 3D printers, and both consumers and producers are developing printing materials based on natural polymers, driven by a need for more sustainable practices [3]. A comprehensive overview of 3D printing technology and available materials is beyond the scope of this work but is available for example in the work of [1] or [3].

Bioprinting as a subsection of 3D printing has also seen rapid and astonishing developments, and has been a major contributor towards the development of 3D printable natural materials and biopolymers, mostly in the form of hydrogels. In this case, the motivation for material development stems from a need for swellable, non-toxic, biocompatible and possible biodegradable materials also known as bio-ink materials, as the main application of bioprinting has been in tissue engineering and regenerative medicine [4–7]. A key aspect of bio-ink material development is the so-called biofabrication window, an attempt to describe the compromises that need to be made to achieve the highest print fidelity without compromising the ability of the printed material to function as a tissue scaffold [8,9]. This window is not only specific to the used bio-ink material but also depends on the used hardware including the used print setting, for example extrusion pressure, print speed or line height and width [10,11].

Polysaccharides such as alginate, hyaluronic acid or chitosan and proteins in the form of collagen and gelatine have been of particular interest and have been successfully printed into complex three-dimensional structures [12]. Most bioprinters rely on extrusion-based 3D printing, meaning the printing material needs to undergo a phase change from liquid to solid during the printing process. In bioprinting this often involves either a gelation process of a solution or, more commonly, thixotropic of a gel [13]. Cellulose is the most abundant biopolymer in the world and has been demonstrated to be a suitable polymer for applications in, for example, tissue engineering, filtration or separation [14]. However, cellulose in its native form is not water soluble and thus has proven challenging to process on bioprinters, either requiring the use of nanocellulose to form thixotropic and thus extrudable gels or suitable solvents to create cellulose solutions within a viscosity range that allows extrusion [15].

Only a few aqueous and non-aqueous solvents have been proven to be suitable for the processing of cellulose, but for most, the viscosity of the resulting solutions is not suitable for bioprinting or the solutions require a solvent exchange process to solidify the cellulose out of the solvent. Cellulose dissolved in the ionic liquid 1-ethyl-3-methylimidazolium acetate exhibited a shear thinning behaviour that made it suitable for extrusion printing [15] and a solution of cellulose in N-methylmorpholine-N-oxide (NMMO) has been successfully extruded on a bioplotter [16]. However, no other non-derivative cellulose solution has been reported to be printable by similar methods and thus little is known about the biofabrication window of cellulose.

Cellulose can be readily dissolved in aqueous solutions of sodium hydroxide (NaOH) and urea, and while this solvent system is often described as attractive due its relatively low cost and low-toxicity, solubility is limited to no more than 6–8 wt.% depending on the type of cellulose used [17]. However, both, the solution viscosity and the properties of cellulose hydrogels made from this solvent can be adjusted by introducing additional amounts of undissolved cellulose particles as a physical cross-linker [18]. Cellulose solutions based on NaOH and urea also show a unique, irreversible gelation behaviour when heated [19,20]. This behaviour has been explored previously to 3D print non-derivatized cellulose using a focused laser beam to locally heat the cellulose solution beyond its gelation point [21]. While there is a significant amount of work published on the cellulose solution, little is known about the behaviour of the gelled solution [17]. In this work we will explore if the cellulose gel can be 3D printed using an extrusion based bioprinter to create three-dimensional objects and determine print related parameters of the biofabrication window.

2. Materials and Methods

2.1. Chemicals

NaOH (99% purity) was purchased in pellets from The Sourcery (Christchurch, NZ). Urea (BioUltra, purity of 99.5%) and cellulose powder (Sigmacell (type 20, 20 μm) were purchased from Sigma-Aldrich (St. Louis, MO, USA). All chemicals were used as received.

2.2. Preparation of Cellulose Formulation

For each print session, deionized water (81 wt.%), urea (12 wt.%) and NaOH (7 wt.%) were measured and mixed using a Velp Scientifica (Usmate Velate, Italy) DLS overhead stirrer at 100 rpm [17,19,20,22]. After mixing for approximately 5 min, the mixture was heated to 50 °C under light stirring on a hot plate (Heidolph Instruments. Schwabach, Germany). After heating,

20 wt.% of cellulose was added to solvent mixture and stirred using a Velp Scientifica DLS overhead stirrer on 100–120 rpm for 2 minutes. The formulation was then cooled in a freezer set to −18 °C for 90 minutes to dissolve the cellulose. Subsequently, the solution was stored in a refrigerator at 5 °C until used for printing or rheometry measurements. This formulation was decided upon from previous trial and error experimentation, altering the weight percentage of the cellulose added to solution. 15, 20 and 25 wt.% cellulose solutions were prepared and extruded along a line from a 10 mL disposable syringe with 0.425 mm diameter and 0.5 mm length metal tip. The extruded gels were graded based on how well they held their shape and how easily they were extruded. From this basic test 20 wt.% cellulose was decided upon for further investigation.

2.3. Rheometry

Oscillatory measurements and rotational measurements were carried out using the Anton Paar Modular Compact Rheometer 302 (Anton Paar, Graz, Austria). For both measurements, the measuring system used was a parallel plate 50 (PP50) and the cell used was a Peltier plate. The temperature was set at 20 °C for all measurements. In total, 12 tests were carried out for the oscillatory measurements and 5 tests were carried out for the rotational tests. Each test was conducted with a cellulose formulation made no older than 5 days before being tested.

2.4. Oscillatory Measurements

For the oscillatory measurements, 4 different angular frequencies were tested 3 times each, the frequencies were 0.5 rad/s, 5 rad/s, 50 rad/s and 500 rad/s with a shear ramp of 0.1–1000%. For each test there were 41 data points plotted. Once the fresh sample was measured, the same sample was then sheared at constant shear strain of 100%, at 50rad/s for an interval of 120s. After the sample was sheared, the initial settings were used to measure the sheared sample.

2.5. Rotational Measurements

1 test was conducted and repeated 5 times with a shear rate of 0.001–1000 1/s which plotted 91 data points with no set duration. The fresh sample was measured and then sheared using a constant shear rate of 100 1/s at a constant interval of 120 s. Once the sample was sheared, the settings from the initial measurements were used to measure the sheared sample.

2.6. Printing Parameters

A Taguchi L9 orthogonal array [23] was used to reach the optimum 3D printing setting of cellulose gel with the minimum number of trials at a minimum cost. The optimum level of the process could be gained by applying a standard analysis. Since the aim of the experiments was to 3D print the nearly cubic shape of samples, a standard analysis was chosen to be *"nominal is the best"*. The experiment was conducted with multi-response results.

Analysis of variance was done on the mean and Signal to Noise (S/N) ratio (Equation (1)). The collected measurements from the experiments were analysed using Minitab software (Minitab LLC, State College, PA, USA). The two important factors studied based on the preliminary tests were print pressure and print speed. The trialled settings are displayed in Table 1. The printing acceleration was kept constant to 4 [mm/s^2] and printing line-width and line-height were set to 0.8 mm. The structure of Taguchi's orthogonal robust design and the results of the measurement are shown in Table 2. The measured parameters were top surface area [mm^2], side height [mm], side width [mm] and side view angles [°].

$$S/N = -10 \times \log s^2 \tag{1}$$

where s is the standard deviation of the responses for all noise factors for the given factor level combination.

Table 1. List of the conducted print trials at varying settings for extrusion pressure and print speed.

Trial	Pressure [psi]	Speed [mm/s]	Trial	Pressure [psi]	Speed [mm/s]
1	8	3	10	14	3
2	8	3	11	14	3
3	8	3	12	14	3
4	8	5	13	14	5
5	8	5	14	14	5
6	8	5	15	14	5
7	8	6	16	14	6
8	8	6	17	14	6
9	8	6	18	14	6
19	20	3	23	20	5
20	20	3	24	20	5
21	20	3	25	20	6
22	20	5	26	20	6
			27	20	6

Table 2. Results of the L9 orthogonal array tests and measurements.

Pressure [psi]	Speed [mm/s]	Top Area [mm^2]	Side Height [mm]	Side Width [mm]	Angle 1 [°]	Angle 2 [°]
8	3	80.75	8.63	9.38	121.75	96.34
8	5	72.65	7.26	7.86	113.58	107.66
8	6	60.12	6.89	7.34	110.01	114.15
14	3	86.24	8.92	9.44	92.12	95.79
14	5	87.69	9.63	9.34	101.60	102.16
14	6	74.45	8.39	9.86	116.77	95.59
20	3	156.09	11.40	11.97	135.07	131.64
20	5	143.216	11.362	12.384	127.77	123.962
20	6	141.6	11.005	12.496	130.728	137.068
	Max	156.09	11.4	12.49	135.07	137.06
	Min	60.12	6.89	7.34	92.12	95.59

Cellulose formulations were prepared and loaded into a 16 mm diameter, 92 mm length barrel then centrifuged in an Eppendorf Centrifuge 5430 (Eppendorf, Hamburg, Germany) at 3000 rpm for 3 minutes to remove any air bubbles. After centrifugation, the barrel was then fitted with a 0.425 mm diameter and 0.5 mm length metal tip. The barrel was loaded into the Advanced Solutions Biobot (Life Sciences–Advanced Solutions, Louisville, KY, USA). For each trial, a cube with a side length of 10 mm was printed. All printing trials were carried out using Tissue Structure Information Modeling (TSIM®) software (Life Sciences– Advanced Solutions).

After a cube was printed, it was removed from the print stage and placed in 60 mL of deionized water for 24 hours to regenerate the cellulose. The water was changed regularly until full regeneration and solvent removal had occurred, as checked by measuring the pH of the water. After regeneration, the cubes were stored in a closed bottle of 60 mL deionized water and a spray of ethanol.

After the 27 samples were completed, pictures were taken of the top view and side view of each cube using a Toupcam UCMOS01300KPA camera and ToupView software (Touptek, Zheijang, China) attached to a Nikon SMZ-1B confocal microscope (Nikon, Tokyo, Japan) for analysing. ImageJ Fiji software (National Institutes of Health, Bethesda, MD, USA) was used to analyse the cubes. The top surface area, width and length of the side view and all 4 angles of both top view and side view were measured and recorded.

3. Results

3.1. Rheometry

To understand the behaviour of the created gels under shear, the viscosity of the gels was measured over a large range of shear stresses, to simulate stresses that occur during extrusion printing (Figure 1).

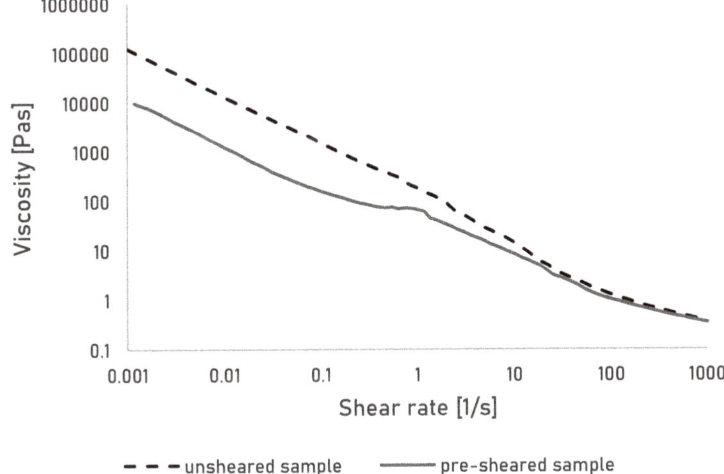

Figure 1. Viscosity of the cellulose gel measured over shear rate. Shown are results of a gel in unsheared and presheared state.

The gels show a continuous shear thinning behaviour, demonstrating a continuous alignment of the polymer chain in the gel. A behaviour that is also seen in ungelled solutions of cellulose [24]. Cellulose gels made from aqueous solutions of NaOH and urea are typically formed by the disruption of a so-called inclusion complex upon heating. In the solution state, the inclusion complex has been described as a sleevelike layer of solvent that surrounds the polymer chains and prevents interchain hydrogen-boding. When heated, those complexes appear to break down, allowing new hydrogen bonds to form resulting in rapid and irreversible gelation of the solution [25,26]. The added cellulose powder acts as a physical cross-linker promoting further hydrogen bonding and in turn stronger gels [18]. The shear thinning behaviour can be explained by the constant breakage and reformation of those hydrogen bonds between aligning polymer chains and the suspended cellulose particles. Similar behaviour was for example seen and deemed preferable for a bio-ink material based on carrageenan and suspended nanosilicates [27].

It is assumed that upon removal of the shear stress, a new gel will be formed and no permanent damage to the gel has occurred. This was further investigated by shearing the sample at a constant rate, followed by a shear sweep (Figure 1). While the initial viscosity of the presheared gel is lower than the original gel, likely due to chain alignment, the shear thinning behaviour is identical in the presheared gel, meaning no permanent damage to the gel has occurred. In turn, this means the viscosity of the gels can be easily modified to allow printing over a large range of shear stresses.

To further investigate the effect of shear on the gel properties, oscillatory measurement over a range of shear strain were performed at different angular frequencies on unsheared and presheared samples (Figure 2). The sample showed a viscoelastic behaviour with the storage modulus G′ being larger than the loss modulus G″ by approximately a factor of 10 at strains below approximately 10% for both tested gels as exemplarily shown for a frequency of 0.5 rad/s. At higher strains the gels start to yield leading to a cross over G′ and G″ at higher strains. This yielding behaviour is common for colloidal gels and indicates initial gel rupture followed by a further break down of gel fragments into finely dispersed particles that allow the gel to flow [28]. An overall reduction between of G′ and G″ is observable between unsheared and presheared gels, as well as a reduction of the loss factor (the ratio of G′: G″) by approximately 50% in the viscoelastic range. This supports the assumption that upon shearing, the gel is broken down into fine fragments, but upon removal of the strain, it is able to form a new, secondary gel through the extensively available hydrogen groups in the dissolved and suspended portions of cellulose, as well as interparticle forces between the dispersed cellulose particles and potentially gel fragments [29]. This in turn means that although the gel is likely to yield under the shear stresses during printing, the formation of a secondary gel will still allow a rapid solidification of the gel fragments into a secondary gel, and thus, allow for an extrusion of the used cellulose gels.

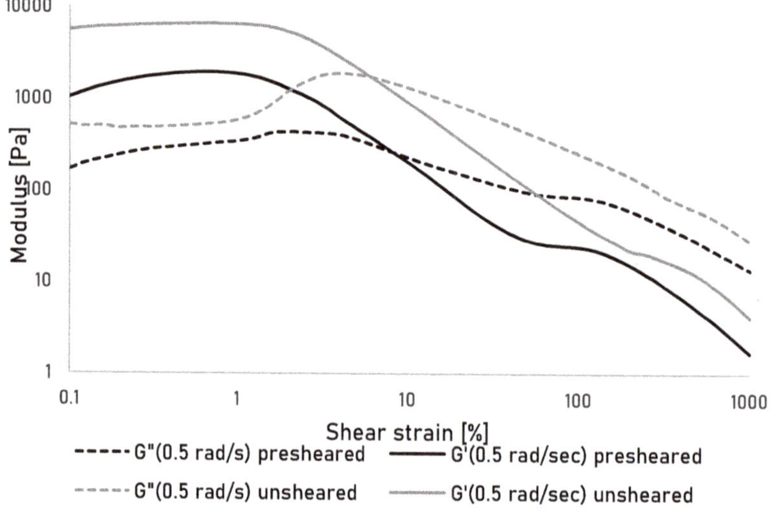

Figure 2. Storage (G′) and Loss (G″) modulus of unsheared and presheared cellulose gels measured over shear strain at an angular frequency of 0.5 rad/sec.

There appears to be a small frequency dependence of the secondary gel regarding the cross-over point of G′ and G″ (Figure 3) which can possibly be explained by a frequency dependence of the inter-particle interactions [28]. However, given the complexity of interaction in the tested systems, further studies that go beyond exploration of the printability of the used cellulose gels will be necessary to understand those interactions with more certainty.

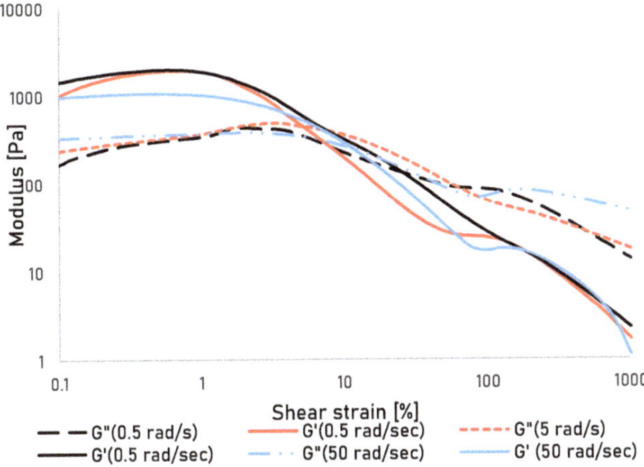

Figure 3. Storage (G′) and Loss (G″) modulus of a presheared cellulose gel over shear strain measured at angular frequencies of 0.5, 5 and 50 rad/sec.

3.2. Taguchi Analysis of Printing Parameters

Not all used combination of printing factors yielded printed cubes suitable for an analysis of print fidelity. Overall, the variance of printing factors created prints with a large range of desirable and poor print fidelity (Figure 4).

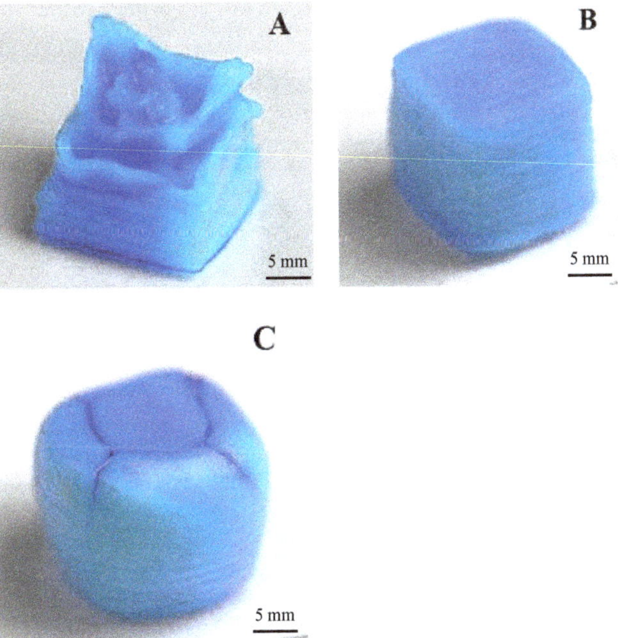

Figure 4. Examples of printed gel cubes with an approximate side length of 10 mm. Shown are a cube printed with insufficient extrusion of gel (**A**), overflowing amounts of extruded gel (**C**) and close to ideal print settings (**B**). Cubes have been dyed blue using food colouring for better contrast in the displayed photos.

The best 3D printed gel cube should have a flat surface with the closest surface area to 10 mm^2, corners of 90°, side view width of 10 mm, side view height of 10 mm. By changing the print settings, the printed samples top area varied between 156.09 mm^2 to 60.12 mm^2 (Table 2).

The analysis of variance for S/N ratio (Table 3) and analysis of variance for means (Table 4) show that the pressure had a significant effect on 3D printing process with a contribution percentage of 99.9%, whereas speed has the minimum effect on the print quality with 25.7% contribution to the print quality. The R^2 was calculated to 96.12%.

Table 3. Analysis of variance for SN ratios.

Source	DOF	Sum of Square	Variance	F-ratio	P
Pressure	2	16.2566	16.2566	61.89	0.001
Speed	2	0.0839	0.0839	0.32	0.743
Residual Error	4	0.5253	0.5253		
Total	8	16.8659			

Table 4. Analysis of variance for means.

Source	DOF	Sum of Square	Variance	F-ratio	P
Pressure	2	1295.73	1295.73	100.88	0.000
Speed	2	2.98	2.98	0.23	0.803
Residual Error	4	25.69	25.69		
Total	8	1324.39			

The F-ratio displayed in Table 2 represents the confidence in the collected data. Experimentation of the F-ratio in Tables 3 and 4 show the control factor 'pressure' with an F-ratio value of 61.89 and 100.88, for S/N ratio and variance of means respectively, have a very low experimental error.

The results from Taguchi analysis tests (mean and S/N ratio) are shown in Figures 5 and 6. A higher S/N ratio, measured in decibels (dB), is preferred because a high value of S/N implies that the signal is much higher than the uncontrollable noise parameters, such as inconsistency in the data measurement.

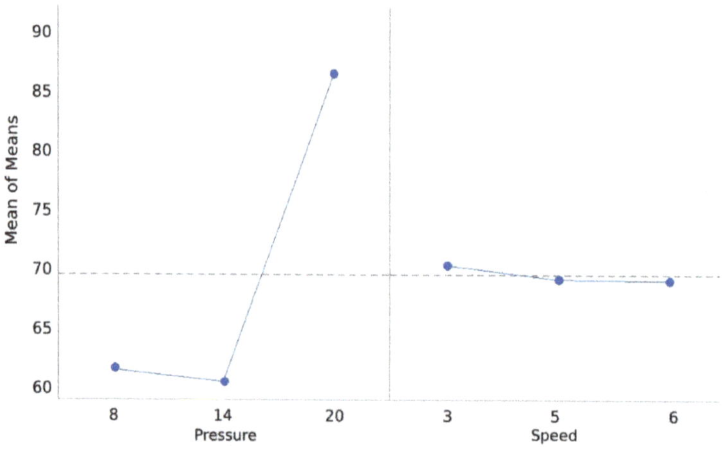

Figure 5. Main effect of the plot for mean values.

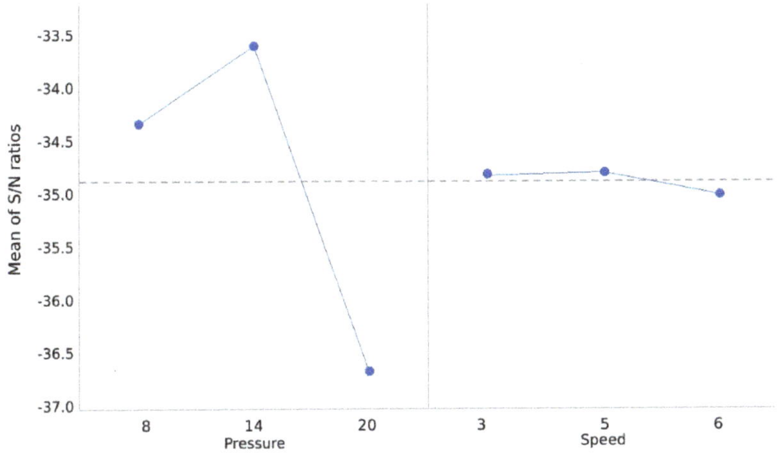

Figure 6. Main effects plot for S/N ratios.

The main effect graph is plotted by determining the means for each value of a categorical variable. The analysis of the mean value shows that pressure has a significant effect on 3D print quality. However, the response mean for speed is almost the same across all parameter levels (Table 5).

Table 5. Response table for S/N ratio.

Level	Pressure	Speed
1	−34.33	−34.82
2	−33.57	−34.80
3	−36.72	−35.01
Delta	3.15	0.21
Rank	1	2

Based on the analysed results shown in Table 4, printing pressure has the highest impact on 3D printing a high-quality part. Speed has the lowest parameter impact on print quality. Thus, unlike trial-and-error experiments that implicitly rely on the experimenter's judgment, Taguchi design of experiment has shown the best combination of controllable parameter levels. The best setting combination from the analysis of the results for 3D printing an object can be seen from the main effect of the plot for mean values (Figure 5) and the main effects plot for S/N ratio (Figure 6). These settings include pressure of 14 psi and speed of 5 mm/s.

To validate the print settings found as optimal a complex three-dimensional shape in the form of an ear was printed including overhanging structures, typically difficult to achieve without the use of support material. The results (Figure 7) clearly indicate the found settings allow the printing of more complex parts with high-fidelity, and thus, confirm the success of the applied method to find optimal print settings for a newly developed bio-ink material.

Figure 7. Test prints of a model of a human ear using optimal settings for extrusion pressure and print speed derived from the Taguchi analysis of the print parameters. (**A**) End view of test print; (**B**) Side view of test print.

4. Conclusions

An investigation into the printability of a novel cellulose-based bio-gel formed from dissolving excess cellulose in a solution of urea and sodium hydroxide has been conducted. Rheological testing has demonstrated that the gel has shown two important properties needed by all extrusion-based inks. The gel is shear thinning and additional to this, forms of a secondary gel structure post shearing, allowing the gel to regain most of its structural integrity. The gel can therefore be easily extruded during printing, while post extrusion retaining its shape. The secondary gel formed may be weaker due to the initial yielding, alignment of polymer chains and dispersion of the gel fragments. This could impact the overall resolution of the printed gel due to slumping or reduced structural integrity of the gel. Further analysis into the gel's behaviour is required to fully understand this complex system. Analyses of some of the many variables involved in bioprinting by the Taguchi DoE method allowed for optimization of the gel's printability. Alteration of pressure was shown to have the greatest effect on the outcome of the print. This is not surprising as the extrusion rate is directly related to the pressure i.e., at higher pressure, more gel is extruded. Therefore, finer tuning of the pressure setting will be the most beneficial to improving the print quality. A more detailed study involving other variables, such as nozzle diameter and acceleration, may further improve printing accuracy and resolution. Optimal settings for pressure and speed were found and used to print a complex organic shape (human ear). The printing of this structure shows the cellulose gel was able to handle slight overhangs without support material, while maintaining shape fidelity.

Prior to the development of this work cellulose was only able to be printed using solutions of ionic liquids or organic compounds [15,30]. This new bio-ink material offers a cheaper less toxic alternative to bioprinting using one of the most abundant materials found on earth. Overall, the printability of this new bio-ink material is incredibly promising as an inexpensive alternative to current cellulose based bio-ink materials. As with all new bio-ink materials, optimization of not only the formulation but also printer settings are required to ensure the best possible print. Here, only one formulation was tested; however, many variations are possible and will be tested in future works.

Author Contributions: Conceptualization, T.H.; methodology, T.H., H.N.Z., S.F.; validation, T.H., H.N.Z., S.F. and T.R.; formal analysis, T.H. and H.N.Z.; investigation, T.H., H.N.Z., S.F. and T.R.; data curation, T.H. and H.N.Z.; writing—original draft preparation, T.H., H.N.Z., S.F. and T.R.; writing—review and editing, T.H. and S.F.; supervision, T.H. and C.F.; project administration, C.F.; funding acquisition, C.F., T.H. All authors have read and agreed to the published version of the manuscript.

Funding: This research was partially funded by the Summer Research Scholarship program, grant number 2019-27 and through the New Zealand Ministry of Business, Innovation and Employment Endeavour Fund, project title: "3-D printed porous media for process engineering".

Acknowledgments: The authors acknowledge the financial support for T.R. through the University of Canterbury Summer Scholarship program. The authors are grateful for the technical assistance provided by Khoa Tran and

Garrick Thorne, and kindly acknowledge access to the rheometer provided by Associate Professor Ken Morrison of the Chemical and Process Engineering department at the University of Canterbury.

Conflicts of Interest: The authors declare no conflict of interest.

References

1. Bikas, H.; Stavropoulos, P.; Chryssolouris, G. Additive manufacturing methods and modelling approaches: A critical review. *Int. J. Adv. Manuf. Technol.* **2016**, *83*, 389–405. [CrossRef]
2. Bhushan, B.; Caspers, M. An overview of additive manufacturing (3D printing) for microfabrication. *Microsyst. Technol.* **2017**, *23*, 1117–1124. [CrossRef]
3. Ngo, T.D.; Kashani, A.; Imbalzano, G.; Nguyen, K.T.; Hui, D. Additive manufacturing (3D printing): A review of materials, methods, applications and challenges. *Compos. Part B Eng.* **2018**, *143*, 172–196. [CrossRef]
4. Ozbolat, I.; Gudapati, H. A review on design for bioprinting. *Bioprinting* **2016**, *3*, 1–14. [CrossRef]
5. Gudupati, H.; Dey, M.; Ozbolat, I. A Comprehensive Review on Droplet-based Bioprinting: Past, Present and Future. *Biomaterials* **2016**, *102*, 20–42. [CrossRef] [PubMed]
6. Jian, H.; Wang, M.; Wang, S.; Wang, A.; Bai, S. 3D bioprinting for cell culture and tissue fabrication. *Bio-Des. Manuf.* **2018**, *1*, 45–61. [CrossRef]
7. Richards, D.; Jia, J.; Yost, M.; Markwald, R.; Mei, Y. 3D bioprinting for vascularized tissue fabrication. *Ann. Biomed. Eng.* **2017**, *45*, 132–147. [CrossRef]
8. Yin, J.; Zhao, D.; Liu, J. Trends on physical understanding of bioink printability. *Bio-Des. Manuf.* **2019**, *2*, 50–54. [CrossRef]
9. Gillispie, G.; Prim, P.; Copus, J.; Fisher, J.; Mikos, A.G.; Yoo, J.J.; Atala, A.; Lee, S.J. Assessment methodologies for extrusion-based bioink printability. *Biofabrication* **2020**, *12*, 022003. [CrossRef]
10. Chimene, D.; Lennox, K.K.; Kaunas, R.R.; Gaharwar, A.K. Advanced bioinks for 3D printing: A materials science perspective. *Ann. Biomed. Eng.* **2016**, *44*, 2090–2102. [CrossRef]
11. Gungor-Ozkerim, P.S.; Inci, I.; Zhang, Y.S.; Khademhosseini, A.; Dokmeci, M.R. Bioinks for 3D bioprinting: An overview. *Biomater. Sci.* **2018**, *6*, 915–946. [CrossRef]
12. Arslan-Yildiz, A.; El Assal, R.; Chen, P.; Guven, S.; Inci, F.; Demirci, U. Towards artificial tissue models: Past, present, and future of 3D bioprinting. *Biofabrication* **2016**, *8*, 014103. [CrossRef] [PubMed]
13. Zhang, B.; Luo, Y.; Ma, L.; Gao, L.; Li, Y.; Xue, Q.; Yang, H.; Cui, Z. 3D bioprinting: An emerging technology full of opportunities and challenges. *Bio-Des. Manuf.* **2018**, *1*, 2–13. [CrossRef]
14. Kang, H.; Liu, R.; Huang, Y. Cellulose-Based Gels. *Macromol. Chem. Phys.* **2016**, *217*, 1322–1334. [CrossRef]
15. Wang, Q.; Sun, J.; Yao, Q.; Ji, C.; Liu, J.; Zhu, Q. 3D printing with cellulose materials. *Cellulose* **2018**, *25*, 4275–4301.
16. Li, L.; Zhu, Y.; Yang, J. 3D Bioprinting of Cellulose with Controlled Porous Structures from NMMO. *Mater. Lett.* **2018**, *210*, 136–138. [CrossRef]
17. Budtova, T.; Navard, P. Cellulose in NaOH-water based solvents: A review. *Cellulose* **2016**, *23*, 5–55.
18. Huber, T.; Feast, S.; Dimartino, S.; Cen, W.; Fee, C. Analysis of the Effect of Processing Conditions on Physical Properties of Thermally Set Cellulose Hydrogels. *Materials* **2019**, *12*, 1066. [CrossRef]
19. Qin, X.; Lu, A.; Zhang, L. Gelation behavior of cellulose in NaOH/urea aqueous system via cross-linking. *Cellulose* **2013**, *20*, 1669–1677. [CrossRef]
20. Cai, J.; Zhang, L. Unique gelation behavior of cellulose in NaOH/urea aqueous solution. *Biomacromolecules* **2006**, *7*, 183–189. [CrossRef]
21. Huber, T.; Clucas, D.; Vilmay, M.; Pupkes, B.; Stuart, J.; Dimartino, S.; Fee, C. 3D Printing Cellulose Hydrogels Using LASER Induced Thermal Gelation. *J. Manuf. Mater. Process.* **2018**, *2*, 42. [CrossRef]
22. Cai, J.; Zhang, L. Rapid dissolution of cellulose in LiOH/urea and NaOH/urea aqueous solutions. *Macromol. Biosci.* **2005**, *5*, 539–548. [CrossRef]
23. Montgomery, D.C. *Design and Analysis of Experiments*; Wiley: Hoboken, NJ, USA, 1984.
24. Huber, T.; Starling, K.; Cen, W.S.; Fee, C.; Dimartino, S. Effect of Urea Concentration on the Viscosity and Thermal Stability of Aqueous NaOH/Urea Cellulose Solutions. *J. Polym.* **2016**, *2016*, 1–9. [CrossRef]
25. Qin, X.; Lu, A.; Cai, J.; Zhang, L. Stability of inclusion complex formed by cellulose in NaOH/urea aqueous solution at low temperature. *Carbohydr. Polym.* **2013**, *92*, 1315–1320. [CrossRef] [PubMed]

are the most important representative among a wide array of bioinks. A large variety of natural polymers commonly used in bioink formulation exemplify collagen, gelatin, and hyaluronic acid in the animal-derived resource catalogue, chitosan and alginate in the marine-derived resource catalogue, and polysaccharides derived from various plant resources [3]. As highlighted in the most recent years, nanocelluloses have been established as a renewable constituent in formulating bioinks for hydrogel-extrusion 3D bioprinting [5,6]. Within the scope of biomedical hydrogels for tissue engineering, the exploiting interests on nanocelluloses are mainly aroused by their structural similarity to extracellular matrices (ECM) in terms of both porosity and interconnect framework within the structural hydrogel and fibrous topography of cellulose fibrils somehow analogous to collagen and fibronectin in native ECM, as well as in terms of their excellent biocompatibility of supporting crucial cellular activities, as suggested in a great number of in vitro and in vivo studies [5,6]. Above all, the printability of these biomaterial systems in extrusion-based 3D printing is, in principle, supported by the shear-thinning properties of the nanocellulose hydrogels.

In recent years, our research group has been active in the development of woody nanocellulose-based inks for 3D bioprinting, mainly in the context of biomedical hydrogels for soft tissue engineering applications. Based on the knowledge gathered in this rather specific but cross-disciplinary field, we intend to present the key research aspects within this mini-review, from the preparation of different types of nanocellulose and their respective material properties, to important perspectives in the hydrogel extrusion-based 3D printing processes to consider, and to a discussion on integrating therapeutically relevant functionalities into the nanocellulose matrix from the viewpoint of cell–matrix interactions and delivery of bioactive cues. Lastly, we also attempt to address the challenges that the nanocellulose-based bioinks still face in in vivo applications presently. This review aims for a confined readership among colleagues developing nanocelluloses into various advanced functional materials and researchers from different disciplines of biomedical engineering who are interested in engaging nanocellulose-based inks in their 3D bioprinting approaches. In addition, we acknowledge a few other up-to-date reviews that have also extensively summarised the usage of cellulose and its derivatives in a wider range of scenarios, e.g., the selection of cellulosic materials, the adapted 3D printing techniques, and the underlined applications, to which readers are referred for potential interests [6–8].

2. Nanocelluloses: Origin, Preparation, and Material Properties on Nano-Scale

In plants, cellulose in the form of para-crystalline microfibrils, as well as nanofibrils, comprises the main load-bearing polymer, which is cross-linked with other macromolecules such as hetero-polysaccharides, lignin, and proteins [9]. The resulting composite confers both strength and flexibility to the plant structure. Keeping in mind the context where a high strength may be offered by the para-crystalline microfibrils and nanofibrils, a large quantity of research work has been laid on the isolation of those fibrils and applying them in various value-added applications [10–12]. Different groups of such cellulose nanostructures are thus in focus but with a consensus term of "nanocellulose", defined as the diameter of the resulted fibril products in one dimension nanoscale. Nanocellulose refers to nanomaterials of three catalogues: bacterial nanocellulose (BNC), cellulose nanofibrils (CNFs), and cellulose nanocrystals (CNCs). CNFs are often noted as nanofibrillated cellulose (NFC) and microfibrillated cellulose (MFC) by researchers in different fields. CNCs referred to nanocrystalline cellulose (NCC) or cellulose nanowhiskers (CNWs) in earlier times. Both CNFs and CNCs can be isolated from biomass resources using top-down approaches that break down the interfibrillated bonds by mechanical disintegration alone or in combination with acidic, enzymatic, or chemical oxidations. However, to note, CNFs contain both crystalline and non-crystalline regions in the fibers, whereas CNCs are generally prepared by hydrolysing the non-crystalline region in acid treatment, thus leaving the crystalline regions. In comparison to CNFs and CNCs, BNCs are prepared in a bottom-up approach via the biosynthesis of cellulose that takes place in a microbial culture through oxidative fermentation. In recent years, cellulosic nanomaterials have shown great potential in biomedical applications owing to their intrinsic characteristics, such as non-cytotoxicity and biocompatibility, high-aspect-ratio material features, strong mechanical properties, and broaden capability for chemical modifications. Furthermore, there is no need to mention the renewable and sustainable nature of nanocelluloses from vast natural resources.

2.1. Bacterial Nanocellulose (BNC)

BNC was first discovered in 1886 by A. J. Brown and can be synthesised in the culture medium of glucose and xylose by bacteria such as *Acetobacter xylinum* [13] and *Gluconacetabacter xylinus* [14], yielding similar structures as plant cellulose. In the culture medium, fibrils are synthesised and secreted as exopolysaccharide and thus generate structural hydrogel with interconnected ribbons of around 100 µm in length and 100 nm diameter (as shown in Figure 1A) [10]. This catalogue of microbiologically derived nanomaterials possesses outstanding features of high purity (free of pectin, hemicellulose, and lignin), degree of polymerisation (up to 10,000), and crystallinity (>85%) of the cellulose microfibrils in BNCs [15]. Importantly, the large hydrophilic surface area originated from cellulose microfibrils endows the BNC hydrogel with excellent water retention capability. Meanwhile, BNCs in the wet state show strong mechanical properties and good flexibility [16]. They are recognised to be highly compatible biomaterials for constructing tissue engineering scaffolds [17] and have been successfully applied for long in a number of biomedical applications such as wound dressing [18], artificial skin [19], vascular and cartilage implants [20,21]. Xylos Corporation (USA) has developed a BNC-based product, XCell, for choric wound dressing, which has been commercialised since 2003 [22]. A series of CELMAT® products that based on BNC are available as facial and eye masks and wound dressing from BOWIL Biotech Ltd. (Poland). However, process factors such as the high cost of glucose as the carbon source and the labor-intensive and low-productivity culture process to yield BNC still restrict the up-scaled production of BNC hydrogels for cost-effective commercialisation [23].

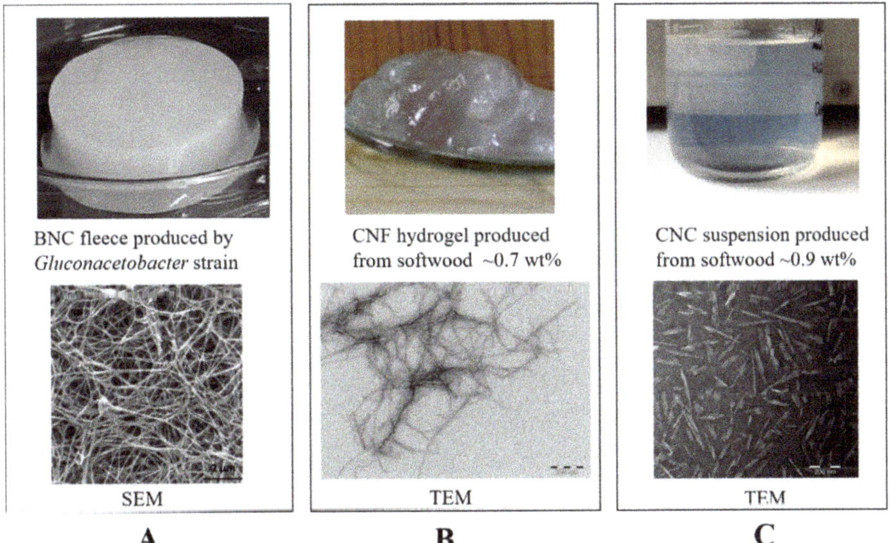

Figure 1. Various nanocellulose products and their microscopic morphology under scanning electron microscopy (SEM) or transmission electron microscopy (TEM) observation: (**A**) BNC (images are reproduced from [10] with the copyright permission from WEILEY-VCH Verlag GmbH & Co.); (**B**) Cellulose nanofibrils (CNFs); (**C**) cellulose nanocrystals (CNCs).

2.2. Cellulose Nanofibrils (CNFs)

Cellulose nanofibrils (CNFs) can be produced from biomasses such as wood pulp. CNFs display as fibrils with diameters of 5–60 nm and lengths of approximately up to a micrometer (as shown in

Figure 1B). Notably, the preparation procedures largely impact on the surface chemistry of the resultant CNFs. At the same time, these surface-modified groups offer an untapped possibility for further functionalisation of CNFs. TEMPO-mediated oxidation, in combination with mechanical defibrillation, produces well-fibrillated CNF with defined surface chemistry of abundant carboxylic groups and a small aldehyde content [24]. Periodate oxidation results in "dialdehyde" CNF by opening the glucose ring on the C2 and C3 sites in the cellulose molecular chain [25]. These flexible CNFs with a high aspect ratio give a gel-like consistency in aqueous suspensions at above a certain dry matter content, depending on the chemical nature and charge density of surface-modified groups on the nanofibrils (as shown in Figure 1B). In the past decade, CNFs have been intensively investigated in versatile applications of cosmetics, pharmaceutics, and biomedical devices [5,26,27]. At present, a CNF product of medical-grade is commercially available from UPM Biomedical (Finland) under the trademark of GrowDex® as a generic 3D cell culture matrix, which is produced with mechanical defibrillation from a sustainable wood resource.

2.3. Cellulose Nanocrystals (CNCs)

Cellulose nanocrystals (CNCs) are prepared by digestion of the cellulosic materials in strong acids (e.g., sulfuric acid or other mild mineral acids) to hydrolyse the amorphous region in cellulose microfibrils, followed by mechanical defibrillation. Hence, wood-derived CNCs that are dispersed in aqueous media display as nanorods of a high aspect ratio (9–50), with a diameter of 3–10 nm and length of 50–500 nm (Figure 1C). More interestingly, CNCs form a chiral nematic liquid crystal phase under the circumstance of controlling the ionic strength of suspension or applying a strong magnetic field [28]. Such features as extremely high stiffness of single nanorod (high crystallinity) and ordered alignment of CNCs in the liquid crystal phase make them attractive nanofillers in preparation of reinforced composite matrices with mechanically anisotropic features [29]. Meanwhile, the surface of CNCs is activated with carboxylates or sulfates, which result from and are dependent on the acidic process adopted. Similar to CNFs, CNCs can also be prepared by comparatively sever periodate oxidation that results in "dialdehyde" on the C2 and C3 sites in glucose units [25]. The surface-modified carboxylates and aldehyde groups facilitate various chemical modification routes to carry out the derivatisation of CNCs. CNCs show great potential for a wide range of applications such as composite reinforcements, flocculants, and certain biomedical areas [30–32]. For instance, nanosized CNCs may pass the cell membrane and thus have found applications in drug delivery and targeted imaging as nanomedicines [33].

These above-described nanocellulose products (BNCs, CNFs, and CNCs) are displayed in Figure 1A–C, along with their representative scanning electron microscopy (SEM) or transmission electron microscopy (TEM) images for microscopic observation.

3. Nanocellulose-Based Bioink: Rheological Properties and Cross-Linking Strategy vs. Ink Fidelity

Direct ink writing (DIW) is the most common extrusion-based additive manufacturing technique used in 3D bioprinting. As illustrated in Figure 2A, this technique employs the cell-laden hydrogel as a feedstock bioink. In printing, the air pressure or the displacement of syringe piston results in stress inside the nozzle on the printer head, where the viscosity of ink decreases and flows through the dispensing nozzle. Once the ink is deposited and the stress disappears, the laid-down hydrogel relaxes and forms a filament of solid gel. Step-by-step, a digitally defined 3D object can be built-up by the layered strand-deposition of the ink.

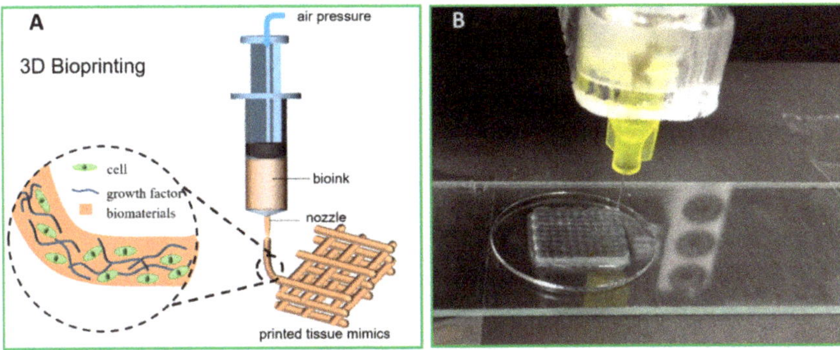

Figure 2. (**A**) Schematic illustration of direct ink writing (DIW) 3D bioprinting and (**B**) DIW printing of TEMPO-oxidised CNF ink at a dry matter content of 1 wt%.

Speaking of appropriate material properties demanded for a bioink, the rheological properties of these biomaterial systems are highly relevant. At first, a shear-thinning behaviour is a must to validate the extrusion-based printing, which enables the ink to pass through the narrow nozzle with low resistance under a certain shear. The yield stress and the shear-thinning response of a bioink are always studied to predict a window of printer operating parameters. Secondly, the viscoelastic property of ink is sufficient to provide ink fidelity when the hydrogel network rapidly recovers elasticity after relaxation, which is technically critical to prevent ink viscous flow and collapse of the wet printed object. In principle, the rheological behaviours of various nanocellulose types largely differ depending on the morphological (size and shape of stiff nanorods in CNCs vs. interconnected nanofibrils in BNC and CNF) and the surface-chemical (charge and other functional groups) attributes of the nanocellulose product. A comprehensive review by M.A. Hubbe and his colleagues is referred to for a detailed understanding of the rheology of nanocellulose-rich aqueous suspensions [34]. For the CNC that is surface-modified with sulfate half-ester groups, its water-like suspensions at low concentrations have a low viscosity (not printable). When such CNCs are added as reinforcing nanofiller in binary systems with other water-soluble polymers, they function as rheological modifiers. For instance, in a binary CNC/alginate bioink system, the addition of CNCs into alginate solution resulted a viscoelastic system with a higher storage modulus (G') than the loss modulus (G″), which consequently assured printability as well as enhancing ink fidelity, compared with the liquid-like alginate solution [35]. The pristine CNC suspensions were shown to be printable only at concentrations above at ~10 wt%, when the shear-thinning behaviour and viscoelastic property of the system meet the rheological demand for a specific DIW setup (e.g., pressure inside syringe and diameter of the extruding nozzle) [36]. These properties are attributed to the interactions between adjacent particles at high concentrations. When disintegrated and suspended in aqueous solution, the BNC fibrils may behave like "flocs" in suspension and can also be oriented under certain shearing conditions to resemble the liquid crystals [37,38]. The dispersions of BNC fibrils show a shear-thinning behaviour and a higher viscosity compared to CNC suspension at the same concentration. In 3D bioprinting, BNC fibrils or oxidised BNC fibrils were printed in blends with alginate for its reinforcing effect [39,40]. The printing of BNC fibrils alone was less practiced, probably due to the lack of cross-linking sites on BNC fibrils to support good performance in terms of ink fidelity. For CNFs, the pretreatments in CNF preparation regulate both the morphology and surface functionality of the CNF products, which in turn determine the interactive forces between nanofibers in the suspensions. The viscosity of carboxymethylated CNF- and TEMPO-oxidised CNF-based inks increases rapidly with increasing the concentration. Both types of CNFs present as a stable hydrogel with strong viscoelastic modulus even at a low concentration of around 1 wt%. This is mainly a result of the electrostatic repulsion among negatively charged carboxylate ($-COO^-$) groups on CNF fibers. For instance, the TEMPO-oxidised

CNFs give a hydrogel-like viscosity when the dry matter content is above a critical concentration in range of 0.5~1 wt%, depending on the charge density on nanofiber resulted from the pretreatment of TEMPO-mediated oxidation. Experimentally, in a dry matter content range of 1~5 wt% in the ink, the shear-thinning rheology and viscoelastic property of these CNFs allow a continuous extrusion of hydrogel filament [41].

More importantly, in situ cross-linking is further relied upon to strengthen the hydrogel network and to generate the adhesion between adjacent layers in order to guarantee the integrity of complex geometry in fabrication. To cross-link the cellulose fiber surface, the strategies can be either physically cross-linked through transient weak interactions (e.g., Van der Waals force, ionic interaction, hydrogen bonding, and hydrophobic interaction) or chemically cross-linked through permanent covalent bonds. Common reaction mechanisms to induce chemical cross-linking in hydrogel systems exemplify condensation, such as the Schiff's base formation and free-radical polymerisation catalysed by enzymes or photo lights [42–44]. These above-mentioned cross-linking strategies have been implemented in the nanocellulose-based inks, aiming to facilitate continuous printability and improve shape fidelity. Physical cross-linking offers a facile and easily applicable approach. Ca^{2+} can cross-link the CNF network by complexing the $-COO^-$ groups on the fiber surface. For instance, a 5-wt% $CaCl_2$ solution was used in droplets during printing the TEMPO-oxidised CNF ink (1 wt%) to gain good shape fidelity via the ionic cross-linking, as shown in Figure 2B [41]. In DIW printing engaging a binary CNF/alginate ink, Ca^{2+} is extensively used as a cross-linker to give good ink performance in terms of the rheological properties, compressive stiffness, and shape fidelity of the printed inks, mainly owing to the presence of large content of $-COO^-$ groups in the alginate molecular [45]. Meanwhile, ionic cross-linking is weak and transient, which is sensitive to a micro-environmental alternation of pH or ionic strength, whereas, chemical cross-linking creates comparatively robust and strong hydrogels with covalent bonds. In this scenario, enzymatic cross-linking and UV-induced cross-linking are engaged when the CNF is in cooperation with an auxiliary biopolymer to formulate compatibly blended binary ink systems. For instance, the intrinsic affinity of cell wall hetero-polysaccharides for cellulose has inspired their use to modify the surface chemistry and mechanical properties of cellulosic materials. As a pioneer for the biomimetic CNF/wood-derived hemicellulose bioinks, K. Markstedt and P. Gatenholm et al. established an enzymatic approach of utilising horseradish peroxidase (HPR) to cross-link the binary hydrogels of CNF/tyramine-modified xylan [46] or CNF/tyramine-modified galactoglucomannan (GGM) [47]. There, HPR catalysed the bond formation between phenolic groups in tyramine-modified xylan or tyramine-modified GGM to result in the cross-linking of the hydrogel network in the printed construct. Photo cross-linking is another easy-to-apply approach as it avoids wet chemistry as well as providing quick gelation, and more importantly, acceptable biocompatibility to the seeded cells. This strategy was applied in binary hydrogels of CNF with a methacrylated biopolymer as the auxiliary polymer, such as in our studies on low-concentration DIW inks of TEMPO-oxidised CNF with methacrylated gelatin (GelMA) [48] and TEMPO-oxidised CNF with methacrylated GGM [49]. In these cases, a UV-LED that is built-in with a DIW printer and moves along with the printing head initialises the radical polymerisation of methacrylates, while the wet filament is laid down to result in an excellent ink fidelity even in the cases of low-concentration inks.

For 3D bioprinting, assessment of ink fidelity is a critical aspect in terms of supporting the layer stacking without the deformation or collapse of overhanging filaments as well as in terms of avoiding compromising of the printing resolution caused by possible fusion between adjacent wet filaments [3,50]. In other words, it reflects how much the printed construct would be able to replicate the digital design. Experimentally, ink fidelity can be determined with regard to the ratio of line width to nozzle diameter (line resolution), the number of layers until collapse, or the curvature of printed lines for constructing complex geometry. For CNFs, the dry matter content in the hydrogel is typically limited up to around 5 wt%, dependent on the fiber dimensions and surface chemistry of the nanofibers. In general, a higher dry matter content used in the CNF ink aids better ink fidelity. Then, higher shear stress has to be applied to yield the ink flow, which might be less desirable from the cell viability viewpoint. When a low-concentration CNF-based ink is printed, an in situ cross-linking strategy has to be integrated to provide strong hydrogel filament in support of a good ink fidelity, as earlier discussed. Meanwhile, the swelling behaviour originating from the hydrophilic nature of cellulose has an impact on the printing resolution. J. Leppiniemi et al. studied the influence of cross-linking on swelling behaviour of 3D printed CNF/alginate grid by soaking the cross-linked and uncross-linked grid into phosphate-buffered saline (PBS buffer). The shape of the cross-linked grid still remained after 24 h (no significant changes on the outer edge), whereas the uncross-linked grid lost its structure in one hour and turned into gel after 24 h [51]. Also, the shape resolution of the cross-linked CNF/alginate grid was sensitive to the variation of ion strength in the soaked media since its swelling behaviour was regulated by ionic cross-linking between Ca^{2+} and the $-COO^-$ groups in alginate [51]. W. Xu et al. studied the influence of cross-linking density on the resolution of the printed GelMA/CNF grid in PBS buffer. The result showed the UV cross-linking could lead to a higher printing resolution and the resolution was enhanced as the increase of cross-linking density [48].

When optimising the ink fidelity, apart from regulating the material properties of the ink itself, as discussed above, other factors should also be taken into consideration, including printing parameters (printing speed, nozzle height, flow rate, and printing path) and the impact of the addition of cells. Actions in optimising printing fidelity of nanocellulose-based inks with respect to printing parameters are still scarce, as challenged by the highly time- and material-consuming lab practice. J. Göhl and his colleagues have adapted the computational fluid dynamics (CFD) tool to stimulate the ink fidelity feature by comparing a 3-wt% binary ink of CNF (non-charged) blended with alginate and a 4-wt% carboxymethylated CNF ink [50]. The CFD simulation helped to understand how various printing parameters affected the line resolution as well as how the viscoelastic stress was distributed throughout the printed filament, which is important for cell viability in 3D bioprinting.

4. Cell–Matrix Interactions and Delivery of Bioactive Cues in Hydrogel Scaffold Fabricated by 3D Bioprinting of Nanocellulose-Based Bioinks

Biocompatibility of cellulose nanomaterials is the primary aspect to investigate as a biomaterial in use. BNC offers excellent biocompatibility. It is a catalogue of well-accepted biomaterials as topical wound healing dressing in treating burns and severe wounds [18,22], as it can offer excellent water retention capability and favorable bioactivities, such as lowering inflammatory response and promoting the fibroblast proliferation [20]. For CNFs and CNCs produced from various resources and preparation methods of different kinds, the non-cytotoxic features of them were verified with a few cell lines in a large number of in vitro cell culture studies. Good biocompatibility of woody nanocellulose was verified with respect to crucial cellular activities of fibroblast growth and proliferation [52]. A study by Y. Lou et al. showed that CNF hydrogels three-dimensionally supported the crucial cellular activities in the culture of human pluripotent stem cells [53]. It was also found in our own studies that the chemical and structural features of CNFs had an impact on the mechanical properties of thus-prepared matrices and further regulated the cell viability in the culture of fibroblasts [54,55]. Furthermore, the binary system of CNF and alginate has been validated to be non-toxic in the culture of various cell lines, e.g., human nasoseptal chondrocytes [45], human-derived induced pluripotent stem cells [56], and fibroblast cells [45]. Collagen and gelatin, the most common biocompatible polymers, are often blended with nanocellulose to create a more biocompatible 3D culture platform, as these biopolymers contain the RGD-moieties that promote cell attachment to the hydrogel matrix [48,57].

Moreover, a bioink system is desired to meet a number of criteria in order for the constructed tissue engineering hydrogel scaffolds to function properly both in vitro and possibly in vivo. In vivo, the ECM provides a microenvironment with proper composition, structure, and stiffness, which is critical for biological processes such as cell adhesion, migration, differentiation, proliferation, and survival [58]. From this perspective, the ECM-mimicking 3D culture platforms would allow us to investigate cell and tissue physiology and pathophysiology in in vitro cell culture to be most relevant and reliable when compared with in vivo conditions [59]. To truly mimic the ECM, the man-made matrices are desired to have innate structural similarities with physiological matrices in the body tissues and to support the most crucial cellular activities, such as cell attachment, proliferation, and subsequent tissue formation [60,61]. First of all, the mechanical characteristics such as ECM rigidity and alignment or organisation play essential roles in various biological processes [62]. For example, A.J. Engler et al. identified that the matrix elasticity for stem cell culture was able to direct the differentiation of human mesenchymal stem cells (hMSCs) toward specific fates [63]. Regarding this perspective, the stiffness control in the ink matrix is a very important biofunctionality to be endowed with. Secondly, cells interact with the biochemical and biophysical cues within their surrounding microenvironment, and such interactions collectively regulate cell behaviour, function, and fate in vivo [58]. Therefore, it is highly demanded to create biofunctionalites of the man-made matrices to spatiotemporally deliver a variety of bioactive cues, aiming to promote the bidirectional crosstalk between the scaffold microenvironment and the resident cells in tissue engineering scaffolds.

4.1. Versatile Cellulose Chemistry to Improve Matrix Reactivty

Various chemical modifications of cellulose are often needed to improve the accessibility of cellulose fibers for adding further functionality to the nanocellulose matrix. Cellulose possesses enormous hydroxyl groups that are amendable for further functionalisation. The chemical modifications are heterogeneous when the nanocellulose presents as BNC, CNF, or CNC. Through reacting with the hydroxyl groups, both cellulose ethers and esters can be prepared by different reaction routes. The functional groups/the tethered moieties are established on the nanocellulose surface, which supports the applicability of nanocellulose hydrogels as ECM-mimicking matrices. In a state-of-the-art study by G. Siqueira et al., the acetylation of CNC with methacrylic anhydride via hydroxyl groups was carried out to result in (hydroxyethyl)methacrylated CNCs, which were established as the foundation of a CNC-reinforced and polyurethane acrylate oligomer-based ink for DIW printing of textured cellular architectures aided by photo cross-linking [36].

Moreover, the active sites such as carboxylates, aldehydes, and sulphates that were introduced by the pretreatments could also be used directly for further functionalisation. Studies on negatively charged, TEMPO-oxidised nanocellulose have been much in focus due to its gelling property, which can be further enhanced by additional multivalent metal ions such as Ca^{2+}. A double cross-linking approach, where TEMPO nanocellulose was cross-linked during printing by addition of an aqueous Ca^{2+} solution, was followed by a post-printing chemical cross-linking with 1, 4-butanediol diglycidyl ether [41]. These printed scaffolds were proven to be stable in PBS buffer for over three months. Carboxylmethylation is another well-established approach to introduce carboxylates to cellulose, which is an effective pretreatment method in producing negatively charged CNFs [64]. The "dialdehyde" resulted in the CNF and CNC from pretreatment of periodate oxidation is actively ready for further chemical derivatisations, e.g., reductive amination. For instance, the dialdehyde-modified CNF reacted with collagen to result in a biocompatible composite platform [65]. In another study, the "dialdehyde"-modified CNCs reinforced the polysaccharide hydrogel of carboxymethyl cellulose–hydrazide (CMC–NHNH$_2$) and dextran–aldehyde (DEX–CHO), being chemically cross-linked within the structural hydrogel network via reductive amination [66].

4.2. Cell–Matrix Interactions

In order to support the cell-matrix interactions better and possibly to add the relevant biofunctionality for a desired therapeutic effect, e.g., osteogenic or angiogenetic effect, biomaterials can be engineered on various aspects, such as surface morphology and chemistry of the biomaterials, micro-engineered mechanical stiffness in the matrix, and stimuli-responsive property of the matrix material. When it comes to the chemical modification to the nanocellulose, it is critical that the reaction medium and reagents selectively used in the chemical modification, as well as the added functionalisation, still guarantees the biocompatibility of biomaterials.

Prof. Paul Gatenholm and his colleagues from Chalmers University of Technology (Sweden) have extensively investigated the interaction of *Gluconacetobacter xylinus*-sourced BNC solely or in binary hydrogel scaffolds with different cell lines such as SH-SY5Y neuroblastoma cells [67], human-derived induced pluripotent stem cells [56,68], and chondrocytes [20,69] towards tissue engineering applications. The mechanical strength of scaffolds was tuned by varying the BNC content to meet the requirement by the targeted tissue [70]. The surface chemistry of the BNC scaffold was also tailored by a coating of collagen to improve cell adhesion, growth, and differentiation [67]. The cell-laden 3D-bioprinting of BNC-based scaffolds has brought up promising solutions for tissue repair [71,72]. A recent review by A. Sionkowska et al. presented recent advances of the most studied medical applications of BNC [73].

For the CNF hydrogels, it is well established that the surface chemistry resulted from various preparation methods has a large impact on cell–matrix interactions. Hence, the CNF hydrogel matrices undergone chemical post-modifications particularly need to be carefully evaluated in cell culture studies prior to formulating bioink. As wound healing and soft tissue engineering are the most discussed end applications for CNF-based bioinks, the fibroblast is the most used cell line in in vitro culture for evaluating the cytotoxicity of CNF hydrogels [52,54]. In addition, cancer cell lines and stem cells were also tested in a number of in vitro studies on evaluating the CNF hydrogels as cell culture platforms [52–54]. L. Alexandrescu et al. created different surface chemistry on cellulose fibers by either enzymatic pretreatment or TEMPO-oxidation in combination with mechanical defibrillation, resulting in CNF free of charge or with anionic charge, respectively [52]. Both types of CNFs showed excellent compatibility and strong cell–matrix interactions in supporting the attachment, proliferation, and growth of 3T3 fibroblasts. The same study also investigated two post-treatment approaches to tune the surface properties, polyethyleneimine (PEI)-cross-linking and sorption of cetyltrimethylammonium bromide (CTAB), both of which induced reduction in cell viability. Thus, when tailoring properties of scaffolds, one needs to keep in mind that cells are sensitive to surface alternation that they will be in contact with.

For the CNF hydrogel produced with the pretreatment of TEMPO-mediated oxidation followed by high-pressure homogenisation, it has demonstrated excellent DIW printability in quite a few studies. The narrow size distribution in terms of nanofiber length in a homogeneous hydrogel phase guarantees a consistent ink flow without clogging the nozzle while being extruded [41]. Speaking of the CNF hydrogel preparation, the charge density of –COO$^-$ introduced by TEMPO-oxidation to the nanofiber plays a decisive role in fiber disintegration to obtain the nano-dimensional fibrils [24]. Meanwhile, the charge density of –COO$^-$ largely impacts the biocompatibility of TEMPO-oxidised CNF with respect to the growth of fibroblasts and Hela cancer cells inside the hydrogel. An intermediate surface charge level of around 1 mmol/g was suggested to provide good biocompatibility in favor of the cell–matrix response [52,54]. Furthermore, the negatively charged COO$^-$ groups on CNF is an important factor to consider in the case of formulating a composite ink with a second biopolymer. When formulating the composite ink of TEMPO-oxidised CNF/GelMA, GelMA had to be kept less than 1 wt% in the binary system containing 1 wt% TEMPO-oxidised CNF in order to avoid the phase separation caused by the ionic interaction between TEMPO-oxidised CNF and GelMA, and the ink homogeneity was thus retained [48].

To meet the requirements of desired matrix stiffness and shape fidelity, a cross-linker is often introduced in ink formulation. Various strategies as dispicted in Figure 3A–D, such as alginate in combination with Ca^{2+} and polymeric methacrylates as UV-curable cross-linkers, could be applied in the nanocellulose-based formulation to provide the control means over the mechanical properties of the bioinks. Then, the cell–matrix response is evaluated in various cell cultures by seeding the cells into 3D-printed porous hydrogel scaffolds or cell-laden bioprinting. Different cell lines were used to investigate the response of cells in those scaffolds with tuned mechanical property: fibroblast, breast cancer cells, human neuroblastoma, different stem cells (seen as in Table 1). In our earlier approach inspired by the intrinsically high affinity between cellulose and hemicellulose in plant cell walls, hemicelluloses (xylan, GGM, and xyloglucan) were engaged as physical cross-linkers to prepare hemicellulose-reinforced TEMPO-oxidised CNF hydrogels [74]. There, it was demonstrated that the incorporation of xyloglucan significantly increased the modulus and yield stress of the aerogels and correspondingly supported the cell functions seeded in the reinforced CNF hydrogel matrix in comparison with the one-component CNF hydrogel [74]. As lately investigated in the double cross-linking approach applied in DIW printing, the compressive Young's modulus of a one-component TEMPO-oxidised CNF hydrogel scaffold resulted in a range of 3 to 8 kPa that suits the attachment of fibroblast cells. Further cell tests confirmed that the hydrogel rigidity had a clear impact on the cell–matrix response: the proliferation of fibroblast was promoted with increased hydrogel stiffness within the studied range [41]. In this study, the stiffness variation was resulted only by chemically cross-linking the hydroxyl groups in a one-component CNF hydrogel, which affects the least the surface chemistry of matrix, and this makes the correlation between the mechanical properties of the hydrogel matrix and cell–matrix response easy to demonstrate. In the binary ink formulations where the methacrylated natural polymers are UV cross-linkers for mechanical stiffness control over the CNF hydrogel, the introduction of a secondary component also alters the chemical nature and surface physiochemical features (surface roughness and porosity) of the hydrogel matrix in different means [48,49]. Together with the mechanical stiffness of the hydrogel matrix, these factors together regulate the cell–matrix response. In the fields of 3D bioprinting, the methacrylated biopolymers (e.g., GelMA) have become popular in formulating various photo cross-linkable bioinks. It is still worth noting that the cytotoxicity of such systems on the resident cells may potentially originate from the free radicals that are generated by the photo-initiator when activating the cross-linking of methacrylate groups. Herein, the selection of biocompatible photoinitiator and degree of substitution (DS) of methacrylate in biopolymer) are important aspects to consider. M.J. Majcher et al. showed that methacrylated starch nanoparticle-based hydrogel showed lower cytotoxicity with a DS lower than 0.10 [75]. Owing to its hydrophilic property, 2-hydroxy-4'-(2-hydroxyethoxy)-2-methylpropiophenone (Irgacure 2959) is widely accepted as a biocompatible photoinitiator for such photo cross-linkable bioink formulations in the concentration range of 0.03%–0.1% w/v [76].

CNCs obtained by acid hydrolysis are characteristics of negatively charged surface groups ($-OSO_3^-$ or $-COO^-$, depending on the treatment acid that is adopted) and nanorod-like morphology. In the context of 3D bioprinting, they are more often seen as reinforcing nanofillers in formulating composite hydrogel bioinks used in bone/cartilage regeneration, owing to their nanorod-like morphology and extraordinarily high stiffness [77]. Importantly, S. Dong et al. performed 3-(4,5-dimethylthiazol-2-yl)-2,5-diphenyltetrazolium bromide (MTT) and lactate dehydrogenase (LDH) assays to study the cytotoxicity of CNCs against nine different cell lines [78], showing no cytotoxic against any of these specific cell lines over the tested concentration range. Furthermore, the unique property of ordered alignment of CNCs in the liquid crystal phase makes them interesting as reinforcing nanomaterials to result in the anisotropically mechanical properties of the matrix [30]. A matrix possessing ordered structure is particularly of interest for tailoring gradients in alignment with or against the order. As reported by K.J. De France et al. [29], the magnetic field-induced alignment of CNCs was successfully translated into a nanocomposite hydrogel based on hydrazone cross-linked poly(oligoethylene glycol methacrylate) (POEGMA) and physically incorporated CNCs after injection, which consequently endowed the hydrogel matrix with anisotropic properties. Meanwhile, the anisotropically mechanical property is directive for the motility and migration of cells, such as skeletal muscle myoblasts in the repair of muscle tissue. The nanocomposite hydrogel of POEGMA/CNCs promoted the differentiation of resident skeletal muscle myoblasts into highly oriented myotubes in situ [29,79]. According to J.M. Dugan et al., oriented surfaces of highly charged CNCs prepared using a spin-coating method also induced contact guidance in skeletal muscle myoblasts [79]. A bit surprisingly, fibroblasts tended not to adhere to the CNC coating in comparison to different types of CNF (no-charge CNF and negatively charged TEMPO-CNF) coatings, as recently reported in our study [80]. This might indicate the effect of material stiffness and nanoscale morphology of nanocelluloses on the cell–matrix interactions for different cell types.

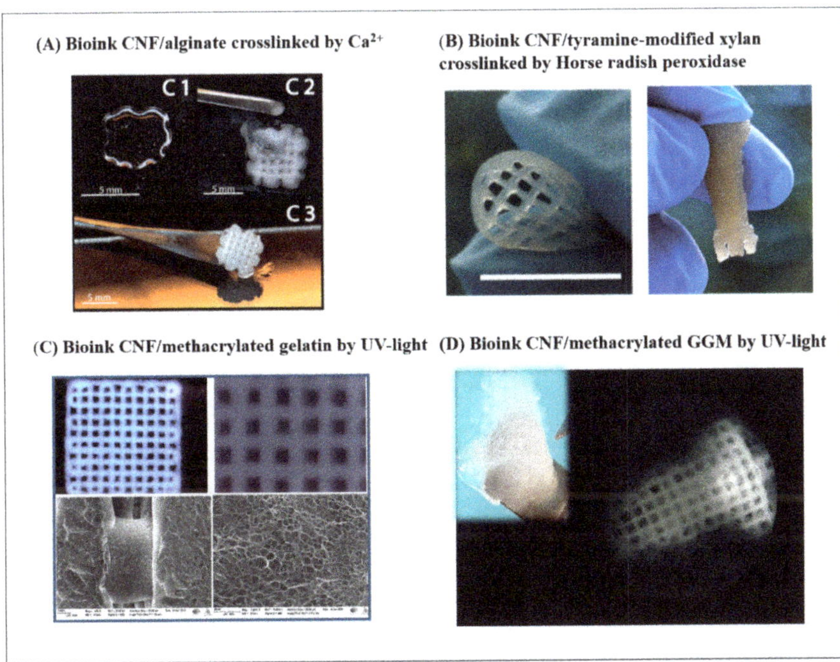

Figure 3. State-of-the-art CNF-based bioinks engaging different cross-linking strategies: (**A**) bioink CNF/alginate cross-linked by Ca^{2+}, as presented in [45] (copyright permission from American Chemistry Society); (**B**) bioink CNF/tyramine-modified xylan cross-linked by horseradish peroxidase, as presented in [46] (copyright permission from American Chemistry Society); (**C**) bioink CNF/methacrylated gelatin by UV light, as presented in [48] (under CC-BY licence); (**D**) bioink CNF/methacrylated gelatin by UV light, as presented in [49] (under CC-BY licence).

4.3. Delivery of Bioactive Cues in the Nanocellulose-Based 3D Bioprinting

As above-mentioned, it is highly demanded to create molecular functionalities of the hydrogel matrix to spatiotemporally deliver a variety of bioactive cues within the tissue engineering scaffolds. More specifically, the hydrogel scaffolds need to be able to act like natural ECM being a carrier for bioactive substances, such as growth factor or bioactive drugs that can regulate the cell behaviour as desired. Owing to the high surface area-to-volume ratio of nanocelluloses, adsorption and entrapment of active substances into their porous structure as such or after surface modifications could enable high levels of drug loading and binding. S. Chatterjee and C. P. Hui recently reviewed the stimuli-responsive polymers, mainly including chitosan, cellulose, and gelatin, imparting sensitivity to act in responding to temperatures and pH conditions for their respective applications in drug delivery [81]. Those hydrogel systems are usually composed of natural polymers and responsive polymers, e.g., poly(N-isopropylacrylamide) (pNIPAAm) by blending or covalent bonding. Another review has summarised the state-of-the-art in chemical modifications of cellulose, lignin, and other wood components via atom transfer radical polymerisation, which broadens their potential applications in medicine and pharmacy as stimuli-responsive micelle delivery systems and gene carriers [82].

The modification of cellulose and other biopolymers often allows better delivery of growth factors and other active drugs with increased binding ability. Thus, those bioactive cues can be directly formulated into the inks. For example, growth factor was blended in alginate/carboxymethyl cellulose formulations for the 3D printed thin films, and significantly improved cell viability was detected for films with incorporated growth factors in the cell tests with the most abundant skin cell types (keratinocytes and fibroblasts) [83].

M. Ojansivu et al. have developed an interesting composite ink containing a polymeric matrix of CNF/alginate/gelatin and bioactive glass (BaG) microparticulates for the fabrication of in vitro tissue equivalents that are proposed for studies of bone/or cartilage regeneration [84]. In their system, CNF regulated the rheological properties of the composite ink to facilitate DIW printing. BaG microparticulates were integrated as a therapeutic functionality carrier as the BaGs have high osteogenic bioactivity attributed to the released therapeutic inorganic ions (Si, P, and Ca) functioning as bioactive cues in vivo [85]. In the printed hMSCs-laden constructs, the BaG microparticulates embedded in the ink were confirmed to stimulate the early osteogenic commitment of the resident hMSCs [84].

With an attempt to summarise the ink formulations that have used nanocelluloses as the major component other than an auxiliary one, the authors present the most recent studies in the literature in Table 1. These studies are discussed with respect to ink composition, printing approach, indications on cell behaviours in cell culture, and biomedical applications, as highlighted.

5. Challenges and Perspectives for Nanocellulose-Based Inks

In the research fields embracing natural polymers as biomaterials, nanocelluloses of various types have gained numerous interests as nanoscaled components for formulating sustainable bioinks. Their potential applications are seen not only in constructing 3D cell culture platform for drug screening and cancer research but also in fabricating skin tissue mimics and composite hydrogel scaffolds used in cartilage/bone tissue reparation via either scaffold-printing or cell-laden 3D bioprinting. On the one hand, as biomaterials are mainly evaluated in in vitro cell culture at the present stage, excellent biocompatibility and strong cell–matrix interactions of nanocelluloses have been highly praised in a number of research studies. As above-mentioned, these outstanding properties of nanocellulose-based bioinks are attributed to their structural similarity in fibrous morphology with the ECM, as well as their nano-sized material features that offer large surface areas and a highly compatible chemistry nature for the accommodated cells to interact with. On the other hand, the popular acceptance of nanocellulose-based bioinks among the potential end-user society, mainly referring to cell biologists and medical surgeons, has been challenged by their concerns on the in vivo biodegradability and validation of in vivo nano-safety for cellulose nanomaterials to date.

Since human beings do not have specific enzymes (e.g., cellulases) that can break down the nanocellulose in vivo, it is not well accepted to engage the nanocellulose itself or the composite material with nanocellulose as the main constituent in implant manufacture where the biomaterials are highly desired to be bioresorbable after tissue healing or organ repair. Some studies showed that oxidised cellulose has the potential to be degraded by the human body, owing to their weak resistance to hydrolysis [26]. A. Rashadet et al. studied the degradation profile of TEMPO-oxidised and carboxymethylated CNF scaffolds in vitro. The result showed 6.7% and 6.5% weight loss of the scaffold after 90 days, respectively [86]. In contrast, periodate-oxidised BNC showed faster degradation kinetics. W. Czaja et al. treated the pressed BNC sheet with pre-γ-irradiation followed by sodium periodate oxidation and the oxidised BNC sheet showed an in vitro degradation rate of 85% in 7 days. The further in vivo test showed the degradation occurring in the first 2 to 4 weeks [87].

A limited number of nanotoxicology studies have addressed the toxicological effect of CNFs and CNCs in in vivo animal models, such as zebrafish [88] and rat [89], although no significant risks were indicated for acute toxicity in small quantities [90]. Meanwhile, long-term nanotoxicity of nanocellulose is another concern that is closely associated with the in vivo degradability problem and it is an important research aspect that still awaits a large number of assessments in in vivo models. Above all, the most intriguing question present for material scientists in basic research is how to chemically modify or engineer the nanocellulose materials to make it self-degrade naturally in the human body.

To date, the most promising applications for CNF-based bioinks can be seen in the fabrication of skin tissue mimics as culture platforms for in vitro studies focusing on cell–cell interactions in elucidating the molecular mechanism of disease or cellular response to the bioactive substances in drug-screening. As one of the frontier players in the commercialisation of medical-grade nanocellulose products (from wood resources), UPM Biomedicals has newly launched a medical-grade CNF hydrogel with the trademark of GrowInk™ as a non-animal-derived bioink, as well as a CNF-based wound dressing product of FibDex® for the European market. As supported in a clinical trial of small group of patients, FibDex® is claimed to provide a favorable environment for the healing of wound to occur [91]. Swedish bioprinter supplier CELLINK is also commercialising CNF/alginate bioink as accessory kits for the use in their bioprinter series. Their CNF/alginate bioink was used to fabricate human cartilage construct with chondrocytes and stem cells co-cultured inside the hydrogel. As evaluated in an in vivo mice model, the matrix supported not only the proliferation of chondrocytes but also the secretion of glycosaminoglycans and collagen II by the chondrocytes [92]. In the near future, with more comprehension of cell–matrix interactions from fundamental studies and validation of nanocellulose products in in vivo models, the nanocellulose-based bioink is anticipated to offer a more sustainable and cost-effective alternative for end-users in biomedical and pharmaceutical fields.

Table 1. Overview of recently developed nanocellulose-incorporated ink formulations and their in-vitro cell culture studies.

Nanocellulose Type	Composition of Inks	Printing Approaches	Cell Lines	Cell Study Results	Potential Applications	References
Bacterial CNF	CNF + silk + gelatin + glycerol	Hydrogel DIW	L929 fibroblasts cells	The in vitro evaluation showed that the composite scaffolds had excellent biocompatibility, while the in vivo results demonstrated that the hierarchical pore structure was beneficial to the ingrowth of tissue	Repair of soft tissues	[93]
CNF	CNF + cross-linkers (CaCl$_2$, Chitosan oligosaccharides, Poly-l-lysine, protamine)	Inkjet spray, cell-laden	Mouse fibroblasts (NIH3T3), human embryonic kidney cells (293A), and human newborn foreskin fibroblasts (Hs68)	cell viability, metabolic activity, and collagen type I secretion were evaluated in the printed objects	Skin tissue mimics	[94]
CNF	CNF + CMC/Alginate	Hydrogel DIW	Human primary pancreatic cells	Promoted cell adhering, aggregation, migration, and support long-term growth of pancreatic cell	Cell culture and disease study	[95]
CNF	CNF + alginate	Hydrogel DIW, cell-laden	Mouse mesenchymal stem cell line C3H10T1/2	The cells accumulate more lipids and have increased gene expression of adipogenic marker genes PPARγ and FABP4 than cells cultured using standard 2D method	3D cell culture of adipocytes	[96]
Enzymatic CNF	CNF + alginate	Hydrogel DIW, cell-laden	L929 fibroblasts, human nasoseptal chondrocytes (hNC; cell-laden)	Biocompatible and a suitable material for cell culture	Cartilage tissue engineering	[45]
Enzymatic CNF	CNF + alginate; CNF + hyaluronic acid	Hydrogel DIW	Pluripotent stem cells	NFC/A bioinks were suitable for bioprinting iPSCs to support cartilage production in co-culture with irradiated chondrocytes	To repair damaged cartilage in joints	[56]
CM-CNF	CNF + Bacterial cellulose (culture medium)	Hydrogel DIW	Fibroblast cells	Healthy growth	Artificial blood vessels and engineered vascular tissue scaffold	[97]
CM-CNF	Methyltrimethoxysilane hydrophobic CNF matrix-assisted	Hydrogel DIW	A549 lung cancer cells	Sustained healthy cell growth	Open cell culture platform and drug test	[98]
CM-CNF	CNF, CNF/carbon nanotubes	Hydrogel DIW	SH-SHY5Y human neuroblastoma cells	Pure CNF materials are not cytotoxic	Neural tissue engineering	[99]

Table 1. *Cont.*

Nanocellulose Type	Composition of Inks	Printing Approaches	Cell Lines	Cell Study Results	Potential Applications	References
TEMPO-CNF	CNF + Alginate/Ca^{2+}	Hydrogel DIW	L929 mouse fibroblasts	The reduction of cytotoxicity as the ash content of the pulps and CNFs was reduced	Wound dressing devices	[100]
TEMPO-CNF	CNF, TEMPO-CNF, Or acetylated TEMPO-CNF	Hydrogel DIW	Cardiac myoblast cells	Enabled the proliferation and attachment of cells	Cellular processes and tissue engineering	[101]
TEMPO-CNF	CNF + galactoglucomannan methacrylate	Hydrogel DIW	Human dermal fibroblast (HDF) cells and pancreatic tumor cell line SW-1990 cells	Support the principal cell behaviours including cell viability, adhesion, and proliferation	Tissue engineering, cancer cell research, and high-throughput drug screening	[49]
TEMPO-CNF	CNF + gelatin methacrylate	Hydrogel DIW	3T3 fibroblasts cells	Promoted proliferative activity of 3T3 fibroblasts	Wound healing	[48]
TEMPO-CNF	CNF + gelatin methacrylamide	Hydrogel DIW, cell-laden	NIH 3T3 fibroblast cell-laden	No cytotoxicity, high cell viability	Biomedical scaffolds	[57]
CNC	CNC + gelatin	Hydrogel DIW	3T3 fibroblast cells	Support the growth and proliferation of 3T3 cells	Tissue engineering	[102]
CNC	CNC-gelatin conjugates	Hydrogel DIW	Human breast cancer MCF-7 cells	Not cytotoxic	Tissue engineering and regenerative medicine	[103]
CNC	CNC + oxidised dextran/gelatin	Hydrogel DIW	3T3, CCK-8 and Hoechst 33342/PI double-staining assays	Support cell growth and proliferation	Tissue repair	[77]
CNC	CNC + yeast cell + binder (PEGDA) + photo initiator	Viscous paste DIW, cell-laden	Yeast cell-laden	Long-term viability	Microbial biocatalysts, bioremediation	[104]

Author Contributions: X.W. and C.X. together conceptualized and outlined the content of this mini-review; X.W. contributed to the major draft of the manuscript and further corresponded the revision in peer-review; Q.W. and C.X. contributed partially to the draft of the manuscript; and Q.W. was responsible for reference formatting. All authors have read and agreed to the published version of the manuscript.

Funding: This research was funded by [Jane and Aatos Erkko Foundation (Finland)] with a grant to Xiaoju Wang at Åbo Akademi University in year of 2019-2022 and by [Academy of Finland Project] with a grant number [298325].

Acknowledgments: Liqiu Hu and Wenyang Xu are acknowledged for the courtesy of their images.

Conflicts of Interest: The authors declare no conflict of interest.

References

1. Murphy, S.V.; Atala, A. 3D Bioprinting of Tissues and Organs. *Nat. Biotechnol.* **2014**, *32*, 773–785. [CrossRef]
2. Vijayavenkataraman, S.; Yan, W.C.; Lu, W.F.; Wang, C.H.; Fuh, J.Y.H. 3D Bioprinting of Tissues and Organs for Regenerative Medicine. *Adv. Drug Deliv. Rev.* **2018**, *132*, 296–332. [CrossRef] [PubMed]
3. Kyle, S.; Jessop, Z.M.; Al-Sabah, A.; Whitaker, I.S. 'Printability' of Candidate Biomaterials for Extrusion Based 3D Printing: State-of-the-Art. *Adv. Healthc. Mater.* **2017**, *6*, 1700264. [CrossRef] [PubMed]
4. Nam, S.Y.; Park, S.H. ECM based bioink for tissue mimetic 3D bioprinting. In *Advances in Experimental Medicine and Biology*; Springer: Singapore, 2018; Volume 1064, pp. 335–353.
5. Chinga-Carrasco, G. Potential and Limitations of Nanocelluloses as Components in Biocomposite Inks for Three-Dimensional Bioprinting and for Biomedical Devices. *Biomacromolecules* **2018**, *19*, 701–711. [CrossRef] [PubMed]
6. Xu, W.; Wang, X.; Sandler, N.; Willför, S.; Xu, C. Three-Dimensional Printing of Wood-Derived Biopolymers: A Review Focused on Biomedical Applications. *ACS Sustain. Chem. Eng.* **2018**, *6*, 5663–5680. [CrossRef] [PubMed]
7. Wang, Q.; Sun, J.; Yao, Q.; Ji, C.; Liu, J.; Zhu, Q. 3D Printing with Cellulose Materials. *Cellulose* **2018**, *25*, 4275–4301. [CrossRef]
8. Dai, L.; Cheng, T.; Duan, C.; Zhao, W.; Zhang, W.; Zou, X.; Aspler, J.; Ni, Y. 3D Printing Using Plant-Derived Cellulose and Its Derivatives: A Review. *Carbohydr. Polym.* **2019**, *203*, 71–86. [CrossRef]
9. Gibson, L.J. The Hierarchical Structure and Mechanics of Plant Materials. *J. R. Soc. Interface* **2012**, *9*, 2749–2766. [CrossRef]
10. Klemm, D.; Kramer, F.; Moritz, S.; Lindström, T.; Ankerfors, M.; Gray, D.; Dorris, A. Nanocelluloses: A New Family of Nature-Based Materials. *Angew. Chemie Int. Ed.* **2011**, *50*, 5438–5466. [CrossRef]
11. Dufresne, A. Nanocellulose: A New Ageless Bionanomaterial. *Mater. Today* **2013**, *16*, 220–227. [CrossRef]
12. Du, X.; Zhang, Z.; Liu, W.; Deng, Y. Nanocellulose-Based Conductive Materials and Their Emerging Applications in Energy Devices—A Review. *Nano Energy* **2017**, *35*, 299–320. [CrossRef]
13. Masaoka, S.; Ohe, T.; Sakota, N. Production of Cellulose from Glucose by Acetobacter Xylinum. *J. Ferment. Bioeng.* **1993**, *75*, 18–22. [CrossRef]
14. Ruka, D.R.; Simon, G.P.; Dean, K.M. Altering the Growth Conditions of Gluconacetobacter Xylinus to Maximize the Yield of Bacterial Cellulose. *Carbohydr. Polym.* **2012**, *89*, 613–622. [CrossRef] [PubMed]
15. Blanco Parte, F.G.; Santoso, S.P.; Chou, C.C.; Verma, V.; Wang, H.T.; Ismadji, S.; Cheng, K.C. Current Progress on the Production, Modification, and Applications of Bacterial Cellulose. *Crit. Rev. Biotechnol.* **2020**, *40*, 397–414. [CrossRef] [PubMed]
16. McKenna, B.A.; Mikkelsen, D.; Wehr, J.B.; Gidley, M.J.; Menzies, N.W. Mechanical and Structural Properties of Native and Alkali-Treated Bacterial Cellulose Produced by Gluconacetobacter Xylinus Strain ATCC 53524. *Cellulose* **2009**, *16*, 1047–1055. [CrossRef]
17. Martínez Ávila, H.; Schwarz, S.; Feldmann, E.M.; Mantas, A.; Von Bomhard, A.; Gatenholm, P.; Rotter, N. Biocompatibility Evaluation of Densified Bacterial Nanocellulose Hydrogel as an Implant Material for Auricular Cartilage Regeneration. *Appl. Microbiol. Biotechnol.* **2014**, *98*, 7423–7435. [CrossRef]
18. Czaja, W.; Krystynowicz, A.; Bielecki, S.; Brown, R.M. Microbial Cellulose—The Natural Power to Heal Wounds. *Biomaterials* **2006**, *27*, 145–151. [CrossRef]
19. Fu, L.; Zhang, J.; Yang, G. Present Status and Applications of Bacterial Cellulose-Based Materials for Skin Tissue Repair. *Carbohydr. Polym.* **2013**, *92*, 1432–1442. [CrossRef]
20. Svensson, A.; Nicklasson, E.; Harrah, T.; Panilaitis, B.; Kaplan, D.L.; Brittberg, M.; Gatenholm, P. Bacterial Cellulose as a Potential Scaffold for Tissue Engineering of Cartilage. *Biomaterials* **2005**, *26*, 419–431. [CrossRef]

21. Schumann, D.A.; Wippermann, J.; Klemm, D.O.; Kramer, F.; Koth, D.; Kosmehl, H.; Wahlers, T.; Salehi-Gelani, S. Artificial Vascular Implants from Bacterial Cellulose: Preliminary Results of Small Arterial Substitutes. *Cellulose* **2009**, *16*, 877–885. [CrossRef]

22. Frankel, V.H.; Serafica, G.C.; Damien, C.J. Development and Testing of a Novel Biosynthesized XCell for Treating Chronic Wounds. *Surg. Technol. Int.* **2004**, *12*, 27–33. [PubMed]

23. Gorgieva, S.; Trček, J. Bacterial cellulose: Production, modification and perspectives in biomedical applications. *Nanomaterials* **2019**, *9*, 1352. [CrossRef] [PubMed]

24. Liu, J.; Korpinen, R.; Mikkonen, K.S.; Willför, S.; Xu, C. Nanofibrillated Cellulose Originated from Birch Sawdust after Sequential Extractions: A Promising Polymeric Material from Waste to Films. *Cellulose* **2014**, *21*, 2587–2598. [CrossRef]

25. Kim, U.J.; Kuga, S.; Wada, M.; Okano, T.; Kondo, T. Periodate Oxidation of Crystalline Cellulose. *Biomacromolecules* **2000**, *1*, 488–492. [CrossRef]

26. Lin, N.; Dufresne, A. Nanocellulose in Biomedicine: Current Status and Future Prospect. *Eur. Polym. J.* **2014**, *59*, 302–325. [CrossRef]

27. Soo Min, K.; Eun Ji, G.; Seung Hwan, J.; Sang Mock, L.; Woo Jong, S.; Jin Sik, K. Toxicity Evaluation of Cellulose Nanofibers (Cnfs) for Cosmetic Industry Application. *J. Toxicol. Risk Assess.* **2019**, *5*. [CrossRef]

28. Gray, D. Recent Advances in Chiral Nematic Structure and Iridescent Color of Cellulose Nanocrystal Films. *Nanomaterials* **2016**, *6*, 213. [CrossRef]

29. De France, K.J.; Yager, K.G.; Chan, K.J.W.; Corbett, B.; Cranston, E.D.; Hoare, T. Injectable Anisotropic Nanocomposite Hydrogels Direct in Situ Growth and Alignment of Myotubes. *Nano Lett.* **2017**, *17*, 6487–6495. [CrossRef]

30. Eichhorn, S.J. Cellulose Nanowhiskers: Promising Materials for Advanced Applications. *Soft Matter* **2011**, *7*, 303–315. [CrossRef]

31. Kan, K.H.M.; Li, J.; Wijesekera, K.; Cranston, E.D. Polymer-Grafted Cellulose Nanocrystals as PH-Responsive Reversible Flocculants. *Biomacromolecules* **2013**, *14*, 3130–3139. [CrossRef]

32. Domingues, R.M.A.; Gomes, M.E.; Reis, R.L. The Potential of Cellulose Nanocrystals in Tissue Engineering Strategies. *Biomacromolecules* **2014**, *15*, 2327–2346. [CrossRef] [PubMed]

33. Hosseinidoust, Z.; Alam, M.N.; Sim, G.; Tufenkji, N.; Van De Ven, T.G.M. Cellulose Nanocrystals with Tunable Surface Charge for Nanomedicine. *Nanoscale* **2015**, *7*, 16647–16657. [CrossRef] [PubMed]

34. Hubbe, M.A.; Tayeb, P.; Joyce, M.; Tyagi, P.; Kehoe, M.; Dimic-Misic, K.; Pal, L. Rheology of Nanocellulose-Rich Aqueous Suspensions: A Review. *BioResources* **2017**, *12*, 9556–9661.

35. Wu, Y.; Lin, Z.Y.; Wenger, A.C.; Tam, K.C.; Tang, X. 3D Bioprinting of Liver-Mimetic Construct with Alginate/Cellulose Nanocrystal Hybrid Bioink. *Bioprinting* **2018**, *9*, 1–6. [CrossRef]

36. Siqueira, G.; Kokkinis, D.; Libanori, R.; Hausmann, M.K.; Gladman, A.S.; Neels, A.; Tingaut, P.; Zimmermann, T.; Lewis, J.A.; Studart, A.R. Cellulose Nanocrystal Inks for 3D Printing of Textured Cellular Architectures. *Adv. Funct. Mater.* **2017**, *27*, 1604619. [CrossRef]

37. Okiyama, A.; Motoki, M.; Yamanaka, S. Bacterial Cellulose III. Development of a New Form of Cellulose. *Top. Catal.* **1993**, *6*, 493–501. [CrossRef]

38. Tsalagkas, D.; Dimic-Misic, K.; Gane, P.; Rojas, O.J.; Maloney, T.; Csoka, L. Rheological behaviour of sonochemically prepared bacterial cellulose aqueous dispersions. In Proceedings of the 6th International Symposium on Industrial Engineering (SIE 2015), Belgrade, Serbia, 24–25 September 2015.

39. Gutierrez, E.; Burdiles, P.A.; Quero, F.; Palma, P.; Olate-Moya, F.; Palza, H. 3D Printing of Antimicrobial Alginate/Bacterial-Cellulose Composite Hydrogels by Incorporating Copper Nanostructures. *ACS Biomater. Sci. Eng.* **2019**, *5*, 6290–6299. [CrossRef]

40. Wei, J.; Wang, B.; Li, Z.; Wu, Z.; Zhang, M.; Sheng, N.; Liang, Q.; Wang, H.; Chen, S. A 3D-Printable TEMPO-Oxidized Bacterial Cellulose/Alginate Hydrogel with Enhanced Stability via Nanoclay Incorporation. *Carbohydr. Polym.* **2020**, *238*, 116207. [CrossRef]

41. Xu, C.; Zhang Molino, B.; Wang, X.; Cheng, F.; Xu, W.; Molino, P.; Bacher, M.; Su, D.; Rosenau, T.; Willför, S.; et al. 3D Printing of Nanocellulose Hydrogel Scaffolds with Tunable Mechanical Strength towards Wound Healing Application. *J. Mater. Chem. B* **2018**, *6*, 7066–7075. [CrossRef]

42. Balakrishnan, B.; Joshi, N.; Jayakrishnan, A.; Banerjee, R. Self-Crosslinked Oxidized Alginate/Gelatin Hydrogel as Injectable, Adhesive Biomimetic Scaffolds for Cartilage Regeneration. *Acta Biomater.* **2014**, *10*, 3650–3663. [CrossRef]

43. Qi, C.; Liu, J.; Jin, Y.; Xu, L.; Wang, G.; Wang, Z.; Wang, L. Photo-Crosslinkable, Injectable Sericin Hydrogel as 3D Biomimetic Extracellular Matrix for Minimally Invasive Repairing Cartilage. *Biomaterials* **2018**, *163*, 89–104. [CrossRef]

44. Yu, F.; Cao, X.; Li, Y.; Zeng, L.; Yuan, B.; Chen, X. An Injectable Hyaluronic Acid/PEG Hydrogel for Cartilage Tissue Engineering Formed by Integrating Enzymatic Crosslinking and Diels-Alder "Click Chemistry". *Polym. Chem.* **2014**, *5*, 1082–1090. [CrossRef]

45. Markstedt, K.; Mantas, A.; Tournier, I.; Martínez Ávila, H.; Hägg, D.; Gatenholm, P. 3D Bioprinting Human Chondrocytes with Nanocellulose-Alginate Bioink for Cartilage Tissue Engineering Applications. *Biomacromolecules* **2015**, *16*, 1489–1496. [CrossRef] [PubMed]

46. Markstedt, K.; Escalante, A.; Toriz, G.; Gatenholm, P. Biomimetic Inks Based on Cellulose Nanofibrils and Cross-Linkable Xylans for 3D Printing. *ACS Appl. Mater. Interfaces* **2017**, *9*, 40878–40886. [CrossRef] [PubMed]

47. Markstedt, K.; Xu, W.; Liu, J.; Xu, C.; Gatenholm, P. Synthesis of Tunable Hydrogels Based on O-Acetyl-Galactoglucomannans from Spruce. *Carbohydr. Polym.* **2017**, *157*, 1349–1357. [CrossRef] [PubMed]

48. Xu, W.; Molino, B.Z.; Cheng, F.; Molino, P.J.; Yue, Z.; Su, D.; Wang, X.; Willför, S.; Xu, C.; Wallace, G.G. On Low-Concentration Inks Formulated by Nanocellulose Assisted with Gelatin Methacrylate (GelMA) for 3D Printing toward Wound Healing Application. *ACS Appl. Mater. Interfaces* **2019**, *11*, 8838–8848. [CrossRef] [PubMed]

49. Xu, W.; Zhang, X.; Yang, P.; Långvik, O.; Wang, X.; Zhang, Y.; Cheng, F.; Österberg, M.; Willför, S.; Xu, C. Surface Engineered Biomimetic Inks Based on UV Cross-Linkable Wood Biopolymers for 3D Printing. *ACS Appl. Mater. Interfaces* **2019**, *11*, 12389–12400. [CrossRef] [PubMed]

50. Göhl, J.; Markstedt, K.; Mark, A.; Håkansson, K.; Gatenholm, P.; Edelvik, F. Simulations of 3D Bioprinting: Predicting Bioprintability of Nanofibrillar Inks. *Biofabrication* **2018**, *10*, 034105. [CrossRef] [PubMed]

51. Leppiniemi, J.; Lahtinen, P.; Paajanen, A.; Mahlberg, R.; Metsä-Kortelainen, S.; Pinomaa, T.; Pajari, H.; Vikholm-Lundin, I.; Pursula, P.; Hytönen, V.P. 3D-Printable Bioactivated Nanocellulose-Alginate Hydrogels. *ACS Appl. Mater. Interfaces* **2017**, *9*, 21959–21970. [CrossRef]

52. Alexandrescu, L.; Syverud, K.; Gatti, A.; Chinga-Carrasco, G. Cytotoxicity Tests of Cellulose Nanofibril-Based Structures. *Cellulose* **2013**, *20*, 1765–1775. [CrossRef]

53. Lou, Y.R.; Kanninen, L.; Kuisma, T.; Niklander, J.; Noon, L.A.; Burks, D.; Urtti, A.; Yliperttula, M. The Use of Nanofibrillar Cellulose Hydrogel as a Flexible Three-Dimensional Model to Culture Human Pluripotent Stem Cells. *Stem Cells Dev.* **2014**, *23*, 380–392. [CrossRef]

54. Liu, J.; Cheng, F.; Grénman, H.; Spoljaric, S.; Seppälä, J.; Eriksson, J.E.; Willför, S.; Xu, C. Development of Nanocellulose Scaffolds with Tunable Structures to Support 3D Cell Culture. *Carbohydr. Polym.* **2016**, *148*, 259–271. [CrossRef] [PubMed]

55. Kummala, R.; Xu, W.; Xu, C.; Toivakka, M. Stiffness and Swelling Characteristics of Nanocellulose Films in Cell Culture Media. *Cellulose* **2018**, *25*, 4969–4978. [CrossRef]

56. Nguyen, D.; Hgg, D.A.; Forsman, A.; Ekholm, J.; Nimkingratana, P.; Brantsing, C.; Kalogeropoulos, T.; Zaunz, S.; Concaro, S.; Brittberg, M.; et al. Cartilage Tissue Engineering by the 3D Bioprinting of IPS Cells in a Nanocellulose/Alginate Bioink. *Sci. Rep.* **2017**, *7*, 1–10. [CrossRef]

57. Shin, S.; Park, S.; Park, M.; Jeong, E.; Na, K.; Youn, H.J.; Hyun, J. Cellulose Nanofibers for the Enhancement of Printability of Low Viscosity Gelatin Derivatives. *BioResources* **2017**, *12*, 2941–2954. [CrossRef]

58. Bonnans, C.; Chou, J.; Werb, Z. Remodelling the Extracellular Matrix in Development and Disease. *Nat. Rev. Mol. Cell Biol.* **2014**, *15*, 786–801. [CrossRef] [PubMed]

59. Fitzgerald, K.A.; Malhotra, M.; Curtin, C.M.; O'Brien, F.J.; O'Driscoll, C.M. Life in 3D Is Never Flat: 3D Models to Optimise Drug Delivery. *J. Control. Release* **2015**, *215*, 39–54. [CrossRef]

60. Moroni, L.; De Wijn, J.R.; Van Blitterswijk, C.A. 3D Fiber-Deposited Scaffolds for Tissue Engineering: Influence of Pores Geometry and Architecture on Dynamic Mechanical Properties. *Biomaterials* **2006**, *27*, 974–985. [CrossRef]

61. Slaughter, B.V.; Khurshid, S.S.; Fisher, O.Z.; Khademhosseini, A.; Peppas, N.A. Hydrogels in Regenerative Medicine. *Adv. Mater.* **2009**, *21*, 3307–3329. [CrossRef]

62. Lu, P.; Takai, K.; Weaver, V.M.; Werb, Z. Extracellular Matrix Degradation and Remodeling in Development and Disease. *Cold Spring Harb. Perspect. Biol.* **2011**, *3*, a005058. [CrossRef]

63. Engler, A.J.; Sen, S.; Sweeney, H.L.; Discher, D.E. Matrix Elasticity Directs Stem Cell Lineage Specification. *Cell* **2006**, *126*, 677–689. [CrossRef] [PubMed]

64. Chinga-Carrasco, G.; Syverud, K. Pretreatment-Dependent Surface Chemistry of Wood Nanocellulose for PH-Sensitive Hydrogels. *J. Biomater. Appl.* **2014**, *29*, 423–432. [CrossRef] [PubMed]

65. Lu, T.; Li, Q.; Chen, W.; Yu, H. Composite Aerogels Based on Dialdehyde Nanocellulose and Collagen for Potential Applications as Wound Dressing and Tissue Engineering Scaffold. *Compos. Sci. Technol.* **2014**, *94*, 132–138. [CrossRef]

66. Yang, X.; Bakaic, E.; Hoare, T.; Cranston, E.D. Injectable Polysaccharide Hydrogels Reinforced with Cellulose Nanocrystals: Morphology, Rheology, Degradation, and Cytotoxicity. *Biomacromolecules* **2013**, *14*, 4447–4455. [CrossRef]

67. Innala, M.; Riebe, I.; Kuzmenko, V.; Sundberg, J.; Gatenholm, P.; Hanse, E.; Johannesson, S. 3D Culturing and Differentiation of SH-SY5Y Neuroblastoma Cells on Bacterial Nanocellulose Scaffolds. *Artif. Cells Nanomed. Biotechnol.* **2014**, *42*, 302–308. [CrossRef]

68. Krontiras, P.; Gatenholm, P.; Hägg, D.A. Adipogenic Differentiation of Stem Cells in Three-Dimensional Porous Bacterial Nanocellulose Scaffolds. *J. Biomed. Mater. Res. Part B Appl. Biomater.* **2015**, *103*, 195–203. [CrossRef]

69. Feldmann, E.M.; Sundberg, J.F.; Bobbili, B.; Schwarz, S.; Gatenholm, P.; Rotter, N. Description of a Novel Approach to Engineer Cartilage with Porous Bacterial Nanocellulose for Reconstruction of a Human Auricle. *J. Biomater. Appl.* **2013**, *28*, 626–640. [CrossRef]

70. Nimeskern, L.; Martínez Ávila, H.; Sundberg, J.; Gatenholm, P.; Müller, R.; Stok, K.S. Mechanical Evaluation of Bacterial Nanocellulose as an Implant Material for Ear Cartilage Replacement. *J. Mech. Behav. Biomed. Mater.* **2013**, *22*, 12–21. [CrossRef]

71. Apelgren, P.; Karabulut, E.; Amoroso, M.; Mantas, A.; Martínez Ávila, H.; Kölby, L.; Kondo, T.; Toriz, G.; Gatenholm, P. In Vivo Human Cartilage Formation in Three-Dimensional Bioprinted Constructs with a Novel Bacterial Nanocellulose Bioink. *ACS Biomater. Sci. Eng.* **2019**, *5*, 2482–2490. [CrossRef]

72. Sämfors, S.; Karlsson, K.; Sundberg, J.; Markstedt, K.; Gatenholm, P. Biofabrication of Bacterial Nanocellulose Scaffolds with Complex Vascular Structure. *Biofabrication* **2019**, *11*, 045010. [CrossRef]

73. Sionkowska, A.; Mężykowska, O.; Piątek, J. Bacterial Nanocelullose in Biomedical Applications: A Review. *Polym. Int.* **2019**, *68*, 1841–1847. [CrossRef]

74. Liu, J.; Chinga-Carrasco, G.; Cheng, F.; Xu, W.; Willför, S.; Syverud, K.; Xu, C. Hemicellulose-Reinforced Nanocellulose Hydrogels for Wound Healing Application. *Cellulose* **2016**, *23*, 3129–3143. [CrossRef]

75. Majcher, M.J.; McInnis, C.L.; Himbert, S.; Alsop, R.J.; Kinio, D.; Bleuel, M.; Rheinstädter, M.C.; Smeets, N.M.B.; Hoare, T. Photopolymerized Starchstarch Nanoparticle (SNP) Network Hydrogels. *Carbohydr. Polym.* **2020**, *236*, 115998. [CrossRef] [PubMed]

76. Fujisawa, S.; Atsumi, T.; Kadoma, Y.; Sakagami, H. Antioxidant and Prooxidant Action of Eugenol-Related Compounds and Their Cytotoxicity. *Toxicology* **2002**, *177*, 39–54. [CrossRef]

77. Jiang, Y.; Zhou, J.; Shi, H.; Zhao, G.; Zhang, Q.; Feng, C.; Xv, X. Preparation of Cellulose Nanocrystal/Oxidized Dextran/Gelatin (CNC/OD/GEL) Hydrogels and Fabrication of a CNC/OD/GEL Scaffold by 3D Printing. *J. Mater. Sci.* **2020**, *55*, 2618–2635. [CrossRef]

78. Dong, S.; Hirani, A.A.; Colacino, K.R.; Lee, Y.W.; Roman, M. Cytotoxicity and Cellular Uptake of Cellulose Nanocrystals. *Nano Life* **2012**, *2*, 1241006. [CrossRef]

79. Dugan, J.M.; Collins, R.F.; Gough, J.E.; Eichhorn, S.J. Oriented Surfaces of Adsorbed Cellulose Nanowhiskers Promote Skeletal Muscle Myogenesis. *Acta Biomater.* **2013**, *9*, 4707–4715. [CrossRef]

80. Kummala, R.; Soto Véliz, D.; Fang, Z.; Xu, W.; Abitbol, T.; Xu, C.; Toivakka, M. Human Dermal Fibroblast Viability and Adhesion on Cellulose Nanomaterial Coatings: Influence of Surface Characteristics. *Biomacromolecules* **2020**, *21*, 1560–1567. [CrossRef]

81. Chatterjee, S.; Chi-Leung Hui, P. Review of Stimuli-Responsive Polymers in Drug Delivery and Textile Application. *Molecules* **2019**, *24*, 2547. [CrossRef]

82. Zaborniak, I.; Chmielarz, P.; Matyjaszewski, K. Modification of Wood-Based Materials by Atom Transfer Radical Polymerization Methods. *Eur. Polym. J.* **2019**, *120*, 109253. [CrossRef]

83. Maver, U.; Gradišnik, L.; Smrke, D.M.; Stana Kleinschek, K.; Maver, T. Impact of Growth Factors on Wound Healing in Polysaccharide Blend Thin Films. *Appl. Surf. Sci.* **2019**, *489*, 485–493. [CrossRef]

84. Ojansivu, M.; Rashad, A.; Ahlinder, A.; Massera, J.; Mishra, A.; Syverud, K.; Finne-Wistrand, A.; Miettinen, S.; Mustafa, K. Wood-Based Nanocellulose and Bioactive Glass Modified Gelatin-Alginate Bioinks for 3D Bioprinting of Bone Cells. *Biofabrication* **2019**, *11*, 035010. [CrossRef]

85. Hupa, L.; Wang, X.; Eqtesadi, S. Bioactive Glasses. In *Springer Handbooks*; Springer: Berlin/Heidelberg, Germany, 2019; pp. 813–849.

86. Rashad, A.; Suliman, S.; Mustafa, M.; Pedersen, T.; Campodoni, E.; Sandri, M.; Syverud, K.; Mustafa, K. Inflammatory Responses and Tissue Reactions to Wood-Based Nanocellulose Scaffolds. *Mater. Sci. Eng. C* **2019**, *97*, 208–221. [CrossRef]

87. Czaja, W.; Kyryliouk, D.; Depaula, C.A.; Buechter, D.D. Oxidation of γ-Irradiated Microbial Cellulose Results in Bioresorbable, Highly Conformable Biomaterial. *J. Appl. Polym. Sci.* **2014**, *131*, 1–12. [CrossRef]

88. Harper, B.J.; Clendaniel, A.; Sinche, F.; Way, D.; Hughes, M.; Schardt, J.; Simonsen, J.; Stefaniak, A.B.; Harper, S.L. Impacts of Chemical Modification on the Toxicity of Diverse Nanocellulose Materials to Developing Zebrafish. *Cellulose* **2016**, *23*, 1763–1775. [CrossRef]

89. Deloid, G.M.; Cao, X.; Molina, R.M.; Silva, D.I.; Bhattacharya, K.; Ng, K.W.; Loo, S.C.J.; Brain, J.D.; Demokritou, P. Toxicological Effects of Ingested Nanocellulose in: In Vitro Intestinal Epithelium and in Vivo Rat Models. *Environ. Sci. Nano* **2019**, *6*, 2105–2115. [CrossRef]

90. Endes, C.; Camarero-Espinosa, S.; Mueller, S.; Foster, E.J.; Petri-Fink, A.; Rothen-Rutishauser, B.; Weder, C.; Clift, M.J.D. A Critical Review of the Current Knowledge Regarding the Biological Impact of Nanocellulose. *J. Nanobiotechnol.* **2016**, *14*, 78. [CrossRef]

91. Koivuniemi, R.; Hakkarainen, T.; Kiiskinen, J.; Kosonen, M.; Vuola, J.; Valtonen, J.; Luukko, K.; Kavola, H.; Yliperttula, M. Clinical Study of Nanofibrillar Cellulose Hydrogel Dressing for Skin Graft Donor Site Treatment. *Adv. Wound Care* **2020**, *9*, 199–210. [CrossRef]

92. Apelgren, P.; Amoroso, M.; Lindahl, A.; Brantsing, C.; Rotter, N.; Gatenholm, P.; Kölby, L. Chondrocytes and Stem Cells in 3D-Bioprinted Structures Create Human Cartilage in Vivo. *PLoS ONE* **2017**, *12*, e0189428. [CrossRef]

93. Huang, L.; Du, X.; Fan, S.; Yang, G.; Shao, H.; Li, D.; Cao, C.; Zhu, Y.; Zhu, M.; Zhang, Y. Bacterial Cellulose Nanofibers Promote Stress and Fidelity of 3D-Printed Silk Based Hydrogel Scaffold with Hierarchical Pores. *Carbohydr. Polym.* **2019**, *221*, 146–156. [CrossRef]

94. Yoon, S.; Park, J.A.; Lee, H.-R.; Yoon, W.H.; Hwang, D.S.; Jung, S. Inkjet-Spray Hybrid Printing for 3D Freeform Fabrication of Multilayered Hydrogel Structures. *Adv. Healthc. Mater.* **2018**, *7*, 1800050. [CrossRef] [PubMed]

95. Milojević, M.; Gradišnik, L.; Stergar, J.; Skelin Klemen, M.; Stožer, A.; Vesenjak, M.; Dobnik Dubrovski, P.; Maver, T.; Mohan, T.; Stana Kleinschek, K.; et al. Development of Multifunctional 3D Printed Bioscaffolds from Polysaccharides and NiCu Nanoparticles and Their Application. *Appl. Surf. Sci.* **2019**, *488*, 836–852. [CrossRef]

96. Henriksson, I.; Gatenholm, P.; Hägg, D.A. Increased Lipid Accumulation and Adipogenic Gene Expression of Adipocytes in 3D Bioprinted Nanocellulose Scaffolds. *Biofabrication* **2017**, *9*, 15022. [CrossRef] [PubMed]

97. Shin, S.; Kwak, H.; Shin, D.; Hyun, J. Solid Matrix-Assisted Printing for Three-Dimensional Structuring of a Viscoelastic Medium Surface. *Nat. Commun.* **2019**, *10*, 1–12. [CrossRef]

98. Shin, S.; Kwak, H.; Hyun, J. Transparent Cellulose Nanofiber Based Open Cell Culture Platform Using Matrix-Assisted 3D Printing. *Carbohydr. Polym.* **2019**, *225*, 115235. [CrossRef]

99. Kuzmenko, V.; Karabulut, E.; Pernevik, E.; Enoksson, P.; Gatenholm, P. Tailor-Made Conductive Inks from Cellulose Nanofibrils for 3D Printing of Neural Guidelines. *Carbohydr. Polym.* **2018**, *189*, 22–30. [CrossRef]

100. Chinga-Carrasco, G.; Ehman, N.V.; Filgueira, D.; Johansson, J.; Vallejos, M.E.; Felissia, F.E.; Håkansson, J.; Area, M.C. Bagasse—A Major Agro-Industrial Residue as Potential Resource for Nanocellulose Inks for 3D Printing of Wound Dressing Devices. *Addit. Manuf.* **2019**, *28*, 267–274. [CrossRef]

101. Ajdary, R.; Huan, S.; Zanjanizadeh Ezazi, N.; Xiang, W.; Grande, R.; Santos, H.A.; Rojas, O.J. Acetylated Nanocellulose for Single-Component Bioinks and Cell Proliferation on 3D-Printed Scaffolds. *Biomacromolecules* **2019**, *20*, 2770–2778. [CrossRef]

102. Xu, X.; Zhou, J.; Jiang, Y.; Zhang, Q.; Shi, H.; Liu, D. 3D Printing Process of Oxidized Nanocellulose and Gelatin Scaffold. *J. Biomater. Sci. Polym. Ed.* **2018**, *29*, 1498–1513. [CrossRef]

103. Prince, E.; Alizadehgiashi, M.; Campbell, M.; Khuu, N.; Albulescu, A.; De France, K.; Ratkov, D.; Li, Y.; Hoare, T.; Kumacheva, E. Patterning of Structurally Anisotropic Composite Hydrogel Sheets. *Biomacromolecules* **2018**, *19*, 1276–1284. [CrossRef]

104. Qian, F.; Zhu, C.; Knipe, J.M.; Ruelas, S.; Stolaroff, J.K.; Deotte, J.R.; Duoss, E.B.; Spadaccini, C.M.; Henard, C.A.; Guarnieri, M.T.; et al. Direct Writing of Tunable Living Inks for Bioprocess Intensification. *Nano Lett.* **2019**, *19*, 5829–5835. [CrossRef]

 *bioengineering*

Article

3D Printing High-Consistency Enzymatic Nanocellulose Obtained from a Soda-Ethanol-O_2 Pine Sawdust Pulp

Heli Kangas [1,*], Fernando E. Felissia [2], Daniel Filgueira [3], Nanci V. Ehman [2],
María E. Vallejos [2], Camila M. Imlauer [2], Panu Lahtinen [1], María C. Area [2,*] and
Gary Chinga-Carrasco [3,*]

[1] VTT Technical Research Centre of Finland Ltd., P.O. Box 1000, FI-02044 VTT, Finland
[2] Instituto de Materiales de Misiones (IMAM), Félix de Azara 1552, 3300 Posadas, Misiones, Argentina
[3] RISE PFI, Høgskoleringen 6b, 7491 Trondheim, Norway
* Correspondence: heli.Kangas@vtt.fi (H.K.); cristinaarea@gmail.com (M.C.A.);
   gary.chinga.carrasco@rise-pfi.no (G.C.-C.)

Received: 13 June 2019; Accepted: 13 July 2019; Published: 16 July 2019

Abstract: Soda-ethanol pulps, prepared from a forestry residue pine sawdust, were treated according to high-consistency enzymatic fibrillation technology to manufacture nanocellulose. The obtained nanocellulose was characterized and used as ink for three-dimensional (3D) printing of various structures. It was also tested for its moisture sorption capacity and cytotoxicity, as preliminary tests for evaluating its suitability for wound dressing and similar applications. During the high-consistency enzymatic treatment it was found that only the treatment of the O_2-delignified pine pulp resulted in fibrillation into nano-scale. For 3D printing trials, the material needed to be fluidized further. By 3D printing, it was possible to fabricate various structures from the high-consistency enzymatic nanocellulose. However, the water sorption capacity of the structures was lower than previously seen with porous nanocellulose structures, indicating that further optimization of the material is needed. The material was found not to be cytotoxic, thus showing potential as material, e.g., for wound dressings and for printing tissue models.

Keywords: pine sawdust; soda ethanol pulping; nanocellulose; 3D printing; cytotoxicity

1. Introduction

Balancing environment, biodiversity, and economic development can be performed through ecosystem services, climate change mitigation strategies, and long-term food security balance. In this context, the possibility of recycling and reusing a resource represents an issue of fundamental importance in the economy of various industrial sectors. The biorefinery of agro and industrial forestry waste implies its integral use and its valorization, being able to satisfy the needs of food, raw materials, and energy, respecting the principles of sustainability. More than 65% of the plantations in Argentina are in two provinces in the Nord-East (Misiones and Corrientes), of which approximately 60% is pine, since they have very favorable conditions for the growth of species such as *P. elliottii* and *P. taeda* (yields above 20 m^3/year). Therefore, pine sawdust is the most important waste of primary wood processing. This waste is not used properly and its accumulation contributes to the pollution of the environment [1].

Fractionation is a sequence of processes, which allow for recovering the different chemical components of a raw material. Different methods have been investigated for pine sawdust fractionation, including hot water, dilute acid, steam explosion, acid organosolv, and traditional alkaline treatments like soda-anthraquinone and kraft [2–8]. This raw material has been proven to be very recalcitrant and acid treatments before the delignification stage produced alterations in the structure that made the fractionation even more difficult [9]. On the contrary, the use of organic solvents combined with bases seems to be promising. Ethanol reduces the surface tension of the pulping liquor favoring the alkali penetration into the material structure [10]. The organic solvent also alters lignin-carbohydrate bonds by hydrolyzing lignin which is then dissolved in the organophilic phase [11]. This processing produces a pulp enriched in cellulose and a less condensed lignin. In studies comparing acid processes for bioethanol production, the organosolv treatment presents higher energy consumption than the diluted acid treatment [12]. Nevertheless, in organosolv processes, a high-quality lignin can be recovered. On the contrary, if in addition to lignin a high-added-value product such as nanocellulose is produced, the process turns auspicious [13]. In addition to the above, the alkaline processes have a much stronger recovery system than the acid ones. Another strategy for further extraction of lignin from the lignocellulosic material after pulping is oxygen delignification, using oxygen and alkali in a pressurized system. Oxygen reactions are generated by radicals which react with lignin removing it in a fraction corresponding to 25% to 65% of the initial kappa number of the pulp [14]. Kappa refers to the lignin content in a pulp, the higher the kappa number the higher the lignin content.

Nanocelluloses are promising bio-based materials for numerous applications, either as replacement of traditional oil-based materials in existing products or in generating completely new materials and products. They can be roughly divided into three categories based on their production methods and properties: Cellulose nanofibrils (CNF), Cellulose nanocrystals (CNC), and bacterial cellulose (BC). Of these, CNF are manufactured by mechanical treatments, often combined with chemical or enzymatic pre-treatment, and different manufacturing methods result in materials with variable properties. Research around different production methods of nanocelluloses has continued actively for over a decade and resulted in pilot, pre-commercial, and even commercial plants all over the world. Despite the rapid development, some challenges still remain with the traditional production technologies. Considering the mechanically manufactured nanocelluloses, the production costs are usually still high, and the resulting material is at low consistency, typically between 1%–3%. The high-water content generates problems, such as difficulties in dewatering and drying, problems in post-treatment, and restricted applicability for certain applications, such as composites or paints and coatings. In addition, long-distance transportation is not feasible leading to limited availability of the material.

To overcome the problems related to high energy consumption of nanocellulose manufacturing and low solids content of the resulting material, a high-consistency enzymatic fibrillation (HefCel) technology has been developed [15]. The benefits of the HefCel nanocellulose are its high-consistency after processing (10%–25%) and lower energy consumption compared to other manufacturing methods [16]. However, similar to other nanocelluloses, HefCel nanocellulose is a potential raw material for many different applications, such as strength additive in the middle ply of board [16] or as a barrier film in packaging materials [17].

During the last years, nanocellulose has been in the focus as a component in bioinks for three-dimensional (3D) printing [18]. The material has shear thinning behavior and consolidates rapidly after deposition on a substrate. These characteristics make it possible to deposit nanocellulose layer-by-layer in order to construct geometrically complex 3D objects. Rees et al. demonstrated the potential of nanocelluloses to be 3D-printed for constructing porous structures with potential as wound dressing materials [19]. Although nanocellulose with appropriate rheological properties are 3D printable, nanocellulose constructs require a post-treatment to be mechanically stable. Nanocellulose inks are usually combined with additional polymers such as alginate and cross-linked with Ca^{2+}, thus yielding a mechanically stable 3D construct [20] with potential as scaffolds for tissue engineering. The composition of inks for 3D printing can be optimized with additional components to tailor the printability and shape fidelity, and to keep the stability at room temperature [21]. Most of the work on nanocellulose for 3D printing has been based on wood nanocellulose from market chemical pulp.

However, recently it has been demonstrated that agro-industrial residues such as bagasse also have potential as resources for nanocellulose production and 3D printing [22].

The purpose of this work was to demonstrate and prove the suitability of a soda-ethanol pulping to yield fibres with adequate composition for production of nanocellulose by high-consistency enzymatic fibrillation, which in turn has appropriate properties for 3D printing. In addition, the cytotoxicity of the produced nanocellulose was tested to evaluate its potential suitability for wound dressing and similar applications.

2. Materials and Methods

2.1. Forestry Residues

Pinus elliotti and *Pinus taeda* sawdust mix was provided by a local sawmill (Forestal Eldorado and Forestal AM, Misiones). The sawdust was air-dried, screened, and maintained in closed plastic bags. The fraction passing 5 mm^2-screen was used.

Before each experiment, sawdust samples were impregnated with water overnight to eliminate bark particles by flotation.

Acid-insoluble lignin (Klason lignin) and pulp structural carbohydrates were measured according to the "Determination of Structural Carbohydrates and Lignin in Biomass" NREL/TP-510-42618. Hydrolysate samples from the aforementioned technique were neutralized with Ba(OH)$_2$ following the methodology proposed by Kaar et al., HPLC with a SHODEX SP810 column was used to determine the carbohydrates content (glucans, xylans, mannans, galactans, and arabinans) [23]. The operational conditions used were water as eluent, 0.6 mL/min, 85 °C, and a refractive index detector. The final chemical composition of the pulp was completed by using an Aminex-HPX87H column (BIO-RAD), operated under the following conditions: 4 mM of H$_2$SO$_4$ as eluent, 0.6 mL/min, 35 °C, and a diode array detector.

2.2. Soda Ethanol Pulping

Soda-ethanol pulping was performed in a 7 L pressurized reactor (M/K Systems, Inc., Peabody, MA, USA), with direct heating and liquor circulation. About 500 g of dry sawdust was cooked. The pulping conditions are shown in Table 1. After the pulping stage, the spent liquor was separated from the pulp by filtration. The pulp was subjected to a 5-cycle washing with water, screened by means of a Somerville device, and properly stored in plastic bags.

Table 1. Pulping conditions.

Parameter	Level
Maximum Temperature (T$_{máx.}$)	170 °C
Time at Maximum Temperature (t$_{máx.}$)	100 min
Time to Maximum Temperature (t$_{heating}$)	60 min
Alkaline load (AL)	23.3% odw
Ethanol:Water Ratio (EtOH:H$_2$O)	35:65% v/v
Liquor:Wood Ratio (L:W)	5.44:1

% odw: oven dried weight on mass of dry wood; *v/v*: percentage in volume.

Residual alkali was measured from the black liquor according to the SCAN-N 33:94 method ("Residual Alkali – Hydroxide Ion Content"), and alkali consumption (%) was calculated. The Kappa number was determined following the TAPPI T236 om-99 procedure and yield was determined.

2.3. Oxygen Stage

The soda-ethanol pulps were treated with oxygen in two stages. Oxygen stages were conducted in a multipurpose reactor, equipped with a high shear rotor which generates the required agitation conditions, an oxygen inlet valve, and a heating system. In both stages the pulp was treated at 100 °C for 60 min, an oxygen pressure of 600 KPa and a consistency of 10%. The alkaline load was 3% on dry pulp in the first stage and 2% on dry pulp in the second stage.

2.4. Chemical Composition of the Pine Sawdust Pulps

The content of acetone extractives was analysed according to standard SCAN-CM 49:03 and the lignin content and carbohydrate composition according to SCAN-CM 71:09. The samples were freeze-dried prior to analysis. The metal (Si, Fe, Mg, Mn, Co, and Ca) contents of the pulps were determined after wet combustion by inductively coupled plasma optical emission spectrometry (ICP-OEP).

2.5. Enzymatic Fibrillation

Two different types of pine sawdust pulps, namely soda-ethanol pulp and oxygen delignified (O_2) soda-ethanol pulp, were used as raw materials for producing nanocellulose according to the HefCel technology. Prior to the treatments, the pulp samples were washed into Na^+ form according to the method described by Lahtinen et al. [24]. The enzymatic treatment with cellulase mixture was carried out at a consistency of 25% dry weight (% odw) for 9 h (soda ethanol) and 5.5 h (O_2 delignified soda-ethanol) at 70 °C and pH 5 using a two-shaft sigma mixer (Jaygo Incorporated, Randolph, NJ, USA) running at 25 rpm. The enzyme dosage in both treatments was 8 mg/g and the pulp batch size 300 g (% odw). After the treatments the enzyme was inactivated by increasing the temperature in the mixer to 90 °C for 30 min. The fibrillated material was diluted with deionised water, filtered, and washed thoroughly with deionised water. Finally, the fibrillated material was dewatered to a consistency of 18.6% (soda ethanol HefCel) and 21% (O_2 soda ethanol HefCel) by filtration. Yield of the fibrillated cellulose material was 91% for the soda ethanol HefCel and 88% for the O_2 delignified soda ethanol HefCel. In order to prevent contamination prior to characterization, the HefCel materials were further autoclaved at 121 °C for 20 min and sealed. The materials were stored at +4 °C until used.

After initial testing by 3D printing, the O_2 delignified soda-ethanol HefCel was further fluidized. Fluidization was done with a Microfluidics microfluidizer type M110-EH. Two passes were run through the chambers having a diameter of 400 μm and 100 μm at 1,800 bar operating pressure. The first pass was at 8% solids and the second pass was at 6% solids.

2.6. Characterisation of Nanocellulose

Preliminary characterization of the HefCel nanocellulose samples included pH, conductivity, dry matter content, optical microscopy, and rheological properties. A standard portable device Metler Toledo SG2 and Inlab 413SG electrode were used for the pH and Jenway 4510 for the conductivity measurement. Dry matter content measurement was based on oven drying.

2.6.1. Apparent Viscosity and Yield Value

The shear viscosities of dilute HefCel nanocellulose samples were measured by a Brookfield rheometer model RVDV-III Ultra using vane-type spindles. The samples were diluted to 5% concentration with Milli-Q water and dispersed with an Ultra-Turrax disperser at 14,000 rpm for approximately 2 min. Viscosity measurements were performed in a 250 mL Pyrex beaker, and each sample was left to settle for a minimum of 30 min at room temperature after the dispersion. This allowed the samples to regain their initial viscosity. The temperature of the samples was adjusted to 20 ± 1 °C. The shear viscosity was measured at 300 measuring points at 0.5 rpm and at 180 measuring points at 10 rpm. The yield stress value was recorded at 0.5 rpm. The apparent viscosities and yield values were measured twice for each sample. Light mixing was performed between the measurements.

2.6.2. Residual Fibre Analysis

Characterization of residual fibres from the HefCel nanocellulose samples was carried out with a Kajaani FibreLab analyser. Each sample was firstly soaked in 0.4 g/L consistency and dispersed using a high shear laboratory blender for 2–3 min. Then the samples were further mixed in 5 L of water at 40 mg/L consistency with an impeller for 10 min. Some 50 mL of dispersed sample or 2 mg in dry weight was pipetted into the analyser and fibre analysis was performed with two repeat measurements. The number of fibres recorded by the analyser was divided by the sample amount, which gives the value for the residual fibres in the sample.

2.6.3. Optical Microscopy Imaging

The HefCel nanocellulose samples were dyed with 1% Congo red solution by mixing nanocellulose and dye in a ratio of 1:1 and further diluting the dyed mixture on the microscope slide (2:1). Optical microscopy was performed with an Olympus BX 61 microscope equipped with WH10X-H eyepieces, fluorite objectives, and a ColorView 12 camera.

2.6.4. Laser Profilometry

Structures of 20 mm × 40 mm were printed directly on microscopy slides (one layer), using a Regemat3D printing unit. The structures were printed with a conical nozzle (size 0.58 mm) and a flow speed of 3.0 mm/s. The structures were allowed to dry for one day at room temperature (23 °C). The microscopy slides were sputtered with a layer of gold for laser profilometry analysis. Ten laser profilometry images (10 × 10 mm, 1.0 µm resolution) were acquired randomly. The laser profilometry images were assessed with the SurfCharJ plugin (v. 1q) for ImageJ (v. 1.50i). The 3D plots were created with the Interactive 3D surface plot plugin (v. 2.4).

2.7. Three-Dimensional (3D) Printing

The nanocellulose gel (fluidized HefCel, 5 wt%) was used as inks for 3D printing. The 3D printing was performed with a Regemat3D bioprinter (version 1.0), equipped with the Regemat3D Designer (version 1.8, Regemat3D, Granada, Spain). Grids having a diameter of 20 mm, a height of 2 mm, and space between the tracks of 2 mm were printed. The inks were kept at room temperature (25 °C) for 24 h before printing. The flow speed was 3 mm/s, using a 0.58 mm conical nozzle. Fluidized HefCel was also combined with alginate (20%) before printing following the same set-up described earlier in this section. After printing, the grids were left in a solution of $CaCl_2$ (50 mmol) for 24 h to cross-link the structure. The grids without and with alginate were freeze-dried with a Telstar LyoQuest-83 for 24 h.

The freeze-dried samples were immersed in MQ water, and the water holding capacity of the 3D-printed grids was estimated according to Equation (1):

$$\text{Moisture sorption capacity (\%)} = (\text{Wt} - \text{Wo})/\text{Wo} \times 100 \tag{1}$$

where, Wt is the weight of the grid at given time points and W0 is the weight of the dry grid.

2.8. Cytotoxicity

The cytotoxicity of extracts from soda-ethanol HefCel, O_2 delignified soda-ethanol HefCel, and fluidized O_2 delignified soda-ethanol HefCel was evaluated according to ISO 10993-5:2009 annex C (3-(4,5-dimethylthiazol-2-yl)-2,5-diphenyltetrazolium bromide (MTT) cytotoxicity test) with extraction according to ISO 10993-12:2012. The HefCel samples were tested at two occasions: first, the unbleached and O_2 delignified HefCel nanocellulose samples and later, the fluidized O_2 delignified HefCel nanocellulose.

The MTT test is based on the measurement of the viability of the cells via metabolic activity. Yellow water-soluble MTT is metabolically reduced in viable cells to a blue-violet insoluble formazan.

The number of viable cells correlates to the colour intensity determined by photometric measurements after dissolving the formazan in alcohol. Extracts from the test item, positive and negative controls, as well as blanks (extraction vehicle not containing the test item but subjected to conditions identical to those to which the test item was subjected to during extraction) were added to a subconfluent monolayer of L929 mouse fibroblast cells and incubated for 24 h at 37 ± 1 °C in $5\% \pm 1\%$ CO_2. After incubation the extracts were removed and MTT solution was added to the cells which were incubated for an additional 2 h at 37 ± 1 °C in $5\% \pm 1\%$ CO_2. Following incubation, the MTT solution was removed, 2-propanol was added, and the plates were shaken rapidly. Finally, the absorbance was measured at 570 nm (reference wavelength 650 nm) and the viability of cells was calculated. Thermanox Plastic Coverslips, Art no 174934 (Thermo Scientific NUNC) were used as the negative control and Latex rubber, Gammex 91-325 (AccuTech Ansell) as the positive control.

3. Results and Discussion

The chemical composition of the raw material, pine sawdust, is shown in Table 2. The amount of total carbohydrates (cellulose and hemicelluloses) were 62.45%, the rest of this material was composed of acid-insoluble lignin (29.16%) and extractives (2.27%).

Table 2. Chemical composition of pine sawdust.

Composition	% odw	StD
Ash	0.04	0.01
Total Extractives	2.27	
Extractives in Ethanol	1.54	0.03
Extractives in Water	0.73	0.01
Acid-Insoluble Lignin	29.16	0.10
Total Carbohydrates	62.45	
Glucans	40.30	0.38
Xylans	6.29	0.10
Galactans	2.18	0.08
Arabinans	0.77	0.02
Mannans	11.69	0.16
Acetyl Groups	1.22	0.02

% odw (% on mass of dry wood); StD (standard deviation).

The yields and kappa numbers of the unbleached and O_2 delignified pulps are shown in Table 3. The O_2 delignification reduced the residual lignin to almost half the kappa number of the unbleached soda-ethanol pulp in Stage 1, and to a third in Stage 2.

Table 3. Yield and kappa number of soda-ethanol pulps.

Pulp	Stage *n*	Yield (%)	Kappa Number
Unbleached Soda-ethanol Pulp		43.6	29.9
O_2 delignified pulps	Stage 1	97.4	13.8
	Stage 2	96.5	9.5

The chemical compositions of the unbleached soda-ethanol and O_2 delignified pulp (from Stage 2) are summarized in Table 4. The O_2 delignification removed almost 70% of total lignin in the unbleached pulp, which slightly increased the composition of carbohydrates in the O_2 delignified pulps. The results show that a significant amount of lignin was removed from the pulp during O_2 delignification, while very little changes were seen in the carbohydrate content or composition. Some metals, such as Si, Fe, and Mn, were removed from the pulp during the delignification treatment, while for some the amount increased.

Table 4. Chemical composition of soda-ethanol pulps.

Chemical Composition	Unbleached Pulp	O_2 Delignified Pulp (Stage 2)
Total Lignin (%)	4.1	1.2
Klason lignin (%)	3.7	0.7
Acid soluble lignin (%)	0.4	0.5
Acetone extract (%)	0.05	0.05
Total carbohydrates (%)	96.1	97.6
Arabinose (%)	0.4	0.4
Galactose (%)	0.5	0.3
Glucose (%)	82.9	84.1
Cellulose (%)	74.6	75.7
Xylose (%)	8.5	8.3
Xylan (%)	7.5	7.3
Mannose (%)	7.7	7.3
Mannan (%)	6.9	6.6
Si, mg/kg	109	60.1
Fe, mg/kg	102	87.8
Mg, mg/kg	438	515
Mn, mg/kg	47.2	35.7
Co, mg/kg	<0,5	1.5
Ca, mg/kg	1080	1180

3.1. Nanocellulose Characterisation

Optical microscopy images of the HefCel nanocellulose samples provide information of their visual appearance, such as fibrillation degree and amount of unfibrillated fibres (Figure 1). Clearly, the degree of fibrillation was better in the case of the O_2 delignified pulp and hardly any intact fibres could be detected. Unbleached material contained many more longer fibres after the enzymatic treatment. The bleached sample was cut into shorter fibre fragments (see Figure 1, middle) and thus, the amount of residual fibre particles was found to be higher according to the fibre analysis (Table 5). In the HefCel process, the accessibility of the enzyme to the cellulose fibre surface is a prerequisite, therefore, the lignin content or the number of adhesives or other impurities cannot be too high in the starting materials. This is probably the reason for the differences observed between the unbleached and O_2 delignified HefCel samples (Table 4, Figure 1, left and middle).

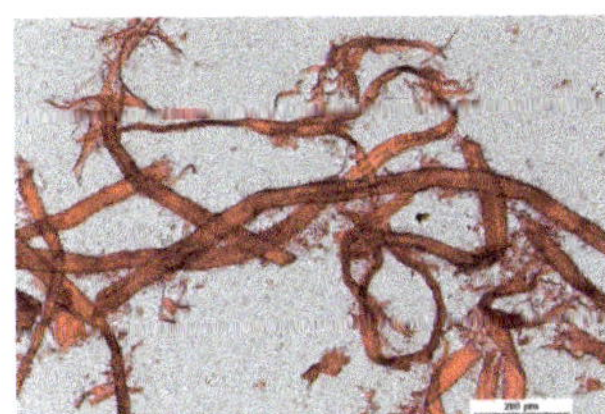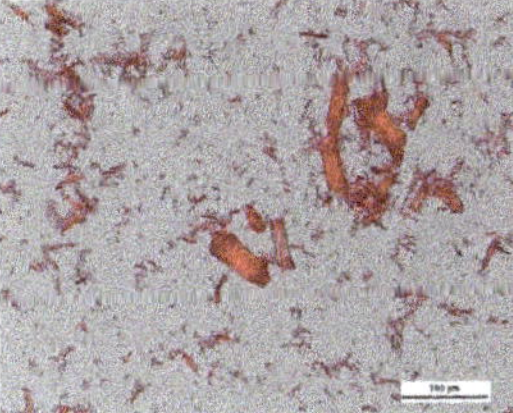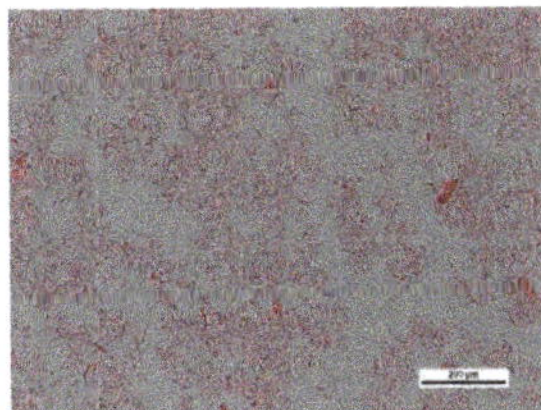

Figure 1. Optical microscopy images. (**left**) Unbleached HefCel, (**middle**) O_2 delignified Hefcel, and (**right**) Fluidized O_2 delignified HefCel. The bars are 200 μm.

Table 5. Characteristics of unbleached and bleached HefCel samples.

Sample	Yield Value (Pa)	StD	Apparent Viscosity, 10 rpm (mPa*s)	StD	Residual Fibres (pcs/mg)	StD
Unbleached HefCel	55	1	5853	337	31,310	34
O_2 delignified HefCel	0.4	0.1	487	14	49,703	2
Fluidized O_2 Delignified HefCel	314	13	121,580	1851	5923	115

Fluidization had a significant effect on the quality of the HefCel nanocellulose. When the O_2 delignified HefCel sample was fluidized twice, the amount of residual fibre particles decreased to 5923 pieces (pcs)/mg from the original value of 49,703 pcs/mg. This change in the quality was also noticed in the microscopy image (Figure 1, right). The sample became more homogeneous and the low shear apparent viscosity increased from 487 mPa*s to 121,580 mPa*s when measured at the fixed 5% consistency.

The laser profilometry analysis revealed the development of the HefCel nanocellulose samples before and after fluidization (Figure 2). The unfibrillated residual fibres observed in Figure 1 and detected by the residual fibre analysis (Table 5) caused a roughening of the surface structure of the films (Figure 2, left). The additional fluidization of the HefCel increased the fibrillation degree, thus reducing the amount of residual fibres. The surface roughness as a function of lateral wavelength clearly showed that the roughness decreased significantly after the fluidization (see wavelengths of 40 and 320 micrometers, Figure 3). This indicates a reduction of residual fibres and an increase of the nanofibril fraction [25].

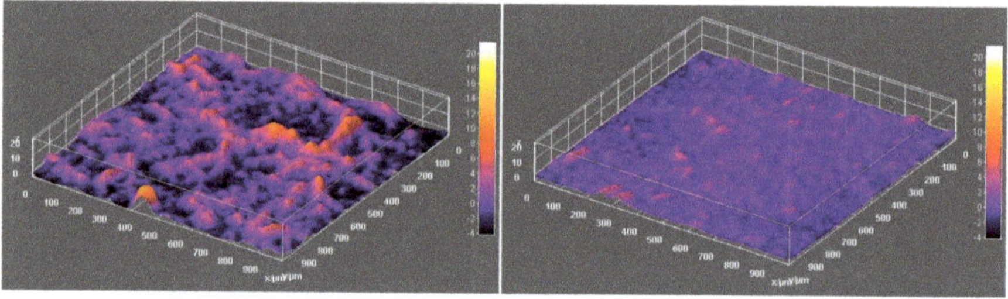

Figure 2. Images from laser profilometry. (**left**) O_2 delignified Hefcel and (**right**) Fluidized O_2 delignified Hefcel. The lateral size of the assessed areas was 1000×1000 μm. The z-direction calibration bar is between −4 and 20 μm.

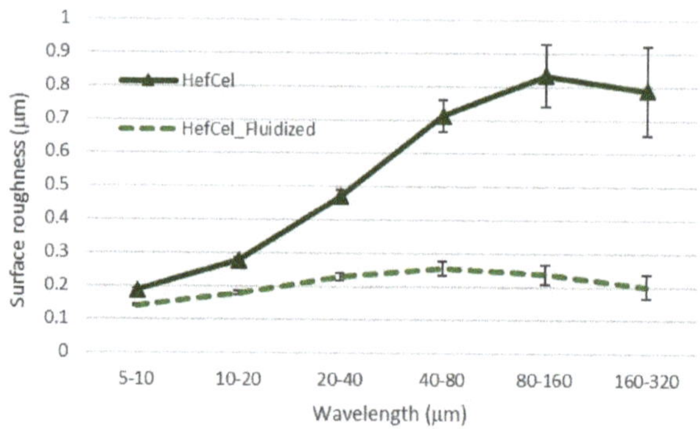

Figure 3. Laser profilometry. Surface roughness as a function of lateral wavelength. Note the large reduction of surface roughness after fluidization.

3.2. 3D Printing

Various 3D structures were printed to verify the suitability of the fibrillated material to be deposited on a substrate layer-by-layer and form a 3D object (Figure 4). The unbleached and O_2 delignified HefCel nanocelluloses were not printable, probably due to a relatively large fraction of residual fibres (Figures 1–3, Table 5) that clogged the printing nozzle. However, after an additional homogenization with a microfluidizer the O_2 delignified Hefcel sample performed well and the material could be extruded through a 0.4 mm nozzle (Figure 4). This behaviour was due to the additional fibrillation of residual fibres into micro- and nano-scale objects (Figure 3).

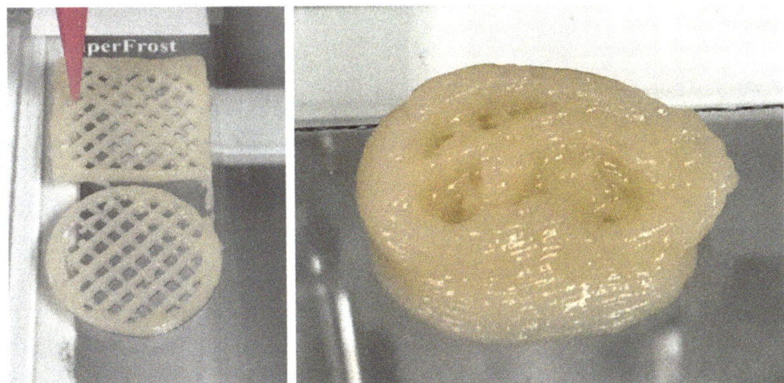

Figure 4. Three-dimensional (3D) printing of fluidized O_2 delignified HefCel. (**left**) Circle (radius: 10 mm) and (**right**) squares (size: 20 mm × 20 mm) exemplifying the good print resolution and print fidelity. Right) Exemplification of a 3D-printed structure, i.e., a self-standing ear.

Fluidized O_2 delignified HefCel was also combined with alginate and cross-linked with calcium as a post-fixing step to solidify the structures (Figure 5). The 3D-printed grids were freeze-dried and immersed in water to assess the stability of the material and their water absorption capacity. The HefCel grid without alginate dissolved rapidly in water and it was not possible to quantify the water absorption capacity. The HefCel combined with alginate (20%) and cross-linked with Ca^{2+} performed well and the structure was stable even after several cycles of water immersion and lifting up to quantify the water absorption. The water absorption capacity reached a 450% weight increase after 78 h in water (Figure 6). This value is considerably lower than the previously reported values for porous nanocellulose structures, which reached levels of up to 10,000–16,000% water uptake [26,27] The relatively low water absorption capacity of the cross-linked HefCel sample was most probably due to two factors: (i) the relatively low fibrillation degree of HefCel and (ii) the cross-linking of the material which limits the water absorption. It is important to note that the water absorption capacity is one of the tests that are recommended for characterizing wound dressings [28].

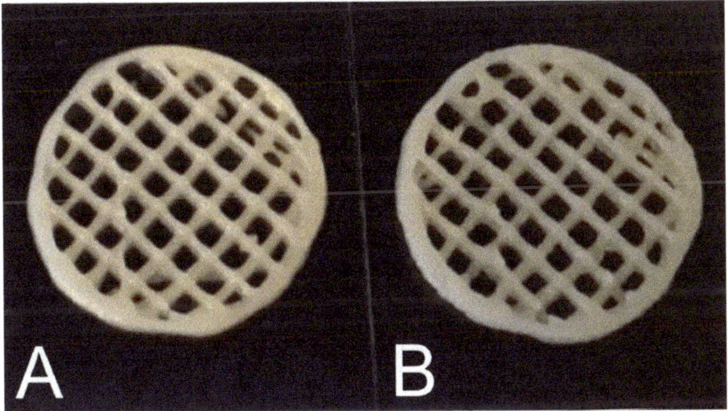

Figure 5. (**A**) Fluidized HefCel material. (**B**) Fluidized HefCel material and 20 wt% alginate. The radius of the circles is 10 mm.

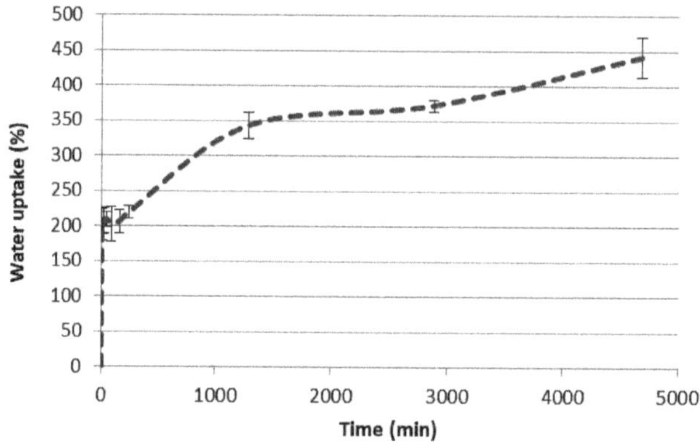

Figure 6. Water uptake of 3D-printed constructs containing HefCel, 20 wt% alginate and cross-linked with CaCl$_2$. The water uptake was measured over 78 h.

3.3. Cytotoxicity

The tested samples (unbleached HefCel, O$_2$ delignified HefCel, and fluidized O$_2$ delignified HefCel nanocellulose) were found not to have cytotoxic potential according to the criteria given in the testing standard. The calculated cell viabilities are presented in Table 6. It is worth noting that the unbleached HefCel sample is just slightly above the limit to be considered not cytotoxic (70% cell viability). This indicates that this sample has a significantly lower cell viability compared to the O$_2$ delignified and fluidized samples. The relatively lower cell viability caused by the unbleached HefCel sample may be caused by impurities and contamination that have not been removed by the pulping and nanocellulose production processes. The O$_2$ delignified samples are subjected to more steps of processing and washing, which may have contributed to removing the potential impurities encountered in the raw material.

Table 6. Calculated average cell viabilities and cytotoxicity grading according to ISO10993-5:2009.

Sample	Viability, % (Average)	STD	Cytotoxicity Grading
Unbleached HefCel	72.9	2.8	Not cytotoxic
O$_2$ delignified HefCel	95.1	6.3	Not cytotoxic
Fluidized O$_2$ delignified HefCel	96.9	1.4	Not cytotoxic
Positive control	1.1	0.1	Cytotoxic
Negative control	106.3	4.8	Not cytotoxic

4. Conclusions

We provided an extensive characterization of pine sawdust pulp and produced high-consistency nanocellulose for 3D printing. O$_2$ delignified soda-ethanol pulp was successfully fibrillated using high-consistency enzymatic fibrillation (HefCel) technology. However, a further homogenisation step by microfluidizer was needed to meet the requirements of 3D printing. After this step, various 3D structures of HefCel nanocellulose were successfully printed, with or without alginate. The structures containing alginate and cross-linked with CaCl$_2$ were more stable against water. However, their water absorption capacity was lower compared to similar structures prepared from other types of nanocelluloses, possibly due to their limited fibrillation degree and use of a cross-linking agent. No cytotoxicity was observed for the HefCel materials studied, which is the first indication of the material suitability for wound dressing or similar applications.

Author Contributions: Conceptualization, H.K., M.C.A. and G.C.-C.; methodology, F.E.F, D.F., N.V.E, M.E.V., C.M.I., P.L. G.C.-C.; formal analysis, P.L., M.C.A. G.C.-C.; investigation, F.E.F, D.F., N.V.E, M.E.V., C.M.I., P.L.

G.C.-C.; resources, H.K., M.C.A. and G.C.-C.; writing—original draft preparation, H.K., M.C.A. and G.C.-C.; writing—review and editing, H.K., D.F. M.E.V. P.L., M.C.A. and G.C.-C.; supervision, H.K., M.C.A. G.C.-C.; project administration, H.K., M.C.A. G.C.-C.; funding acquisition, H.K., M.C.A. and G.C.-C.

Funding: This work has been funded by the ValBio-3D project (Grant ELAC2015/T03-0715 Valorization of residual biomass for advanced 3D materials). The authors acknowledge the Consejo Nacional de Investigaciones Científicas y Técnicas (CONICET), the Universidad Nacional de Misiones (Argentina), the Research Council of Norway (Grant no. 271054), and Academy of Finland (Agreement No. 311973, VTT) for the financial support.

Conflicts of Interest: The authors declare no conflict of interest.

References

1. Area, M.C.; Vallejos, M.E. La Biorrefinería Forestal, Chapter 1. In *Biorrefinería a Partir de Residuos Lignocelulosicos*; Area, M.C., Vallejos, M.E., Eds.; Editorial Academica Española: Saarbrücken, Germany, 2012; ISBN 978-3-659-05295-8.
2. Stoffel, R.B.; Felissia, F.E.; Silva Curvelo, A.A.; Gassa, L.M.; Area, M.C. Optimization of sequential alkaline-acid fractionation of pine sawdust for a biorefinery. *Ind. Crops Prod.* **2014**, *61*, 160–168. [CrossRef]
3. Stoffel, R.B.; Neves, P.V.; Felissia, F.E.; Ramos, L.P.; Gassa, L.M.; Area, M.C. Hemicellulose extraction from slash pine sawdust by steam explosion with sulfuric acid. *Biomass Bioenergy* **2017**, *107*, 93–101. [CrossRef]
4. Imlauer, C.; Vallejos, M.E.; Area, M.C.; Felissia, F.E.; Ramankutty, N.; da Silva Curvelo, A.A. Hydrothermal treatment and organosolv pulping of softwood assisted by carbon dioxide. In Proceedings of the 51° Congresso Internacional de Celulose e Papel—X Congresso Iberoamericano de Pesquisa em Celulose e Papel, São Paulo, Brazil, 12 March 2018.
5. Imlauer, C.; Vergara, P.; Area, M.C.; Revilla, E.; Felissia, F.; Villar, J.C. Fractionation of Pinus radiata wood by combination of steam explosion and organosolv delignification. *Maderas Cienc. y Tecnol.* **2018**, *21*. (Accept).
6. Imlauer, C.M.; Ehman, N.V.; Area, M.C.; Felissia, F. Soda/ethanol-oxygen delignification of pine sawdust for a biorefinery. In Proceedings of the 3er Congreso Iberoamericano sobre Biorrefinerías (CIAB), 4to Congreso Latinoamericano sobre Biorrefinerías, y 2do Simposio Internacional sobre Materiales Lignocelulósicos, Concepción, Chile, 23–25 November 2015.
7. Imlauer, C.M.; Kruyeniski, J.; Area, M.C.; Felissia, F.E. Fraccionamiento a la soda-AQ de aserrín de pino para la biorrefinería forestal. In Proceedings of the VIII Congreso Iberoamericano de Investigación en Celulosa y Papel, Medellìn, Colombia, 26–28 November 2014.
8. Kruyeniski, J.; Felissia, F.E.; Area, M.C. Pretreatment soda-ethanol of pine and its influence on enzymatic hydrolysis. *Rev. Cienc. y Tecnol.* **2017**, *28*, 38–42.
9. Das, P.; Stoffel, R.B.; Area, M.C.; Ragauskas, A.J. Effects of one-step alkaline and two-step alkaline/dilute acid and alkaline/steam explosion pretreatments on the structure of isolated pine lignin. *Biomass Bioenergy* **2019**, *120*, 350–358. [CrossRef]
10. Muurinen, E. *Organosolv Pulping. A Review and Distillation Study Related to Peroxyacid Pulping*; Oulun Yliopisto: Oulu, Finland, 2000.
11. Pan, X.; Arato, C.; Gilkes, N.; Gregg, D.; Mabee, W.; Pye, K.; Xiao, Z.; Zhang, X.; Saddler, J. Biorefining of softwoods using ethanol organosolv pulping: Preliminary evaluation of process streams for manufacture of fuel-grade ethanol and co-products. *Biotechnol. Bioeng.* **2005**, *90*, 473–481. [CrossRef] [PubMed]
12. Kautto, J.; Realff, M.J.; Ragauskas, A.J.; Kässi, T. Economic analysis of an organosolv process for bioethanol production. *BioResources* **2014**, *9*, 6041–6072. [CrossRef]
13. Albarelli, J.; Paidosh, A.; Santos, D.T.; Maréchal, F.; Meireles, M.A.A. Environmental, energetic and economic evaluation of implementing a supercritical fluid-based Nanocellulose production process in a sugarcane biorefinery. *Chem. Eng. Trans.* **2016**, *47*, 49–54.
14. Colodette, J.L.; Martino, D.C.; Seção, V. Capítulo 1. Deslignificação com oxigênio. In *Branqueamento de Polpa Celulósica. Da Produção da Polpa Marrom ao Produto Acabado*; Colodette, J.L., Borges Gomes, F.J., Eds.; Editora UFV: Viçosa, Brazil, 2015; pp. 267–312.
15. Hiltunen, J.; Kemppainen, K.; Pere, J. Process for Producing Fibrillated Cellulose Material. U.S. Patent Application No. 15/104,991, 2015.
16. Lehmonen, J.; Pere, J.; Hytönen, E.; Kangas, H. Effect of cellulose microfibril (CMF) addition on strength properties of middle ply of board. *Cellulose* **2017**, *24*, 1041–1055. [CrossRef]
17. Vartiainen, J. Development towards all-cellulosic packaging films. *Packag. Technol.* **2017**, *1*, 6–13.

18. Chinga-Carrasco, G. Potential and limitations of Nanocelluloses as components in biocomposite inks for three-dimensional bioprinting and for biomedical devices. *Biomacromolecules* **2018**, *19*, 701–711. [CrossRef] [PubMed]

19. Rees, A.; Powell, L.C.; Chinga-Carrasco, G.; Gethin, D.T.; Syverud, K.; Hill, K.E.; Thomas, D.W. 3D Bioprinting of Carboxymethylated-Periodate Oxidized Nanocellulose Constructs for Wound Dressing Applications. *Biomed Res. Int.* **2015**, *2015*, 1–7. [CrossRef] [PubMed]

20. Markstedt, K.; Mantas, A.; Tournier, I.; Martínez Ávila, H.; Hägg, D.; Gatenholm, P. 3D Bioprinting Human Chondrocytes with Nanocellulose–Alginate Bioink for Cartilage Tissue Engineering Applications. *Biomacromolecules* **2015**, *16*, 1489–1496. [CrossRef] [PubMed]

21. Leppiniemi, J.; Lahtinen, P.; Paajanen, A.; Mahlberg, R.; Metsä-Kortelainen, S.; Pinomaa, T.; Pajari, H.; Vikholm-Lundin, I.; Pursula, P.; Hytönen, V.P. 3D-Printable Bioactivated Nanocellulose–Alginate Hydrogels. *ACS Appl. Mater. Interfaces* **2017**, *9*, 21959–21970. [CrossRef] [PubMed]

22. Chinga-Carrasco, G.; Ehman, N.V.; Pettersson, J.; Vallejos, M.E.; Brodin, M.W.; Felissia, F.E.; Håkansson, J.; Area, M.C. Pulping and Pretreatment Affect the Characteristics of Bagasse Inks for Three-dimensional Printing. *ACS Sustain. Chem. Eng.* **2018**, *6*, 4068–4075. [CrossRef]

23. Kaar, W.E.; Cool, L.G.; Merriman, M.M.; Brink, D.L. The Complete Analysis of Wood Polysaccharides Using HPLC. *J. Wood Chem. Technol.* **1991**, *11*, 447–463. [CrossRef]

24. Lahtinen, P.; Liukkonen, S.; Pere, J.; Sneck, A.; Kangas, H. A Comparative Study of Fibrillated Fibres from Different Mechanical and Chemical Pulps. *BioResources* **2014**, *9*, 2115–2127. [CrossRef]

25. Chinga-Carrasco, G.; Averianova, N.; Kondalenko, O.; Garaeva, M.; Petrov, V.; Leinsvang, B.; Karlsen, T. The effect of residual fibres on the micro-topography of cellulose nanopaper. *Micron* **2014**, *56*, 80–84. [CrossRef] [PubMed]

26. Chinga-Carrasco, G.; Syverud, K. Pretreatment-dependent surface chemistry of wood nanocellulose for pH-sensitive hydrogels. *J. Biomater. Appl.* **2014**, *29*, 423–432. [CrossRef]

27. Nordli, H.R.; Chinga-Carrasco, G.; Rokstad, A.M.; Pukstad, B. Producing ultrapure wood cellulose nanofibrils and evaluating the cytotoxicity using human skin cells. *Carbohydr. Polym.* **2016**, *150*, 65–73. [CrossRef] [PubMed]

28. Boateng, J.S.; Matthews, K.H.; Stevens, H.N.E.; Eccleston, G.M. Wound Healing Dressings and Drug Delivery Systems: A Review. *J. Pharm. Sci.* **2008**, *97*, 2892–2923. [CrossRef] [PubMed]

bioengineering

Communication

Nanocellulose-Based Inks—Effect of Alginate Content on the Water Absorption of 3D Printed Constructs

Eduardo Espinosa [1], Daniel Filgueira [2], Alejandro Rodríguez [1] and Gary Chinga-Carrasco [2,*]

1   Chemical Engineering Department, Faculty of Science, Universidad de Córdoba, Building Marie-Curie, Campus de Rabanales, 14014 Córdoba, Spain
2   RISE PFI, Høgskoleringen 6b, 7491 Trondheim, Norway
*   Correspondence: gary.chinga.carrasco@rise-pfi.no

Received: 27 May 2019; Accepted: 27 July 2019; Published: 30 July 2019

Abstract: 2,2,6,6-tetramethylpyperidine-1-oxyl (TEMPO) oxidized cellulose nanofibrils (CNF) were used as ink for three-dimensional (3D) printing of porous structures with potential as wound dressings. Alginate (10, 20, 30 and 40 wt%) was incorporated into the formulation to facilitate the ionic cross-linking with calcium chloride ($CaCl_2$). The effect of two different concentrations of $CaCl_2$ (50 and 100 mM) was studied. The 3D printed hydrogels were freeze-dried to produce aerogels which were tested for water absorption. Scanning Electronic Microscopy (SEM) pictures demonstrated that the higher the concentration of the cross-linker the higher the definition of the printed tracks. CNF-based aerogels showed a remarkable water absorption capability. Although the incorporation of alginate and the cross-linking with $CaCl_2$ led to shrinkage of the 3D printed constructs, the approach yielded suitable porous structures for water and moisture absorption. It is concluded that the 3D printed biocomposite structures developed in this study have characteristics that are promising for wound dressings devices.

Keywords: nanocellulose; 3D printing; absorption; wound dressings

---

1. Introduction

The biomedical area is constantly developing new biomaterials with optimized properties for a given application. Wound management is a particularly demanding area with constant challenges, which motivate continuous efforts to develop tailor-made materials for specific wounds, e.g., chronic and burn wounds. In such cases, foams and hydrogels can be applied to absorb exudates, provide moisture, reduce pain and limit bacterial growth [1,2]. During the last years, cellulose nanofibrils (CNF) have appeared as a promising material for wound dressings. Importantly, CNF have several beneficial characteristics for wound dressing applications, including: absorption of large quantities of liquid, capability to form highly translucent structures with adequate mechanical properties, ability to inhibit bacterial growth and is generally cytocompatible with interesting immunogenic properties [3–11].

2,2,6,6-tetramethylpyperidine-1-oxyl (TEMPO) oxidation is one of the most commonly used methods for the production of carboxylated CNF [12]. TEMPO pre-treatment consists on the oxidation of the C6 primary hydroxyls groups of the glucose units to carboxyl groups. The higher amount of negatively charged groups increases the electrostatic repulsion between the nanofibrils, which facilitates the fibrillation and reduces the energy consumption during homogenization. Interestingly, the presence of negatively charged carboxyl groups in the CNF may provide new properties such as ionic cross-linking capability.

TEMPO CNF-based structures have the capability to absorb large quantity of water and maintain a moistening environment suitable for wound healing [3,13]. Additionally, we have previously demonstrated that TEMPO CNF inhibits growth of *P. aeruginosa*, an opportunistic pathogen commonly occurring in infected wounds [5,14]. These characteristics suggest that TEMPO CNF could be directly used as moistening dressings for, e.g., treatment of burns. Moreover, such properties could be improved by manufacturing tailor-made porous structures by three-dimensional (3D) printing [15].

It is worth to mention that the hydrophilic nature of CNF, due to its large number of hydroxyl groups on their surface, and the electrical sensitivity of cellulose to water vapor, allows CNF structures to be used as moisture sensor [16,17], which was recently demonstrated for TEMPO CNF [18]. Moreover, moisture balance is especially critical in wound dressings, as excessively moist tissue can lead to maceration and insufficient moisture can lead to drying of the wound, affecting the healing process. Thus, proper wound moisture monitoring can reduce wound healing time as well as the number of dressing changes [19]. In this respect, it is also valuable to assess the moisture and water absorption capability of CNF-based materials. This will in addition provide the necessary data for clinicians to evaluate the specific wound dressing for a particular wound and wound management.

Three-dimensional (3D) printing is a layer-by-layer manufacturing process, which enables the rapid fabrication of model objects with complex structures and geometries. Depending on the technology, different raw materials (i.e., plastic, metal, ceramic, glass or biocomposites) can be used for 3D printing [20,21]. For instance, hydrogels can be 3D printed by Direct Ink Writing (DIW), which basically consists on the cold extrusion of a hydrogel through a syringe [22]. Common hydrogels used in 3D printing are biopolymers such as collagen, hyaluronic acid, chitosan or alginate [23]. A major challenge for the 3D printing of hydrogels is their collapse after 3D printing, which dramatically affects the shape fidelity of the printed structure. Due to its shear thinning behavior and rapid consolidation after deposition, CNF has excellent properties to be used in 3D printing applications [24]. Additionally, CNF can be combined with other biopolymers such as alginates to tailor the mechanical properties [25,26]. There are various types of alginates, which can be obtained from different algae and with varying composition of β-D-mannuronic acid (M) and α-L-guluronic acid (G). The composition of M- and G-blocks affects the mechanical properties of the alginates. For a good description of the effect of different alginates on the mechanical properties of CNF/alginates biocomposite gels, see Aarstad et al. [25].

CNF provides a rheology suitable for the extrusion process and alginate potentiates the cross-linking with divalent cations such as calcium [27]. CNF and alginates can thus be applied as inks for the controlled structuring of porous materials. Therefore, 3D printing of CNF/alginate biocomposite hydrogels is a promising pathway for the manufacturing of biobased porous structures with adequate mechanical and liquid absorption properties.

According to Boateng et al. [28], water uptake is one of the characteristics relevant for wound dressings. However, little information is available in the literature about water absorption of 3D printed constructs for wound dressing applications. Hence, in the present study, 3D printing of constructs based on CNF and varying amounts of alginate was demonstrated and the corresponding moisture and water absorption was quantified.

2. Materials and Methods

2.1. CNF Preparation

Pinus radiata kraft pulp fibers (CMPC, Chile) were chemically pre-treated with (2,2,6,6-tetramethylpiperidinyl-1-oxyl (TEMPO), using 3 mmol of sodium hypochlorite (NaClO) per gram of cellulose. The kraft pulp fibers had a concentration of 1 wt% and were homogenized with a Rannie 15 type 12.56X homogenizer (operated at 1000 bar pressure). The CNF was collected after three passes through the homogenizer. The carboxyl acid content has been previously quantified to 982 ± 7.6 μmol/g for the same CNF grade used in this study [29].

2.2. Ink Composition

A series of CNF-alginate compositions (inks) were prepared for 3D printing (Table 1). Alginate (PROTANAL LF 10/60, FMC corporation) and CNF were mixed using mechanical stirring until obtaining a homogeneous mixture. The amounts of alginate added were based on the dry weight of the CNF.

Table 1. Composition of the inks for three-dimensional (3D) printing. $CaCl_2$: calcium chloride.

Series	Alginate (wt%)	$CaCl_2$ (mmol)
CNF *	-	-
CNF_C50	-	50
CNF_C100	-	100
CNF_A10_C50	10	50
CNF_A20_C50	20	50
CNF_A30_C50	30	50
CNF_A40_C50	40	50
CNF_A10_C100	10	100
CNF_A20_C100	20	100
CNF_A30_C100	30	100
CNF_A40_C100	40	100

* The CNF concentration was 1 wt%.

2.3. Viscosity

The viscosities of the inks CNF, CNF_A20 and CNF_A40 were assessed using a Brookfield viscometer (Brookfield DV2TRV, John Morris Group, Sydney, Australia). The assessed volume was 20 mL at a temperature of 23 ± 1 °C and speeds of 1, 2, 6 and 10 RPM, using a spindle V-73.

2.4. 3D Printing

The inks were 3D printed using a Regemat3D bioprinter (version 1.0), equipped with the Regemat3D Designer, version 1.8, Regemat3D (Granada, Spain). The target length, width and height of the 3D printed structures were 40 mm, 20 mm and 2 mm, respectively. The structures were printed directly on microscopy slides. The target width of the printed tracks was 0.41 mm. The space between the tracks was 2 mm. The flow speed was 2 mm/s, using a 0.58 mm conical nozzle. The inks were kept at room temperature (25 °C) for 24 h before printing. The 3D printed structures were cross-linked immersing the samples in calcium chloride ($CaCl_2$) solution (either at concentration of 50 mM or 100 mM) for 24 h. A blank sample without cross-linking was also prepared.

The area of the printed 3D constructs was quantified by image analysis using the ImageJ program.

After the cross-linking the samples were flushed with distilled water to remove the excess of $CaCl_2$ and freeze-dried for 24 h in a Telstar LyoQuest at −83 °C.

2.5. Water Absorption Capacity

The water absorption capability of the 3D printed structures was measured by immersing a pre-weighed dry sample in distilled water for 24 h and weighed at different specific times. The excess surface water was removed with filter paper before weighing. The water absorption capacity was calculated using Equation (1):

$$Water\ sorption\ capacity\ (\%) = \frac{W_t - W_o}{W_o} \cdot 100 \tag{1}$$

where W_t is the weight of the sample at a specific time and W_o is the weight of dry sample.

For measurement of moisture absorption, the prepared aerogels were placed in a climate chamber (23 °C, 90% relative humidity) during 24 h and weighed periodically. The moisture content was calculated as the difference of mass measurements at different times and the initial dry state weight using Equation (1). This analysis was carried out in triplicate and the mean value was provided.

2.6. SEM and Porosity

3D printed samples were prepared for scanning electron microscopy analysis (SEM, Hitachi SU3500 Scanning Electron Microscope, Hitachi High-Technologies Co., Tokyo, Japan). The freeze-dried samples were coated with a layer of gold to make the surface conductive. The equipment for gold coating was an Agar Auto Sputter Coater (Agar Scientific, Essex, UK). The images were acquired in secondary electron imaging (SEI), using 5 kV and 6 mm acceleration voltage and working distance, respectively.

3. Results

Figure 1 shows the viscosity of the inks CNF, CNF_A20 and CNF_A40. These compositions were used for comparison purposes. As expected, increasing the amount of alginate from 0, to 20 and 40 wt% decreases the viscosity of the inks, confirming also previous results [26]. Figure 1 exemplifies also the reduction of the viscosity as the speed increases, i.e., the inks have clear shear thinning behavior. The contribution of CNF to the rheological properties of the ink is a clear advantage for 3D printing operations.

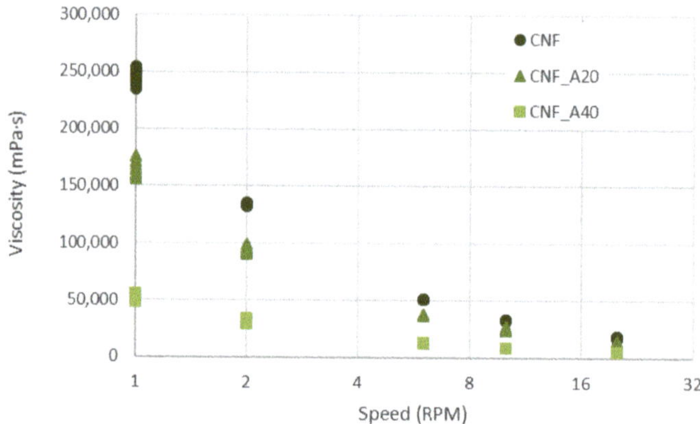

Figure 1. Viscosity of the inks CNF, CNF_A20 and CNF_A40. Ten single measurements are included for each speed interval and for each sample.

Structures composed of four layers (height = 2 mm) and a size of 40 × 20 mm were 3D printed (Figure 2) and freeze-dried. Morphological aspects of the aerogels made of CNF (with and without $CaCl_2$ as crosslinker) and samples with 40% of alginate (with $CaCl_2$ as crosslinker) were investigated by SEM images (Figure 3). The microscopy assessment reveals that an increase in the alginate content leads to a lower resolution of the 3D printed constructs. Likely, the presence of alginate in the formulation increased the lateral flow of the inks, which reduced the shape fidelity of the 3D printed objects (Figures 2 and 3). The alginate applied in this study had a higher flowability compared to CNF (Figure S1: Alginate and CNF inks for 3D printing). Hence, alginate does not provide good printability and shape fidelity (Figure S2: 3D printing with alginate and CNF inks).

For the neat CNF, the spaces between the printed tracks are clearly visible (Figure 3). When the cross-linking was performed with $CaCl_2$ (CNF_C50 and CNF_C100), the shrinkage of the hydrogel caused a greater definition of the printed tracks and a greater detail in the porous structure. The SEM analysis also confirmed that the inks containing alginate had a larger lateral flow, since the spaces between the tracks were not clearly visible. Such effect was lower when a higher concentration of $CaCl_2$ was used. Apparently, constriction of the ink occurred during the cross-linking, as shown in sample CNF_A40_C100.

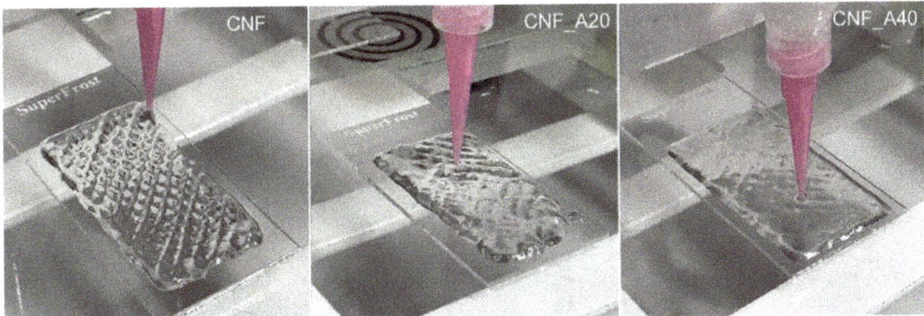

Figure 2. 3D printed gels. Note the relatively large lateral flow of inks CNF_A20 and CNF_A40. The target dimensions of the 3D printed structures were 20 mm × 40 mm.

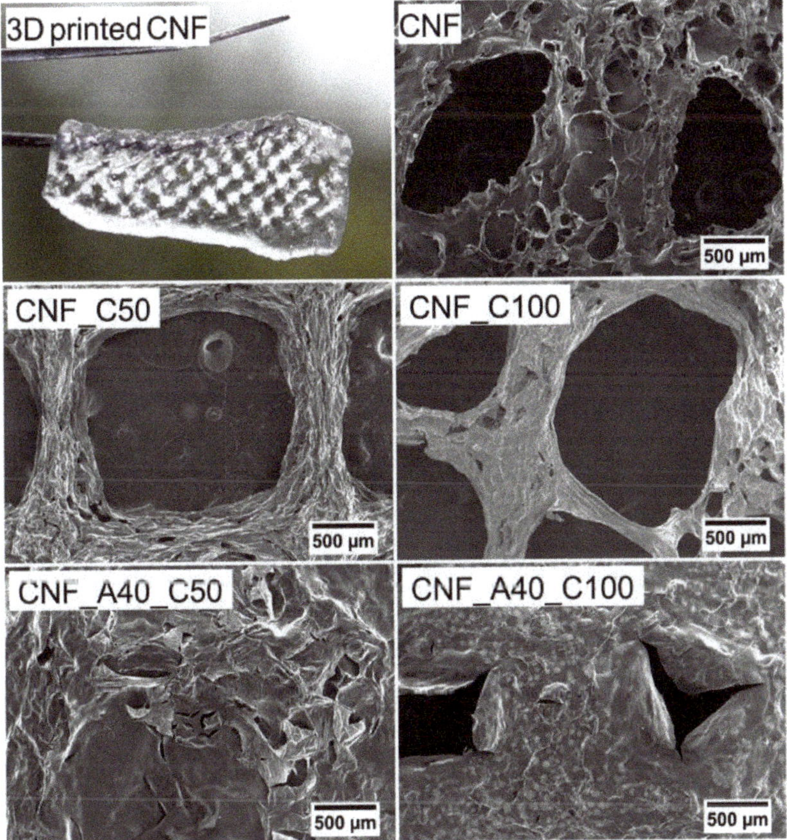

Figure 3. A 3D printed CNF wound dressing cross-linked with $CaCl_2$ (CNF_C50), and Scanning Electron Microscope (SEM) images of a region of five freeze-dried 3D printed constructs. The target dimension of the 3D printed CNF structure (upper left) was 20 mm × 40 mm.

The hydroxyl, carboxyl and other polar groups found in the chemical structure of polysaccharides have the capability to form intermolecular hydrogen bonds, which have a remarkable effect on the moisture and water absorption. Moreover, the chains of polysaccharides can form networks between themselves, which keep the moisture content [30].

Figure 4 shows that neat CNF, cross-linked with CaCl$_2$ had the highest moisture absorption in comparison with the inks based on the combination of CNF with alginate. The use of a higher CaCl$_2$ concentration produced an increase in the moisture absorption capacity of the samples due to the hygroscopic behavior of the salt. The 3D printed structures developed in the present study showed a holding capacity of up to 165% (1.65 g of water vapor per g material) when CNF was cross-linked with CaCl$_2$. Thus, the combination of both CNF and CaCl$_2$ seems to be an effective material as moisture absorber. In general, the addition of alginate reduced the moisture absorption capacity of CNF. Nonetheless, the moisture absorption of the CNF-alginate structures was between 30 and 100% (0.3–1 g of water vapor per g material).

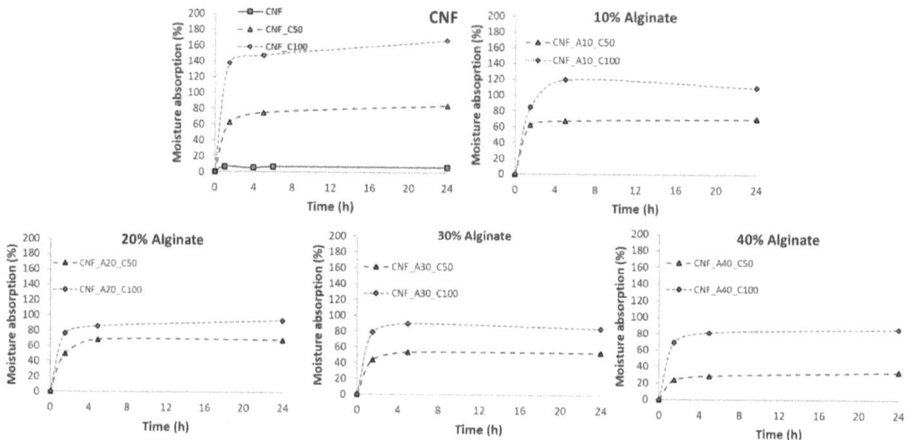

Figure 4. Moisture absorption curves for the different aerogels.

In Figure 5, the water absorption isotherms are presented for the different samples. The graphs show that the greatest capacity of water absorption is presented by neat CNF (1800% weight gain). However, the addition of alginate reduced the capacity of water absorption up to 1200% weight gain over the initial dry weight. This behavior has been reported in previous studies, where the addition of oxidized CNF improves the absorption and retention properties of alginate sponges, reaching values of 1400% of water absorbance [31]. At the same time, CNF presented better results than the use of cellulose nanocrystals. This is due to the reduction of porosity, as the internal microstructure network is modified by the interconnection of alginate with CNF. The pore network allows water molecules to pass through and fully permeate the full structure. In addition, a uniform porous structure will "lock" the water inside the structure and limit the runoff of the water [31]. It was also observed that the use of CaCl$_2$ as cross-linker reduced the water absorption capacity. Hence, the higher the concentration of CaCl$_2$, the lower the water absorption of neat CNF. The reduction of water absorption for the samples containing alginate was detected only for the samples cross-linked with 100 mmol CaCl$_2$. It has been demonstrated that inks containing increasing alginate content led to stiffer structures after the cross-linking with Ca^{2+} [26]. During the cationic cross-linking, the ionic links between COO– groups of the oxidized CNF and the Ca^{2+} increased the cross-linking density and reduced the swelling capacity of the aerogels [32]. The cross-linking thus decreases the dimensional changes of the 3D printed structure compared with the non-cross-linked samples. This limits the deformation of the structure and thus hinders further absorption of water.

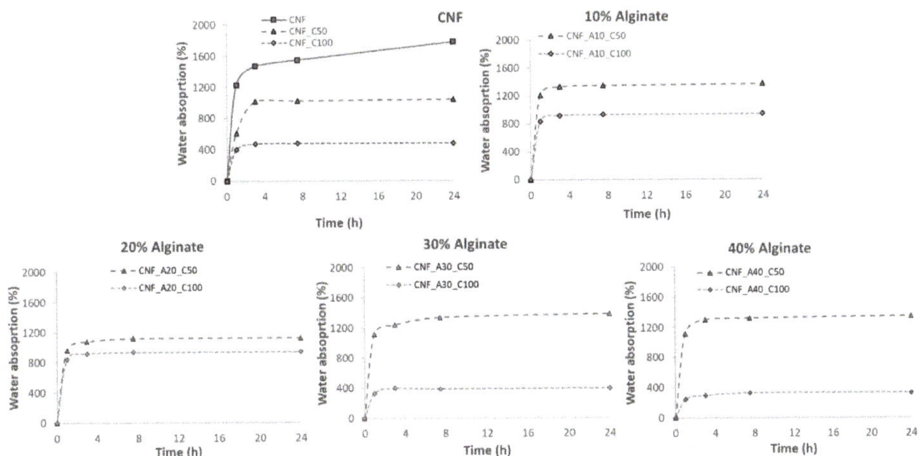

Figure 5. Water absorption curves for the different aerogels.

As we have demonstrated, the wound dressings developed in this study can hold a large fraction of liquid, yet the structures were sufficiently solid to be applied as dressings, attaching and conforming easily to the surface of the skin without disintegrating (Figure 6). The quantified areas of the cross-linked dressings were 667, 612 and 519 mm^2 for the dressings CNF, CNF_A20 and CNF_A40, respectively. Although, the alginate-containing dressings are more robust, these dressings had a larger tendency to shrink caused by the cross-linking with Ca^{2+}. Based on the obtained results regarding moisture and water absorption, we propose an alginate content of 20 wt%, relative to the CNF content. This level seems to provide an adequate stability of the construct and also a good water and moisture absorption (Figures 4–6). The concentration of the CNF used in this study was 1 wt%. Less shrinkage can be expected when using CNF/alginate inks with higher concentration, which may be beneficial in the design of tailor-made wound dressing devices.

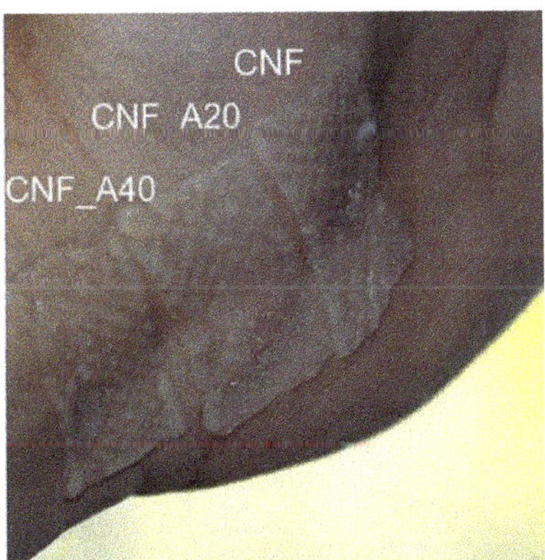

Figure 6. Wound dressings cross-linked with CaCl$_2$ (50 mM). Note the relatively large shrinkage of the samples containing alginate. The target dimensions of the 3D printed structures were 20 mm × 40 mm.

It is also worth to emphasize that TEMPO CNF from the same pulp fibers applied in this study (bleached kraft softwood pulp) has been proven to be non-cytotoxic [3], with interesting dose-dependent inhibition of bacterial growth [5] and has been extensively evaluated as a potential wound dressing material [4]. Additionally, TEMPO CNF from bleached sulfite softwood pulp in combination with Ca^{2+}

(as cross-linking agent) has been evaluated in in vitro and in vivo testing, demonstrating the wound healing ability of CNF hydrogels [6,7,33].

4. Conclusions

Wound dressings with a great capacity to maintain large amounts of water can be manufactured by 3D printing of CNF-based inks. The addition of alginate to the CNF and the cross-linking with $CaCl_2$ consolidated remarkably the structure of the porous constructs. The highest water absorption was measured in the structures composed of neat CNF. The ionic cross-linking reduced the water absorption of CNF from roughly 1800% to 400% depending on the alginate content and Ca^{2+} cross-linking. This study demonstrates the suitability of carboxylated CNF in combination with alginate as ink for 3D printing of porous constructs for wound dressing devices.

Supplementary Materials: The following are available online at http://www.mdpi.com/2306-5354/6/3/65/s1, Figure S1: Alginate and CNF inks for 3D printing. Note that the alginates have a higher flowability compared to the CNF. Alginates with a concentration of 1–4 wt% flow easily to the bottom of the vial (lower panel). Alginate with 8 wt% concentration has a higher viscosity as exemplified in the lower panel, when the vials are placed upside-down. Due to the high zero-shear viscosity of the CNF sample (concentration: 0.9 wt%) the material keeps its shape even when the vial is place upside-down. Figure S2: 3D printing with alginate and CNF inks. (Left) 4 wt% alginate in water. (Middle) 8% Alginate in water. (Right) 0.9% CNF in water. Note that 8 wt% alginate is required to print with some resolution, compared with CNF that prints adequately with only 0.9 wt%. Photos acquired during the printing of the 4th layer. The size of the squares is 20 mm × 20 mm.

Author Contributions: Idea and supervision: G.C.-C.; Investigation: E.E., D.F. and G.C.-C., Draft preparation, reviewing and editing: G.C.-C., E.E. and A.R.

Funding: The authors acknowledge the European Commission for funding part of this work through the MANUNET III program (Project No. MNET17/NMCS-1204), MedIn Project Grant No. 283895, "New functionalized medical devices for surgical interventions in the pelvic cavity". The COST Action (European Cooperation in Science and Technology) FP1405 is acknowledged for the STSMs granted to Eduardo Espinosa and Daniel Filgueira. The work was performed at the RISE PFI 3D printing lab.

Acknowledgments: Mirjana Filipovic and Johnny Kvakland Melbø (RISE PFI) are acknowledged for valuable laboratory assistance.

Conflicts of Interest: The authors declare no conflict of interest.

References

1. Kamoun, E.A.; Kenawy, E.-R.S.; Chen, X. A review on polymeric hydrogel membranes for wound dressing applications: PVA-based hydrogel dressings. *J. Adv. Res.* **2017**, *8*, 217–233. [CrossRef]

2. Sood, A.; Granick, M.S.; Tomaselli, N.L. Wound Dressings and Comparative Effectiveness Data. *Adv. Wound Care* **2014**, *3*, 511–529. [CrossRef]

3. Nordli, H.R.; Chinga-Carrasco, G.; Rokstad, A.M.; Pukstad, B. Producing ultrapure wood cellulose nanofibrils and evaluating the cytotoxicity using human skin cells. *Carbohydr. Polym.* **2016**, *150*, 65–73. [CrossRef]

4. Nordli, H.R.; Pukstad, B.; Chinga-Carrasco, G.; Rokstad, A.M. Ultrapure Wood Nanocellulose—Assessments of Coagulation and Initial Inflammation Potential. *ACS Appl. Bio Mater.* **2019**, *2*, 1107–1118. [CrossRef]

5. Jack, A.A.; Nordli, H.R.; Powell, L.C.; Powell, K.A.; Kishnani, H.; Johnsen, P.O.; Pukstad, B.; Thomas, D.W.; Chinga-Carrasco, G.; Hill, K.E. The interaction of wood nanocellulose dressings and the wound pathogen *P. aeruginosa*. *Carbohydr. Polym.* **2017**, *157*, 1955–1962. [CrossRef]

6. Basu, A.; Hong, J.; Ferraz, N. Hemocompatibility of Ca^{2+}-Crosslinked Nanocellulose Hydrogels: Toward Efficient Management of Hemostasis. *Macromol. Biosci.* **2017**, *17*, 1700236. [CrossRef]

7. Basu, A.; Heitz, K.; Strømme, M.; Welch, K.; Ferraz, N. Ion-crosslinked wood-derived nanocellulose hydrogels with tunable antibacterial properties: Candidate materials for advanced wound care applications. *Carbohydr. Polym.* **2018**, *181*, 345–350. [CrossRef]

8. Liu, Y.; Sui, Y.; Liu, C.; Liu, C.; Wu, M.; Li, B.; Li, Y. A physically crosslinked polydopamine/nanocellulose hydrogel as potential versatile vehicles for drug delivery and wound healing. *Carbohydr. Polym.* **2018**, *188*, 27–36. [CrossRef]

9. Shefa, A.A.; Amirian, J.; Kang, H.J.; Bae, S.H.; Jung, H.-I.; Choi, H.-j.; Lee, S.Y.; Lee, B.-T. In vitro and in vivo evaluation of effectiveness of a novel TEMPO-oxidized cellulose nanofiber-silk fibroin scaffold in wound healing. *Carbohydr. Polym.* **2017**, *177*, 284–296. [CrossRef]

10. Souza, S.F.; Mariano, M.; Reis, D.; Lombello, C.B.; Ferreira, M.; Sain, M. Cell interactions and cytotoxic studies of cellulose nanofibers from Curauá natural fibers. *Carbohydr. Polym.* **2018**, *201*, 87–95. [CrossRef]

11. Bacakova, L.; Pajorova, J.; Bacakova, M.; Skogberg, A.; Kallio, P.; Kolarova, K.; Svorcik, V. Versatile Application of Nanocellulose: From Industry to Skin Tissue Engineering and Wound Healing. *Nanomaterials* **2019**, *9*, 164. [CrossRef]

12. Saito, T.; Kimura, S.; Nishiyama, Y.; Isogai, A. Cellulose Nanofibers Prepared by TEMPO-Mediated Oxidation of Native Cellulose. *Biomacromolecules* **2007**, *8*, 2485–2491. [CrossRef]

13. Sun, F.; Nordli, H.R.; Pukstad, B.; Kristofer Gamstedt, E.; Chinga-Carrasco, G. Mechanical characteristics of nanocellulose-PEG bionanocomposite wound dressings in wet conditions. *J. Mech. Behav. Biomed. Mater.* **2017**, *69*, 377–384. [CrossRef]

14. Powell, L.C.; Khan, S.; Chinga-Carrasco, G.; Wright, C.J.; Hill, K.E.; Thomas, D.W. An investigation of Pseudomonas aeruginosa biofilm growth on novel nanocellulose fibre dressings. *Carbohydr. Polym.* **2016**, *137*, 191–197. [CrossRef]

15. Rees, A.; Powell, L.C.; Chinga-Carrasco, G.; Gethin, D.T.; Syverud, K.; Hill, K.E.; Thomas, D.W. 3D Bioprinting of Carboxymethylated-Periodate Oxidized Nanocellulose Constructs for Wound Dressing Applications. *Biomed. Res. Int.* **2015**. [CrossRef]

16. Bethke, K.; Palantöken, S.; Andrei, V.; Roß, M.; Raghuwanshi, V.S.; Kettemann, F.; Greis, K.; Ingber, T.T.K.; Stückrath, J.B.; Valiyaveettil, S.; et al. Functionalized Cellulose for Water Purification, Antimicrobial Applications, and Sensors. *Adv. Funct. Mater.* **2018**, *28*, 1800409. [CrossRef]

17. Safari, S.; van de Ven, T.G.M. Effect of Water Vapor Adsorption on Electrical Properties of Carbon Nanotube/Nanocrystalline Cellulose Composites. *ACS Appl. Mater. Interfaces* **2016**, *8*, 9483–9489. [CrossRef]

18. Syrový, T.; Maronová, S.; Kuberský, P.; Ehman, N.V.; Vallejos, M.E.; Pretl, S.; Felissia, F.E.; Area, M.C.; Chinga-Carrasco, G. Wide range humidity sensors printed on biocomposite films of cellulose nanofibril and poly(ethylene glycol). *J. Appl. Polym. Sci.* **2019**, *136*, 47920. [CrossRef]

19. Milne, S.D.; Seoudi, I.; Al Hamad, H.; Talal, T.K.; Anoop, A.A.; Allahverdi, N.; Zakaria, Z.; Menzies, R.; Connolly, P. A wearable wound moisture sensor as an indicator for wound dressing change: An observational study of wound moisture and status. *Int. Wound J.* **2016**, *13*, 1309–1314. [CrossRef]

20. Ford, S.; Despeisse, M. Additive manufacturing and sustainability: an exploratory study of the advantages and challenges. *J. Clean. Prod.* **2016**, *137*, 1573–1587. [CrossRef]

21. Filgueira, D.; Holmen, S.; Melbø, J.K.; Moldes, D.; Echtermeyer, A.T.; Chinga-Carrasco, G. Enzymatic-Assisted Modification of Thermomechanical Pulp Fibers To Improve the Interfacial Adhesion with Poly(lactic acid) for 3D Printing. *ACS Sustain. Chem. Eng.* **2017**, *5*, 9338–9346. [CrossRef]

22. Xu, W.; Wang, X.; Sandler, N.; Willför, S.; Xu, C. Three-Dimensional Printing of Wood-Derived Biopolymers: A Review Focused on Biomedical Applications. *ACS Sustain. Chem. Eng.* **2018**, *6*, 5663–5680. [CrossRef]

23. Gopinathan, J.; Noh, I. Recent trends in bioinks for 3D printing. *Biomater. Res.* **2018**, *22*, 11. [CrossRef]

24. Chinga-Carrasco, G. Potential and Limitations of Nanocelluloses as Components in Biocomposite Inks for Three-Dimensional Bioprinting and for Biomedical Devices. *Biomacromolecules* **2018**, *19*, 701–711. [CrossRef]

25. Aarstad, O.; Heggset, E.B.; Pedersen, I.S.; Bjørnøy, S.H.; Syverud, K.; Strand, B.L. Mechanical Properties of Composite Hydrogels of Alginate and Cellulose Nanofibrils. *Polymers* **2017**, *9*, 378. [CrossRef]

26. Heggset, E.B.; Strand, B.L.; Sundby, K.W.; Simon, S.; Chinga-Carrasco, G.; Syverud, K. Viscoelastic properties of nanocellulose based inks for 3D printing and mechanical properties of CNF/alginate biocomposite gels. *Cellulose* **2019**, *26*, 581–595. [CrossRef]

27. Markstedt, K.; Mantas, A.; Tournier, I.; Martínez Ávila, H.; Hägg, D.; Gatenholm, P. 3D Bioprinting Human Chondrocytes with Nanocellulose–Alginate Bioink for Cartilage Tissue Engineering Applications. *Biomacromolecules* **2015**, *16*, 1489–1496. [CrossRef]

28. Boateng, J.S.; Matthews, K.H.; Stevens, H.N.E.; Eccleston, G.M. Wound Healing Dressings and Drug Delivery Systems: A Review. *J. Pharm. Sci.* **2008**, *97*, 2892–2923. [CrossRef]

29. Silva, F.; Gracia, N.; McDonagh, B.H.; Domingues, F.C.; Nerín, C.; Chinga-Carrasco, G. Antimicrobial activity of biocomposite films containing cellulose nanofibrils and ethyl lauroyl arginate. *J. Mater. Sci.* **2019**, *54*, 12159–12170. [CrossRef]

30. Zhang, Z.-S.; Wang, X.-M.; Han, Z.-P.; Zhao, M.-X.; Yin, L. Purification, antioxidant and moisture-preserving activities of polysaccharides from papaya. *Carbohydr. Polym.* **2012**, *87*, 2332–2337. [CrossRef]

31. Lin, N.; Bruzzese, C.; Dufresne, A. TEMPO-Oxidized Nanocellulose Participating as Crosslinking Aid for Alginate-Based Sponges. *ACS Appl. Mater. Interfaces* **2012**, *4*, 4948–4959. [CrossRef] [PubMed]

32. De Vos, P.; De Haan, B.; Wolters, G.H.J.; Van Schilfgaarde, R. Factors influencing the adequacy of microencapsulation of rat pancreatic islets. *Transplantation* **1996**, *62*, 888–893. [CrossRef] [PubMed]

33. Basu, A.; Celma, G.; Strømme, M.; Ferraz, N. In Vitro and in Vivo Evaluation of the Wound Healing Properties of Nanofibrillated Cellulose Hydrogels. *ACS Appl. Bio Mater.* **2018**, *1*, 1853–1863. [CrossRef]

 *bioengineering*

Article

3D Printed Nanocellulose Scaffolds as a Cancer Cell Culture Model System

Jennifer Rosendahl [1,†], Andreas Svanström [2,†], Mattias Berglin [1], Sarunas Petronis [1], Yalda Bogestål [1], Patrik Stenlund [1], Simon Standoft [1], Anders Ståhlberg [2,3,4], Göran Landberg [2,5], Gary Chinga-Carrasco [6,*] and Joakim Håkansson [1,7,*]

1. Unit of Biological Function, Division Materials and Production, RISE Research Institutes of Sweden, Box 857, SE-50115 Borås, Sweden; jennifer.rosendahl@ri.se (J.R.); Mattias.Berglin@ri.se (M.B.); sarunas.petronis@ri.se (S.P.); Yalda.bogestal@ri.se (Y.B.); patrik.stenlund@ri.se (P.S.); Simon.standoft@ri.se (S.S.)
2. Sahlgrenska Center for Cancer Research, Department of Laboratory Medicine, Institute of Biomedicine, Sahlgrenska Academy, University of Gothenburg, Box 425, Medicinaregatan 1G, SE-41390 Gothenburg, Sweden; andreas.svanstroem@gmail.com (A.S.); anders.stahlberg@gu.se (A.S.); goran.landberg@gu.se (G.L.)
3. Wallenberg Centre for Molecular and Translational Medicine, University of Gothenburg, SE-40530 Gothenburg, Sweden
4. Department of Clinical Genetics and Genomics, Region Västra Götaland, Sahlgrenska University Hospital, SE-40530 Gothenburg, Sweden
5. Department of Clinical Pathology, Sahlgrenska University Hospital, SE-41345 Gothenburg, Sweden
6. RISE PFI AS, Høgskoleringen 6b, NO-7491 Trondheim, Norway
7. Department of Laboratory Medicine, Institute of Biomedicine, University of Gothenburg, P.O. Box 440, SE-40530 Gothenburg, Sweden
* Correspondence: Gary.Chinga.Carrasco@rise-pfi.no (G.C.-C.); Joakim.hakansson@ri.se (J.H.); Tel.: +47-45412304 (G.C.-C.); +46-702172197 (J.H.)
† Equal contribution.

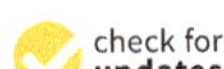

check for updates

Citation: Rosendahl, J.; Svanström, A.; Berglin, M.; Petronis, S.; Bogestål, Y.; Stenlund, P.; Standoft, S.; Ståhlberg, A.; Landberg, G.; Chinga-Carrasco, G.; et al. 3D Printed Nanocellulose Scaffolds as a Cancer Cell Culture Model System. *Bioengineering* **2021**, *8*, 97. https://doi.org/10.3390/bioengineering8070097

Academic Editor: Joaquim M. S. Cabral

Received: 14 May 2021
Accepted: 28 June 2021
Published: 10 July 2021

Publisher's Note: MDPI stays neutral with regard to jurisdictional claims in published maps and institutional affiliations.

Abstract: Current conventional cancer drug screening models based on two-dimensional (2D) cell culture have several flaws and there is a large need of more in vivo mimicking preclinical drug screening platforms. The microenvironment is crucial for the cells to adapt relevant in vivo characteristics and here we introduce a new cell culture system based on three-dimensional (3D) printed scaffolds using cellulose nanofibrils (CNF) pre-treated with 2,2,6,6-tetramethylpyperidine-1-oxyl (TEMPO) as the structural material component. Breast cancer cell lines, MCF7 and MDA-MB-231, were cultured in 3D TEMPO-CNF scaffolds and were shown by scanning electron microscopy (SEM) and histochemistry to grow in multiple layers as a heterogenous cell population with different morphologies, contrasting 2D cultured mono-layered cells with a morphologically homogenous cell population. Gene expression analysis demonstrated that 3D TEMPO-CNF scaffolds induced elevation of the stemness marker *CD44* and the migration markers *VIM* and *SNAI1* in MCF7 cells relative to 2D control. T47D cells confirmed the increased level of the stemness marker *CD44* and migration marker *VIM* which was further supported by increased capacity of holoclone formation for 3D cultured cells. Therefore, TEMPO-CNF was shown to represent a promising material for 3D cell culture model systems for cancer cell applications such as drug screening.

Keywords: nanocellulose; 3D printing; cancer; 3D cell culture; CNF; cancer stemness

1. Introduction

Cancer is one of the most abundant diseases worldwide and there is a clear need for developing new and effective drugs and therapies [1]. Such development requires relevant pre-clinical in vitro models that adequately represent the human situation. Conventional cell cultures in plastics have been applied to fill this gap [2]; however, in most situations these cell cultures do not represent a relevant tissue mimicking its natural environment and consequently only 5% of new drug candidates that enter clinical trial reach the market with a proper safety profile and shown effect in humans [3–5]. Hence, the scientific community needs improved tumor model systems.

There is increasing evidence that cancer stem cells (CSC) which possess the capacity for self-renewal, unlimited proliferation and multidrug- and radiotherapy resistance are the driving force in many cancer types [6,7]. Two-dimensional (2D) cell cultures are usually grown in a mono layer and have a homogenous cell population with limited CSC properties. Three-dimensional models represent the in vivo situation more closely by inducing the cells to grow in multiple layers, stimulate CSC properties, communicate in all directions and enable representative cellular heterogeneity [2,8,9]. Three-dimensional culturing platforms represent promising predictive models for in vivo tumorigenesis [9–11] as the microenvironment, both structurally, biochemically and biomechanically, plays a vital role in cell fate, differentiation, organ development, signal transduction and countless other biological processes [12,13]. A cell culture tumor model system would therefore benefit from a microenvironment that stimulates the formation of a heterogeneous cell population with CSC traits in order to represent a physiological tumor microenvironment.

Various biopolymers have been applied for 3D cell culture including polysaccharides and proteins, e.g., fibrin, alginate and collagen [14,15]. These have been reported to be biocompatible and to alter cellular response [16], which is relevant for mimicking complex tissues. Preclinical cell culture models should be able to mimic the tumor microenvironment and may also benefit from physiologically non-biodegradable structures in order to provide support and stability to the cell cultures. In addition, the ability to 3D print biopolymers is beneficial for constructing optimized 3D models based on architectural preference. Wood nanocellulose has proven to be a suitable biobased material for various biomedical applications. Cellulose nanofibrils (CNF) has been extensively studied as gels for wound healing, drug carriers and scaffolds for tissue engineering [17–21]. This is due to the versatility of CNF grades regarding properties such as biocompatibility and the capability to form gels at low concentration with adequate viscoelastic behavior, the latter being most interesting from a 3D bioprinting perspective [22]. Various types of CNF grades have an appropriate viscosity at low concentrations and good shear thinning behavior [23–25]. The gels can easily be extruded through a nozzle, deposited on a substrate and maintain a defined 3D shape post printing due to the rapid zero-shear viscosity recovery [26–29]. Carboxylated CNF, produced through 2,2,6,6-tetramethylpiperidinyl-1-oxyl (TEMPO)-mediated oxidation [30], has been used in various studies concerning biomedical applications demonstrating benefits for cell viability, proliferation, migration, biocompatibility and cytotoxicity [18,19]. In addition, TEMPO-CNF has been shown to have a positive effect on fibroblast, HeLA and Jurkat tumor cell growth and proliferation [31], indicating the potential of TEMPO-CNF as a printing ink for 3D cell culturing applications.

The aim of this study was to evaluate TEMPO-CNF as a 3D printable ink to form scaffolds for tumor cell culturing and to study the effect of the 3D model on cell behavior. We demonstrate how 3D printed TEMPO-CNF may induce cancer stemness and affect cellular morphology and heterogeneity, which suggest TEMPO-CNF to be a promising material for cell culturing in 3D.

2. Materials and Methods

2.1. Nanocellulose

Bleached *Pinus radiata* kraft pulp fibers were kindly provided by CMPC (Planta Pacifico, Chile). The chemical composition of the pulp fibers has been quantified to 87% cellulose, 12.2% hemicellulose and 0.8% lignin [32]. The pulp fibers were chemically pre-treated with 2,2,6,6-tetramethylpiperidinyl-1-oxyl (TEMPO) mediated oxidation [33], applying 3.0 mmol/g hypochlorite (NaClO, 9%) with the reaction kept at pH 10. The pulp fibers were washed with deionized water to achieve a conductivity of less than 5 µS/cm to secure removal of unreacted reactants. The pre-treated pulp fibers in deionized water (1% w/v) were homogenized using a Rannie 15 type 12.56X homogenizer operating at 1000 bar pressure. Conductometric titration was applied to measure the carboxylic acid content of the produced TEMPO-CNF and was quantified to 982 ± 8 µmol/g [34]. The nanocellulose was stored at 2–8 °C until 3D-printing.

2.2. Atomic Force Microscopy and Nano-Mechanical Assessment

Atomic force microscopy analysis (AFM) was performed on the TEMPO-CNF to reveal the morphology of single cellulose nanofibrils. AFM (Veeco multimode) was performed using tips with a spring constant of $\sim$0.4 N m^{-1} (Bruker AFM probes) on areas of 2×2 µm (1024×1024 pixels). The stiffness of the scaffolds was analyzed using nano-mechanical assessment. The freeze-dried samples were immersed in either DMEM- or RPMI-medium (Thermo Fisher Scientific, Gothenburg, Sweden) for 1 h prior to measurement and the following nanoindentation parameters were applied: Berkovich tip; displacement controlled at peak indentation depth of 2000 nm; 0.125 s loading, 0.4 s holding, 0.125 s unloading (total testing time 0.65 s for one indent). More than 20 indents on random areas on each sample were performed and two replicate samples were assessed in solution.

2.3. 3D Printing

TEMPO-CNF hydrogel with a concentration of 1% (w/v) in water was used as ink for 3D printing. Previously, viscosity data of the same CNF gel and the corresponding printability was reported [35]. In the present study, the CNF was 3D printed in a laminar airflow hood as square box scaffolds with the dimension $30 \times 30 \times 5$ mm; grid distance 1.5 mm, 90°, with an outer frame in 4 layers to produce a mesh like scaffold (3DPS) using an EnvisionTEC 4th Gen 3D-Bioplotter (EnvisionTEC, Gladbeck, Germany). The 3D print was designed using CAD (computer aided design) software, Autodesk Fusion 360 2019 (Autodesk Inc., San Rafael, CA, USA) and exported to the printer as an STL-file. The printing parameters for TEMPO-CNF were: temperature (20 °C), pressure (0.2 bar), speed (70 mm/s) and needle offset (0.50 mm). The printing was controlled using Visual machine software (EnvisonTEC, Gladbeck, Germany) and a tapered needle with 410 µm internal diameter. The strand width of the printed nanocellulose was measured on images acquired directly after printing with the bioprinter EnvisionTEC 4th Gen 3D-Bioplotter (EnvisionTEC, Gladbeck, Germany) using ImageJ software (1.52a). The data was plotted in GraphPad Prism v9 (GraphPad). The scaffolds were frozen at -20 °C, transferred to -80 °C and then freeze-dried for at least 48 h, or to stable weight, using VirTis Sentry 2.0 Benchtop Freeze Dryer (SP Scientific, Warminster, PA, USA). Each scaffold was cut into 9 pieces using a scalpel.

2.4. 2D Cell Culture

Conventional polystyrene culture plates were used for 2D cell culture (6-well plate, Sarstedt, Helsingborg, Sweden). The breast cancer cell line MCF7 (ATCC) was cultured in 1X DMEM (Thermo Fisher Scientific, Gothenburg, Sweden) supplemented with 10% v/v fetal bovine serum (Thermo Fisher Scientific, Gothenburg, Sweden), 1% v/v L-glutamine (Thermo Fisher Scientific, Gothenburg, Sweden), 1% v/v penicillin/streptomycin (Thermo Fisher Scientific, Gothenburg, Sweden) and 1% v/v MEM non-essential amino acid solution (100X) (Sigma-Aldrich, Stockholm, Sweden). The breast cancer cell lines MDA-MB-231 (ATCC) and T47D (ATCC) were cultured in 1X RPMI (Thermo Fisher Scientific, Gothenburg, Sweden) supplemented with 10% v/v FBS, 1% v/v L-glutamine, 1% v/v penicillin/streptomycin and 1% v/v 100 mM sodium pyruvate (Thermo Fisher Scientific, Gothenburg, Sweden). Cells were seeded with a density of 2.0–2.5×10^5 cells per well in 6-well plates and cultured for 48 h for holoclone formation and gene expression analysis, respectively, or 1×10^5 cells per well in 6-well plates for up to 8 d for cell growth assessment. Longer culturing time was not used to avoid confluence and cell arrest, which would otherwise make the cells incomparable with growth in 3D-printed scaffolds. All cell lines were cultured at 5% CO_2 at 37 °C.

2.5. 3D Cell Culture

To grow cells in 3D printed scaffolds, cells were detached from 2D culture plates using 0.25% Trypsin-EDTA (Thermo Fisher Scientific, Gothenburg, Sweden), centrifuged at $300 \times g$ for 3 min and the cell pellet was resuspended in cell medium. The freeze-dried scaffolds were pre-wetted in in cell culture medium for 15 min before adding cells in desired density on top of the scaffolds. Cells were counted using MOXI, (Orlflo, Ketchum, ID, USA), seeded with a density of 3×10^5 cells (for cell viability assay) or 5×10^4 cells (for quantitative PCR (qPCR) and scanning electron microscopy analysis) per well in a 48 well plate. Three-dimensional prints were transferred to 6-well plates (Polystyrene, Thermo Fisher Scientific, Gothenburg, Sweden) after 24 h and thereafter transferred to new medium every 3–4 d until final analysis. Gene expression analysis with qPCR and imaging with SEM was performed after 14 d, and holoclone formation was analyzed after 21 d of culture.

2.6. Cell Viability Assessment

Cells were detached as described above, centrifuged at $300 \times g$ for 3 min and resuspended in cell medium and trypan blue 1:2 (Thermo Fisher Scientific, Gothenburg, Sweden). The number of live and dead cells were counted using a dispensable hemocytometer (C-Chip, NanoEnTek, Seoul, Korea).

2.7. Gene Expression Analysis

Scaffolds were gently washed twice in PBS, transferred to 2 mL tubes, lysed with 600 μL QIAzol lysing reagent (Qiagen, Kista, Sweden) and 50 μL lysis buffer (5 μL RNase out (Thermo scientific, Gothenburg, Sweden), 2.5 μL BSA 20 μg/μL (Thermo scientific, Gothenburg, Sweden) and 42.5 μL RNase free water), snap-frozen in liquid nitrogen and stored at -80 °C. The samples were thawed on ice for 30 min and disrupted using 5 mm stainless steel beads (Qiagen, Kista, Sweden) and TissueLyser II (Qiagen, Kista, Sweden) for 2×2.5 min at 25 Hz. Total RNA isolation was extracted using miRNeasy Micro kit (Qiagen, Kista, Sweden) with on-column DNase digestion according to the manufacturer's instructions and RNA concentration was determined using a NanoDrop (ND-1000, Saveen Werner AB, Malmo, Sweden). Complementary DNA (cDNA) was produced using the GrandScript cDNA synthesis kit (TATAA Biocenter, Gothenburg, Sweden) and a T100 Thermal Cycler (Bio-Rad, Solna, Sweden). Reverse transcription was performed in 10 μL reactions at 25 °C for 5 min, 42 °C for 30 min and 85 °C for 5 min followed by cooling to 4 °C. The cDNA was diluted 1:10 in RNase free water (Invitrogen) and pre-amplified in 50 μL reactions with $1 \times$ SYBR GrandMaster Mix (TATAA Biocenter, Gothenburg, Sweden), 0.04 μM of each primer in a primerpool (Supplementary Table S1) and 1 μg/mL of BSA (Thermo Scientific, Gothenburg, Sweden) using a CFX96 Touch Real Time Cycler (Bio-Rad) and the following temperature profile: 95 °C for 3 min, 15 cycles at 95 °C for 20 s, 60 °C for 3 min and 72 °C for 20 s. The samples were snap frozen on dry ice, thawed on ice and diluted 20 times in TE-buffer (Thermo scientific, Gothenburg, Sweden). For qPCR, all samples were diluted 1:3 and run in 6 μL reactions with 400 nM primers (Supplementary Table S1) and $1 \times$ SYBR GrandMaster Mix (TATAA Biocenter, Gothenburg, Sweden). qPCR was run using a CFX384 Touch Real Time Cycler (Bio-Rad, Solna, Stockholm) with the following temperature profile: 95 °C for 2 min, 39 cycles at 95 °C for 5 s, 60 °C for 20 s and 70 °C for 20 s followed by a melting curve analysis at 65 °C to 95 °C with 0.5 °C per 5 s increments. CFX Manager Software version 3.1 (Bio-Rad, Solna, Stockholm) regression method was used to determine the cycles of quantification (Cq) values. Data was analyzed using GenEx software (MultiD). Samples with >25% missing values were removed followed by the same criteria for genes. Missing values were imputed based on replicates, cut-off and missing Cq-values were set to 35. Data was normalized to reference genes identified with NormFinder algorithm, transformed to relative values and log2 scale. For each experiment optimal reference genes were evaluated. Minimum information for publication of quantitative real-time PCR experiments guidelines was followed for qPCR [36].

2.8. Scanning Electron Microscopy (SEM)

Cells cultured for 48 h in 2D culture plates, or for 14 d in 3DPS, were gently washed once in cell medium and fixated for 1 h at room temperature in 2.5% glutaraldehyde (Product ref G7651-10ML, Sigma-Aldrich Sweden AB, Stockholm, Sweden) diluted in cell medium. Scaffolds were washed once in TBS (50 mM Trizma-HCl, Sigma-Aldrich, Stockholm, Sweden; 150 mM NaCl, Merck, Stockholm, Sweden; pH 7.5) with 0.1 mM $CaCl_2$ (Merck, Stockholm, Sweden), fixated for 1 h at room temperature in 1% osmium tetroxide (Sigma-Aldrich, Stockholm, Sweden) in TBS (50 mM Trizma-HCl (Sigma-Aldrich, Stockholm, Sweden); 150 mM NaCl, (Merck, Stockholm, Sweden); pH 7.5) with 0.1 mM $CaCl_2$ (Merck, Stockholm, Sweden), rinsed in deionized water, plunge-frozen in liquid propane (EMS-002 Rapid Immersion Freezer, Electron Microscopy Sciences) and freeze-dried overnight (VirTis Sentry 2.0 Benchtop Freeze Dryer, SP Scientific, Warminster, PA, USA). Samples were mounted on aluminum stubs by adhesive carbon tabs (Agar Scientific) and coated with a 10 nm Au/Pd conducting thin film by sputter-coater (PECS Mod 682, Gatan Inc., Pleasanton, CA, USA). Samples were imaged by Zeiss SUPRA 40VP scanning electron microscope operated in secondary electron mode at 3.0–4.1 kV acceleration voltage, 9–17 mm working distance and 100–250,000× magnification range.

2.9. Hematoxylin and Eosin Staining

TEMPO-CNF scaffolds with cells cultured for 14 d were gently washed once in cell medium and fixated for 1 h at room temperature in 4% formaldehyde (Histolab, Gothenburg, Sweden) diluted in PBS (Medicago, Uppsala, Sweden). The formaldehyde liquid was removed and replaced with 70% ethanol for 1 h at room temperature, followed by 95% ethanol for 1 h at room temperature, 99% ethanol for 1 h at room temperature and finally xylene (Univar) for 1 h at room temperature. Scaffolds were paraffin embedded, sectioned to 4.5 µm thickness using a retracting microtome (Rotary one, LKB Stockholm, Sweden), dried overnight at 60 °C, de-paraffinized, counterstained with hematoxylin and eosin, de-hydrated and mounted with pertex (Histolab, Gothenburg, Sweden). Sections were imaged with Leica SCN400 Slide Scanner (Meyer instruments, Houston, TX, USA). Representative images were selected and exported using Leica SlidePath Gateway software and analyzed with ImageJ software [37].

2.10. Holoclone Assay

Cells cultured for 48 h in 2D or for 14 d on TEMPO-CNF were detached as described above and seeded in 6 well plates at a density of 50 cells/cm^2, cultured for 6–7 d to form colonies and stained with crystal violet. Holoclones were identified as colonies ($\geq$32 cells) with small clustered as well as differentiated cells and quantified using Gelcount software (Oxford Optronix, Oxford, UK).

2.11. Statistics

All statistical data analyses were performed using GraphPad Prism v8 (GraphPad). The values in the graphs are presented as average ± standard error of the mean.

3. Results

3.1. Nanocellulose Characteristics

In order to develop an in vivo-like 3D cell culture system to be used as a cancer drug screening model, TEMPO-CNF was used as a base material to produce extrusion 3D printed scaffolds. After 3D printing, the TEMPO-CNF colloidally stable gel material was transformed to a porous TEMPO-CNF structure by a freeze-drying procedure and the nanomorphology confirmed the formation of close-packed nanofibrillar networks (nanofibril diameter <20 nm) (Figure 1A). Mechanical properties of the TEMPO-CNF hydrogel were analyzed with nanoindentation, and the quantified elastic modulus was comparable with previously published results, i.e., lower MPa regime [15], but varied with growth medium. Different cells lines need different culture condition and in the present study, DMEM and RPMI cell growth medium was used. It was found that the different cell growth media affected the elastic modulus of the material where RPMI statistically significantly increased the material stiffness with 50% compared with DMEM with elastic modulus' of approximately 4 MPa for DMEM and 6 MPa for RPMI (Figure 1B).

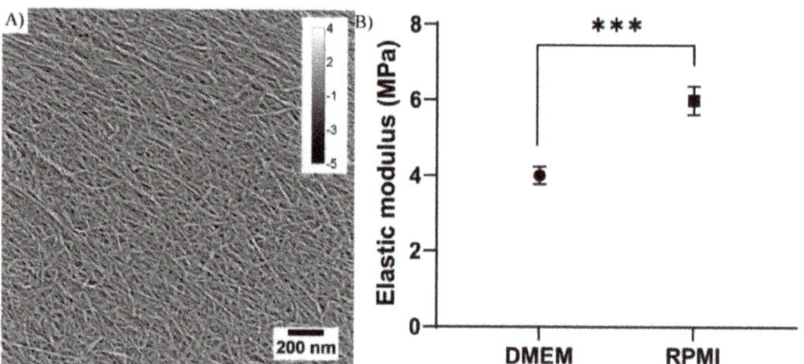

Figure 1. Structural morphology and stiffness of TEMPOCNF. (**A**) AFM nanoscale image of nanofibrils with diameter <20 nm. The scale bar (lower right) and calibration (upper right) indicate the scales in nanometers, in the x, y and z-directions, respectively. (**B**) Elastic modulus of analysis on TEMPO-CNF in DMEM and RPMI medium, respectively. The dots represent average elastic modulus and error bars represent standard error of the mean (n = 42–43); unpaired t-test. *** p-value < 0.001.

3.2. Printability Evaluation

Nanocellulose was printed in a 4-layer grid structure to create a porous mesh in which the cells and nutrients could access the whole scaffold (Figure 2A). The printability of the TEMPO-CNF material was evaluated by analyzing the flow out of the printed strands. This was performed by determining the strand width directly after printing, and was assessed to 0.86 ± 0.07 mm (average ± standard error of the mean) when printed with a 410 μm needle (Figure 2B), which is similar to other materials, e.g., alginate and gelatin mix [38]. The average pore size between strands was 0.52 ± 0.01 mm^2 (average ± standard error of the mean) and the pore size within the material ranged from <1 × 10^{-6} to 0.2 mm^2. The strand and pore size of the 3D printed construct were similar to previous studies where the same CNF ink was used [35].

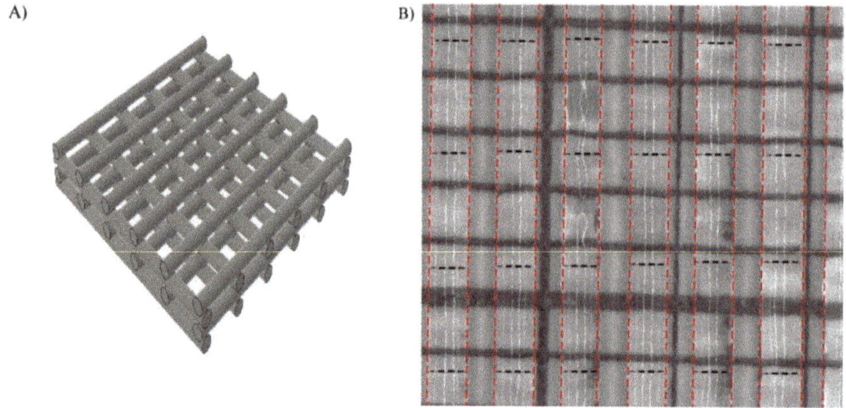

Figure 2. Design of scaffold and printability of TEMPO-CNF (**A**) Schematic illustration of the printed grid model. (**B**) One layer of printed TEMPO-CNF strands on mm-marked paper (black solid lines in the image) to determine the strand width. The strands are marked with red dotted lines and the measurement position is marked with black dotted lines.

3.3. 3D Printed CNF Suitable as Cell Growth Scaffolds

To evaluate how well cells grow in 3D printed TEMPO-CNF structures, the estrogen receptor- and progesterone receptor positive breast cancer cell line MCF7 or the estrogen receptor- and progesterone negative breast cancer cell line MDA-MB-231 were seeded on the scaffolds. Both cell lines attached to the scaffolds to the same extent, as analyzed by the cell number after 1 day of culture, and expanded well during the following 7 days, approximately 90-fold for MCF7 and 24-fold for MDA-MB-231 (Figure 3A). The cell viability was assessed to >96% for both cell lines, at both 1 and 8 days of culture, which further confirmed the material and growth conditions as suitable (Figure 3B). By comparing the cell size between cells grown in 2D cell culture dishes with 3D printed TEMPO-CNF scaffolds, it was shown that both MCF7 and MDA-MB-231 cells were significantly smaller in 3D compared with 2D after 14 days culture, 36% and 38%, respectively. Interestingly, the cell size also decreased from 8 to 14 days culture in the 3D environment, 24% and 30% for MCF7 and MDA-MB231, respectively (Figure 3C).

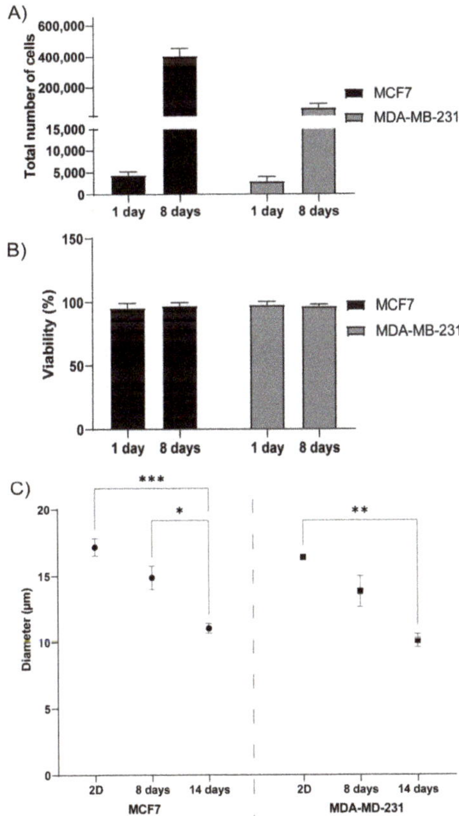

Figure 3. 3D printed TEMPO-CNF as a suitable scaffold for breast cancer cells. (**A**) Total number of MCF7 and MDA-MB-231 breast cancer cells quantified after 1- and 8-days culture, respectively, in 3D printed TEMPO-CNF scaffolds, $n = 6$–9. (**B**) Viability of MCF7 and MDA-MB-231 cells quantified after 1- and 8-days culture, respectively, in 3D printed TEMP-CNF scaffolds, $n = 6$–9. (**C**) Quantification of cell diameter of MCF7 and MDA-MB-231 cells cultured in 2D or in 3D printed TEMPO-CNF scaffolds for 8 and 14 days, respectively, $n = 3$–9. (**A**–**C**) Average ± standard error of the mean. Two-way ANOVA; Sidak's post hoc test for multiple comparison. * p-value < 0.05, ** p-value < 0.01, *** p-value < 0.001.

To analyze the microscopic structure of the 3D printed TEMPO-CNF and evaluate the scaffolds' ability to affect the cellular phenotype, MCF7 and MDA-MB-231 cells were analyzed with SEM and histochemistry after growth on TEMPO-CNF scaffolds for 14 days or in conventional 2D cell cultured for 2 days (Figure 4). The surface of the 3D printed TEMPO-CNF material displayed a porous structure at two distinct scales. Elongated large pores with cavity sizes up to several hundred microns with a sheet-like and rather smooth wall surface for cells to attach to (Figure 4A(i)). In higher magnification, we could see TEMPO-CNF networks forming open pores of sub-micron size enabling cell medium transport (Figure 4A(ii)). In 2D culture, the cells grew in a mono layer as morphologically homogenous populations, (Figure 4B(i),C(i)). In contrast, cells cultured in 3D printed TEMPO-CNF scaffolds grew in multiple layers and developed cellular heterogeneity. This was shown by different morphologies in association with pores and cavities illustrated by low and high magnification images with SEM (Figure 4B(ii,iii),C(ii,iii)) and hematoxylin and eosin-stained paraffin sections (Figure 4B(iv),C(iv)).

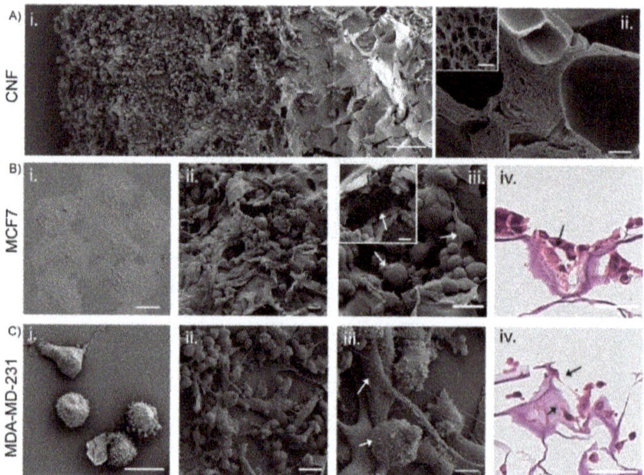

Figure 4. Cells cultured on TEMPO-CNF develop a heterogenous cell population with different cell morphologies. (**A**,**B**(**i–iii**),**C**(**i–iii**)) SEM images of (**A**) TEMPO-CNF material in (i) low magnification overview with MDA-MB-231 cells and (ii) cross section and cavities in the material. Inset in (**A**(**ii**)) higher magnification of the cross section. (**B**) MCF7 and (**C**) MDA-MB-231 cells. (**B**(**i**),**C**(**i**)) Cells grown in 2D cell culture plates in a homogenous mono layer. (**B**(**ii**,**iii**),**C**(**ii**,**iii**)) illustrates cells grown in multiple layers with different morphologies (indicated by white arrows in (**B**(**iii**),**C**(**iii**))). Inset in Biii shows cells filling out holes and cavities in the TEMPO-CNF material. (**B**(**iv**),**C**(**iv**)) Hematoxylin and eosin-stained sections of TEMPO-CNF scaffolds with MCF7 and MDA-MB-231, respectively. Black arrows indicate the cells. Scale bars 200 μm (**A**(**i**)), 20 μm (**B**(**i**),**C**(**i**)), 10 μm (**A**(**ii**),**C**(**iii**)), 1 μm (inset in (**A**(**ii**))), 10 μm (**C**(**iv**)), 30 μm ((**B**(**ii–iv**),**C**(**ii**)), inset (**B**(**iii**))).

3.4. 3D CNF Environment Induce Stem Cell Properties

Cells adjust their gene expression profile depending on the microenvironment and growth conditions. Here, we profiled genes related to stemness, migration and epithelial-mesenchymal transition (EMT), pluripotency, differentiation and proliferation in cells cultured in 3D printed TEMPO-CNF scaffolds compared with 2D plastic cell culture. In addition to MCF7, another estrogen receptor- and progesterone receptor positive breast cancer cell line, T47D, was used for gene expression studies. Relative to the expression in 2D cell culture, the 3D TEMPO-CNF environment induced expression of the stemness marker *CD44* and the migration marker *VIM* in both cell lines. The migration marker *SNAI1* was upregulated in MCF7 while *TWIST* was downregulated in T47D. The pluripotency marker *NANOG* was slightly upregulated in MCF7. *ESR1* and *EPCAM* were downregulated in MCF7, and *CD24* and *PGR* was down- and upregulated in T47D, respectively. The proliferation marker *MKI67* was slightly upregulated in MCF7, while *CCNA2* was more downregulated in T47D. In the estrogen receptor- and progesterone negative breast cancer cell line MDA-MB-231, the only significantly detected gene expression change was upregulation of the differentiation marker *CD24*, (Figure 5).

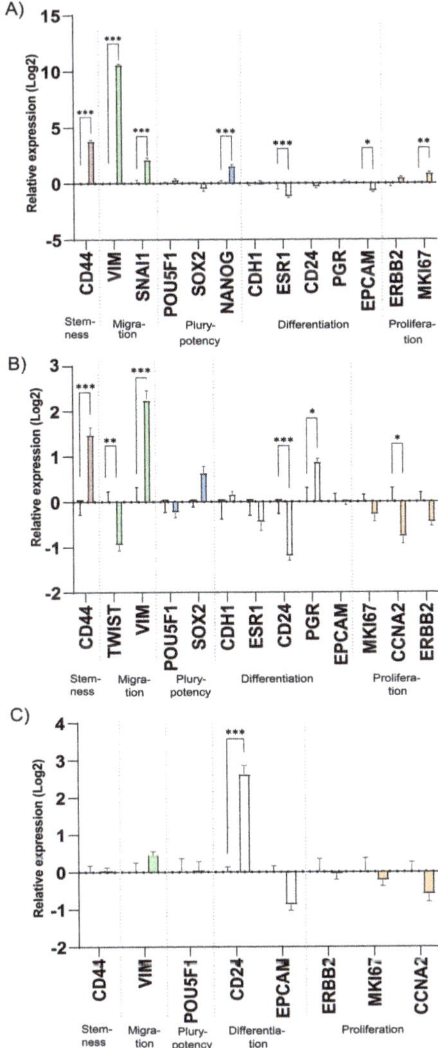

Figure 5. The microenvironment of 3D printed TEMPO-CNF scaffolds alters expression of stemness and migration gene markers. Gene expression levels analyzed by qPCR on cells cultured for 2 weeks in 3D scaffolds and 48 h in 2D for the cell lines (**A**) MCF7, (**B**) T47D and (**C**) MDA-MB-231. The data was normalized against the gene expression of the 2D control, set to zero for each gene. Average ± standard error of the mean ($n = 3$). Two-way ANOVA; Sidak's post hoc test for multiple comparison (2D). * p-value < 0.05, ** p-value < 0.01, *** p-value < 0.001.

To test the stemness properties of the 3D TEMPO-CNF cultured MCF7 and MDA-MB-231 cells at cellular level we performed the holoclone formation assay. The 3D TEMPO-CNF environment significantly increased the ability to form holoclones for both MCF7, with 120%, and MDA-MB-231 cells, with 54%, compared with 2D cultured cells (Figure 6A).

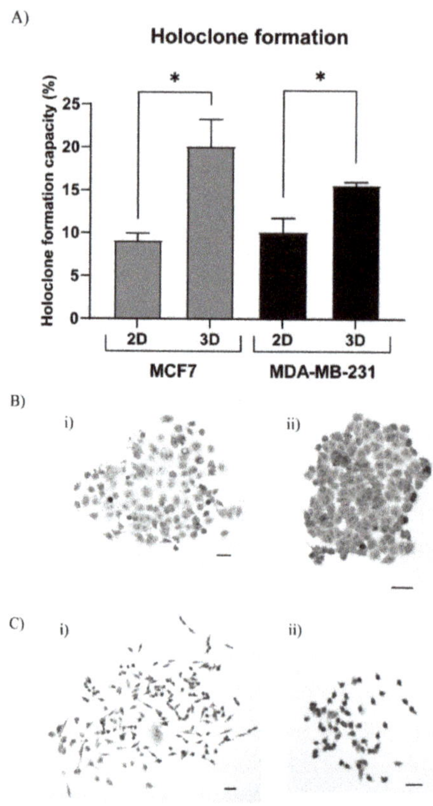

Figure 6. 3D TEMPO-CNF environments increase holoclone formation properties. MCF7 and MDA-MB-231 cells were cultured in 3DPS and 2D for 2 weeks or 48 h, respectively, detached and analyzed for holoclone formation properties. (**A**) The holoclone formation ability was quantified for MCF7 and MDA-MB-231 cells grown in 3D TEMPO-CNF scaffolds compared with 2D culture. Unpaired *t*-test: Gaussian distribution between 2D and 3D within each cell line. Average ± standard error of the mean, * *p*-value < 0.05, *n* = 9. Holoclone formation assay with (**B**) MCF7 and (**C**) MDA-MB-231 cells grown in 3D TEMPO-CNF scaffolds. Images illustrates differentiated cells (**B**(i),**C**(i)) and holoclones (**B**(ii),**C**(ii)). Scale bars 50 μm.

4. Discussion

In attempts to mimic native human tissue, there are several ways to produce scaffolds as a structural base for tissue engineering or in vitro 3D cell culturing models. For 3D printing applications, an extensive material database for bio-inks has been developed during the last decade [39] and the various materials have different limitations and unique properties depending on how the material is alternated, what cell types to be used and the application strategy. One promising option is the various types of nanocelluloses that can be obtained from different raw materials, including wood, of which hard- and softwood chemical pulp fibers are the most commonly used to produce nanocelluloses. Wood pulp fibers are treated chemically or enzymatically, to generate two main types of wood nanocellulose fibers, i.e., CNF and cellulose nanocrystals. In this work, we present a novel approach where TEMPO-CNF [34,40] is 3D printed as scaffolds for cancer cells to develop a 3D cell growth system simulating the human cancer environment. To be able to obtain a good understanding of how cells interact with the material, it is important to perform a thorough material characterization. The chemical modifications on the TEMPO-CNF used in this study introduced carboxylic functionality on the surface of the material. Hence, the interaction with cells could, in addition to hydrogen bonds, act via electrostatic interaction as well as cationic metal chelation. The carboxyl acid content of the TEMPO-CNF was 982 ± 8 μmol/g [34] and similar surface charge of TEMPO-CNF has previously been demonstrated to be suitable for cell growth and proliferation [34,41].

The mechanical properties of the material also have a critical role regulating the cell phenotype and function [42]. The range of elastic modulus for cancer tissue is wide and has generally been found to vary from 800 to 4500 Pa depending on location [43]. However, in calcified regions the elastic modulus can be much higher locally and reach the same as bone. So, even if the TEMPO-CNF hydrogels are considerably stiffer than cancer tissue, they could, from a cellular perspective, belong to the same microenvironmental niche and the biological response can be comparable. The response to elastic modulus could hypothetically be studied by varying the TEMPO-CNF-concentration during the preparation of the TEMPO-CNF hydrogel and should not vary the porosity and nanostructure.

When TEMPO-CNF is dispersed in water to form a gel, the individual nanofibrils interact with each other and, at a sufficient concentration, establishes a percolated network thus increasing the corresponding viscosity. The concentration of nanofibrils used in this study is above the concentration to form network percolation reported for TEMPO-CNF [23]. The TEMPO-CNF at 1% therefore has suitable viscosity for 3D printing [35] as demonstrated in the present study (Figure 2).

Depending on the composition of the cell culture medium used, the material will behave differently, e.g., cations will react with carboxyl groups in the TEMPO-CNF [44,45]. The different cell culture medium used in this study affected the nanomechanical properties of the TEMPO-CNF material significantly, where RPMI resulted in a higher elastic modulus of the TEMPO-CNF compared with DMEM. This difference could arise from differences in composition since 50% of the Ca^{2+} in RPMI are replaced by Mg^{2+} and it also contains the reducing agent glutathione. It is known from crosslinking of alginates that different cations form hydrogels with varying mechanical properties [45]. The reasons for the difference in mechanical properties of the TEMPO-CNF-hydrogels is solely speculative at this stage, but if the differences are large enough to induce differences in cell responses, selection of growth medium could be important.

Cells will respond to roughness, porosity, pore size and microstructure of the material as well as chemical properties such as charge, valency and functional groups [46–48]. For the 3D printed TEMPO-CNF scaffolds used in this study, we show that breast cancer cells can be seeded on the scaffolds, the cells expand as expected (Figure 3A,B) and adopt a smaller cell size compared with cells grown in 2D cell culture dishes (Figure 3C), which is consistent with previous studies [46–48]. The cells adapted to the material surface, grew into cavities in the porous material structure (Figure 4B(ii)) and developed a heterogenous cell population (Figure 4B(ii,iii),C(ii,iii)) mimicking the in vivo situation. This is also consistent with previous work using de-cellularized patient derived scaffolds and 3D-printed alginate hydrogels where seeded cancer cells populate the entire material surface ([12,15] and unpublished data).

By analyzing biomarkers for tumor cell characteristics at mRNA level, 3D TEMPO-CNF scaffolds increased expression of genes related to stemness and migratory properties compared with 2D cultures (Figure 5). To functionally determine if cells derive from stem cells, transit-amplifying cells or differentiated cells, the ability of the cells to form holoclones, meroclones or paraclones, respectively, can be studied [49,50]. Increased ability of MCF7 and MDA-MB-231 breast cancer cells, cultured on scaffolds, to form holoclones supported the gene expression results and the 3D TEMPO-CNF scaffolds can be regarded to enrich for CSC (Figure 6).

Our results confirm our previous data showing that cells cultured in 3D-printed alginate hydrogel scaffolds fill out pores and cavities throughout the scaffold with a heterogenous population of cells in multilayers compared with cells grown in a monolayer with a homogenous cell population in 2D [15]. Previous studies have also shown that fine-tuning of the mesh network, and using different geometries, affect the cell phenotype [51].

5. Conclusions

Carboxylated CNF demonstrated good capability to be applied as ink in a micro-extrusion printer producing 3D scaffolds with micrometer-sized pores. The breast cancer cells lines MCF7, MDA-MB-231 and T47D were successfully cultured on the scaffolds and developed a cellular heterogeneity growing in multiple layers throughout the scaffolds with enhanced CSC features. The data suggests that TEMPO-CNF-based 3D printed scaffolds can be used as cancer cell culture models mimicking the in vivo situation in a more relevant way compared with conventional 2D cell culturing on plastics making them suitable for applications such as drug testing.

Supplementary Materials: The following are available online at https://www.mdpi.com/article/10.3390/bioengineering8070097/s1, Table S1. Primer sequences for qPCR.

Author Contributions: Conceptualization, J.R., A.S. (Andreas Svanström), M.B., Y.B., A.S. (Anders Ståhlberg), G.L., G.C.-C. and J.H.; data curation, J.R., A.S. (Andreas Svanström), S.P., P.S. and S.S.; formal analysis, J.R., A.S. (Andreas Svanström), S.P., M.B., Y.B., A.S. (Anders Ståhlberg), G.L., G.C.-C. and J.H.; funding acquisition, Y.B., A.S. (Anders Ståhlberg), G.L., G.C.-C. and J.H.; investigation, J.R., A.S. (Andreas Svanström), S.P., M.B., Y.B., P.S., S.S., A.S. (Anders Ståhlberg), G.L., G.C.-C. and J.H.; methodology, J.R., A.S. (Andreas Svanström), S.P., M.B., Y.B., P.S., S.S., A.S. (Anders Ståhlberg), G.L., G.C.-C. and J.H.; project administration, J.R., A.S. (Andreas Svanström), A.S. (Anders Ståhlberg), G.L., G.C.-C. and J.H.; resources, Y.B., A.S. (Anders Ståhlberg), G.L., G.C.-C. and J.H.; supervision, M.B., Y.B., A.S. (Anders Ståhlberg), G.L., G.C.-C. and J.H.; validation, J.R., A.S. (Andreas Svanström), S.P., M.B., Y.B., A.S. (Anders Ståhlberg), G.L., G.C.-C. and J.H.; visualization, J.R., A.S. (Andreas Svanström), S.P., P.S. and G.C.-C.; writing—original draft, J.R., A.S. (Andreas Svanström), S.P., M.B., Y.B., P.S., S.S., A.S. (Anders Ståhlberg), G.L., G.C.-C. and J.H.; writing—review and editing, J.R., A.S. (Andreas Svanström), S.P., M.B., Y.B., P.S., S.S., A.S. (Anders Ståhlberg), G.L., G.C.-C. and J.H. All authors have read and agreed to the published version of the manuscript.

Funding: This research was funded by Sweden's Innovation Agency (2017-03737); the Swedish Cancer Society (2019 0306 and 2019-0317); the Swedish Research Council (2019-01273, 2017-01392, 2016-01530 and 2015-03256); the Swedish state under the agreement between the Swedish government and the county councils (ALF-agreement) (716321 and 721091); Region Västra Götaland, Sweden (RUN 2018-00017 and infrastructure support to A.S. (Anders Ståhlberg)), Swedish Foundation for Strategic Research (FID15-0008); Johan Jansson Foundation for Cancer Research; Wilhelm and Martina Lundgrens Foundation; Assar Gabrielsson Foundation and the Sigurd och Elsa Goljes Minne foundation for funding this project.

Institutional Review Board Statement: Not applicable.

Informed Consent Statement: Not applicable.

Data Availability Statement: The data presented in this study are available on request from the corresponding author.

Conflicts of Interest: G.L. declares stock ownership and is also a board member in Iscaff Pharma. A.S. (Anders Ståhlberg) declares stock ownership in TATAA Biocenter, Iscaff Pharma and SiMSen Diagnostics. A.S. (Anders Ståhlberg) is also a board member of Iscaff Pharma and SiMSen Diagnostics.

References

1. Torre, L.A.; Bray, F.; Siegel, R.L.; Ferlay, J.; Lortet-Tieulent, J.; Jemal, A. Global cancer statistics, 2012: Global Cancer Statistics, 2012. *CA Cancer J. Clin.* **2015**, *65*, 87–108. [CrossRef] [PubMed]
2. Yamada, K.M.; Cukierman, E. Modeling Tissue Morphogenesis and Cancer in 3D. *Cell* **2007**, *130*, 601–610. [CrossRef]

3. Zhang, H.; Wang, Z.Z. Mechanisms that mediate stem cell self-renewal and differentiation. *J. Cell. Biochem.* **2008**, *103*, 709–718. [CrossRef] [PubMed]
4. Hutchinson, L.; Kirk, R. High drug attrition rates—Where are we going wrong? *Nat. Rev. Clin. Oncol.* **2011**, *8*, 189–190. [CrossRef] [PubMed]
5. Hay, M.; Thomas, D.W.; Craighead, J.L.; Economides, C.; Rosenthal, J. Clinical development success rates for investigational drugs. *Nat. Biotechnol.* **2014**, *32*, 40. [CrossRef]
6. Qiao, S.-P.; Zhao, Y.-F.; Li, C.-F.; Yin, Y.-B.; Meng, Q.-Y.; Lin, F.-H.; Liu, Y.; Hou, X.-L.; Guo, K.; Chen, X.; et al. An alginate-based platform for cancer stem cell research. *Acta Biomater.* **2016**, *37*, 83–92. [CrossRef] [PubMed]
7. Vlashi, E.; Pajonk, F. Cancer stem cells, cancer cell plasticity and radiation therapy. *Semin. Cancer Biol.* **2015**, *31*, 28–35. [CrossRef]
8. Cavo, M.; Caria, M.; Pulsoni, I.; Beltrame, F.; Fato, M.; Scaglione, S. A new cell-laden 3D Alginate-Matrigel hydrogel resembles human breast cancer cell malignant morphology, spread and invasion capability observed "in vivo". *Sci. Rep.* **2018**, *8*, 5333. [CrossRef]
9. Pampaloni, F.; Reynaud, E.G.; Stelzer, E.H.K. The third dimension bridges the gap between cell culture and live tissue. *Nat. Rev. Mol. Cell Biol.* **2007**, *8*, 839–845. [CrossRef]
10. Edmondson, R.; Broglie, J.J.; Adcock, A.F.; Yang, L. Three-Dimensional Cell Culture Systems and Their Applications in Drug Discovery and Cell-Based Biosensors. *ASSAY Drug Dev. Technol.* **2014**, *12*, 207–218. [CrossRef] [PubMed]
11. Langhans, S.A. Three-Dimensional in Vitro Cell Culture Models in Drug Discovery and Drug Repositioning. *Front. Pharmacol.* **2018**, *9*, 6. [CrossRef] [PubMed]
12. Landberg, G.; Fitzpatrick, P.; Isakson, P.; Jonasson, E.; Karlsson, J.; Larsson, E.; Svanström, A.; Rafnsdottir, S.; Persson, E.; Gustafsson, A.; et al. Patient-derived scaffolds uncover breast cancer promoting properties of the microenvironment. *Biomaterials* **2020**, *235*, 119705. [CrossRef] [PubMed]
13. Le, X.; Poinern, G.E.J.; Ali, N.; Berry, C.M.; Fawcett, D. Engineering a Biocompatible Scaffold with Either Micrometre or Nanometre Scale Surface Topography for Promoting Protein Adsorption and Cellular Response. *Int. J. Biomater.* **2013**, *2013*, 1–16. [CrossRef] [PubMed]
14. Huang, L.; Abdalla, A.M.; Xiao, L.; Yang, G. Biopolymer-Based Microcarriers for Three-Dimensional Cell Culture and Engineered Tissue Formation. *Int. J. Mol. Sci.* **2020**, *21*, 1895. [CrossRef] [PubMed]
15. Svanström, A.; Rosendahl, J.; Salerno, S.; Leiva, M.C.; Gregersson, P.; Berglin, M.; Bogestål, Y.; Lausmaa, J.; Oko, A.; Chinga-Carrasco, G.; et al. Optimized alginate-based 3D printed scaffolds as a model of patient derived breast cancer microenvironments in drug discovery. *Biomed. Mater.* **2021**, *16*, 045046. [CrossRef]
16. Caddeo, S.; Boffito, M.; Sartori, S. Tissue Engineering Approaches in the Design of Healthy and Pathological in vitro Tissue Models. *Front. Bioeng. Biotechnol.* **2017**, *5*, 40. [CrossRef]
17. Basu, A.; Celma, G.; Strømme, M.; Ferraz, N. In Vitro and in vivo Evaluation of the Wound Healing Properties of Nanofibrillated Cellulose Hydrogels. *ACS Appl. Bio Mater.* **2018**, *1*, 1853–1863. [CrossRef]
18. Rashad, A.; Mustafa, K.; Heggset, E.B.; Syverud, K. Cytocompatibility of Wood-Derived Cellulose Nanofibril Hydrogels with Different Surface Chemistry. *Biomacromolecules* **2017**, *18*, 1238–1248. [CrossRef]
19. Nordli, H.R.; Chinga-Carrasco, G.; Rokstad, A.M.; Pukstad, B. Producing ultrapure wood cellulose nanofibrils and evaluating the cytotoxicity using human skin cells. *Carbohydr. Polym.* **2016**, *150*, 65–73. [CrossRef]
20. Mertaniemi, H.; Escobedo-Lucea, C.; Sanz-García, A.; Gandía, C.; Mäkitie, A.; Partanen, J.; Ikkala, O.; Yliperttula, M. Human stem cell decorated nanocellulose threads for biomedical applications. *Biomaterials* **2016**, *82*, 208–220. [CrossRef]
21. Lou, Y.-R.; Kanninen, L.; Kuisma, T.; Niklander, J.; Noon, L.; Burks, D.; Urtti, A.; Yliperttula, M. The Use of Nanofibrillar Cellulose Hydrogel as a Flexible Three-Dimensional Model to Culture Human Pluripotent Stem Cells. *Stem Cells Dev.* **2014**, *23*, 380–392. [CrossRef]
22. Hong, N.; Yang, G.-H.; Lee, J.; Kim, G. 3D bioprinting and its in vivo applications. *J. Biomed. Mater. Res. Part B Appl. Biomater.* **2018**, *106*, 444–459. [CrossRef] [PubMed]
23. Moberg, T.; Sahlin, K.; Yao, K.; Geng, S.; Westman, G.; Zhou, Q.; Oksman, K.; Rigdahl, M. Rheological properties of nanocellulose suspensions: Effects of fibril/particle dimensions and surface characteristics. *Cellulose* **2017**, *24*, 2499–2510. [CrossRef]
24. Naderi, A.; Lindström, T.; Sundström, J. Carboxymethylated nanofibrillated cellulose: Rheological studies. *Cellulose* **2014**, *21*, 1561–1571. [CrossRef]
25. Iotti, M.; Gregersen, Ø.W.; Moe, S.; Lenes, M. Rheological Studies of Microfibrillar Cellulose Water Dispersions. *J. Polym. Environ.* **2011**, *19*, 137–145. [CrossRef]
26. Ojansivu, M.; Rashad, A.; Ahlinder, A.E.; Massera, J.; Mishra, A.; Syverud, K.; Finne-Wistrand, A.; Miettinen, S.; Mustafa, K. Wood-based nanocellulose and bioactive glass modified gelatin–alginate bioinks for 3D bioprinting of bone cells. *Biofabrication* **2019**, *11*, 035010. [CrossRef]
27. Xu, C.; Molino, B.Z.; Wang, X.; Cheng, F.; Xu, W.; Molino, P.; Bacher, M.; Su, D.; Rosenau, T.; Willför, S.; et al. 3D printing of nanocellulose hydrogel scaffolds with tunable mechanical strength towards wound healing application. *J. Mater. Chem. B* **2018**, *6*, 7066–7075. [CrossRef]
28. Chinga-Carrasco, G.; Ehman, N.V.; Pettersson, J.; Vallejos, M.E.; Brodin, M.W.; Felissia, F.E.; Håkansson, J.; Area, M.C. Pulping and Pretreatment Affect the Characteristics of Bagasse Inks for Three-dimensional Printing. *ACS Sustain. Chem. Eng.* **2018**, *6*, 4068–4075. [CrossRef]

29. Markstedt, K.; Mantas, A.; Tournier, I.; Ávila, H.M.; Hägg, D.; Gatenholm, P. 3D Bioprinting Human Chondrocytes with Nanocellulose–Alginate Bioink for Cartilage Tissue Engineering Applications. *Biomacromolecules* **2015**, *16*, 1489–1496. [CrossRef]
30. Saito, T.; Nishiyama, Y.; Putaux, J.-L.; Vignon, M.; Isogai, A. Homogeneous Suspensions of Individualized Microfibrils from TEMPO-Catalyzed Oxidation of Native Cellulose. *Biomacromolecules* **2006**, *7*, 1687–1691. [CrossRef]
31. Liu, J.; Cheng, F.; Grénman, H.; Spoljaric, S.; Seppälä, J.; Eriksson, J.E.; Willför, S.; Xu, C. Development of nanocellulose scaffolds with tunable structures to support 3D cell culture. *Carbohydr. Polym.* **2016**, *148*, 259–271. [CrossRef]
32. Chinga-Carrasco, G.; Kuznetsova, N.; Garaeva, M.; Leirset, I.; Galiullina, G.; Kostochko, A.; Syverud, K. Bleached and unbleached MFC nanobarriers: Properties and hydrophobisation with hexamethyldisilazane. *J. Nanopart. Res.* **2012**, *14*, 1–10. [CrossRef]
33. Saito, T.; Isogai, A. TEMPO-Mediated Oxidation of Native Cellulose. The Effect of Oxidation Conditions on Chemical and Crystal Structures of the Water-Insoluble Fractions. *Biomacromolecules* **2004**, *5*, 1983–1989. [CrossRef]
34. Silva, F.; Gracia, N.; McDonagh, B.H.; Domingues, F.C.; Nerín, C.; Chinga-Carrasco, G. Antimicrobial activity of biocomposite films containing cellulose nanofibrils and ethyl lauroyl arginate. *J. Mater. Sci.* **2019**, *54*, 12159–12170. [CrossRef]
35. Espinosa, E.; Filgueira, D.; Rodríguez, A.; Chinga-Carrasco, G. Nanocellulose-Based Inks—Effect of Alginate Content on the Water Absorption of 3D Printed Constructs. *Bioengineering* **2019**, *6*, 65. [CrossRef]
36. Bustin, S.A.; Benes, V.; Garson, J.A.; Hellemans, J.; Huggett, J.; Kubista, M.; Mueller, R.; Nolan, T.; Pfaffl, M.W.; Shipley, G.L.; et al. The MIQE Guidelines: Minimum Information for Publication of Quantitative Real-Time PCR Experiments. *Clin. Chem.* **2009**, *55*, 611–622. [CrossRef] [PubMed]
37. Schneider, C.A.; Rasband, W.S.; Eliceiri, K.W. NIH Image to ImageJ: 25 years of image analysis. *Nat. Methods* **2012**, *9*, 671–675. [CrossRef] [PubMed]
38. Webb, B.; Doyle, B.J. Parameter optimization for 3D bioprinting of hydrogels. *Bioprinting* **2017**, *8*, 8–12. [CrossRef]
39. Jose, R.R.; Rodriguez, M.J.; Dixon, T.A.; Omenetto, F.; Kaplan, D.L. Evolution of Bioinks and Additive Manufacturing Technologies for 3D Bioprinting. *ACS Biomater. Sci. Eng.* **2016**, *2*, 1662–1678. [CrossRef]
40. Nordli, H.R.; Pukstad, B.; Chinga-Carrasco, G.; Rokstad, A.M. Ultrapure Wood Nanocellulose—Assessments of Coagulation and Initial Inflammation Potential. *ACS Appl. Bio Mater.* **2019**, *2*, 1107–1118. [CrossRef]
41. Liu, J.; Chinga-Carrasco, G.; Cheng, F.; Xu, W.; Willför, S.; Syverud, K.; Xu, C. Hemicellulose-reinforced nanocellulose hydrogels for wound healing application. *Cellulose* **2016**, *23*, 3129–3143. [CrossRef]
42. Radisic, M.; Alsberg, E. Special Issue on Tissue Engineering. *ACS Biomater. Sci. Eng.* **2017**, *3*, 1880–1883. [CrossRef] [PubMed]
43. Butcher, D.T.; Alliston, T.; Weaver, V.M. A tense situation: Forcing tumour progression. *Nat. Rev. Cancer* **2009**, *9*, 108–122. [CrossRef]
44. Engler, A.; Sen, S.; Sweeney, H.L.; Discher, D.E. Matrix Elasticity Directs Stem Cell Lineage Specification. *Cell* **2006**, *126*, 677–689. [CrossRef] [PubMed]
45. Loh, Q.L.; Choong, C. Three-Dimensional Scaffolds for Tissue Engineering Applications: Role of Porosity and Pore Size. *Tissue Eng. Part B Rev.* **2013**, *19*, 485–502. [CrossRef] [PubMed]
46. Mrksich, M.; Chen, C.; Xia, Y.; Dike, L.E.; Ingber, D.E.; Whitesides, G.M. Controlling cell attachment on contoured surfaces with self-assembled monolayers of alkanethiolates on gold. *Proc. Natl. Acad. Sci. USA* **1996**, *93*, 10775–10778. [CrossRef] [PubMed]
47. Arima, Y.; Iwata, H. Effects of surface functional groups on protein adsorption and subsequent cell adhesion using self-assembled monolayers. *J. Mater. Chem.* **2007**, *17*, 4079–4087. [CrossRef]
48. Lee, M.H.; Brass, D.A.; Morris, R.; Composto, R.J.; Ducheyne, P. The effect of non-specific interactions on cellular adhesion using model surfaces. *Biomaterials* **2005**, *26*, 1721–1730. [CrossRef] [PubMed]
49. Pellegrini, G.; Golisano, O.; Paterna, P.; Lambiase, A.; Bonini, S.; Rama, P.; De Luca, M. Location and Clonal Analysis of Stem Cells and Their Differentiated Progeny in the Human Ocular Surface. *J. Cell Biol.* **1999**, *145*, 769–782. [CrossRef]
50. Barrandon, Y.; Green, H. Three clonal types of keratinocyte with different capacities for multiplication. *Proc. Natl. Acad. Sci. USA* **1987**, *84*, 2302–2306. [CrossRef]
51. Callens, S.J.; Uyttendaele, R.J.; Fratila-Apachitei, L.E.; Zadpoor, A.A. Substrate curvature as a cue to guide spatiotemporal cell and tissue organization. *Biomaterials* **2020**, *232*, 119739. [CrossRef] [PubMed]

MDPI

St. Alban-Anlage 66

4052 Basel

Switzerland

Tel. +41 61 683 77 34

Fax +41 61 302 89 18

www.mdpi.com

Bioengineering Editorial Office

E-mail: bioengineering@mdpi.com

www.mdpi.com/journal/bioengineering